Introduction
To
Stochastic
Processes

INTRODUCTION

TO

STOCHASTIC

PROCESSES

ERHAN ÇINLAR

Northwestern University

Prentice-Hall, Inc.

Englewood Cliffs, New Jersey

Library of Congress Cataloging in Publication Data

ÇINLAR, ERHAN, *date*
   Introduction to stochastic processes.

   Bibliography: p.
   1. Stochastic processes. 2. Markov
renewal theory. I. Title.
QA274.C56   519.2   74-5256
ISBN 0-13-498089-1

PRENTICE-HALL INTERNATIONAL, INC., *London*
PRENTICE-HALL OF AUSTRALIA, PTY. LTD., *Sydney*
PRENTICE-HALL OF CANADA, LTD., *Toronto*
PRENTICE-HALL OF INDIA PRIVATE LIMITED, *New Delhi*
PRENTICE-HALL OF JAPAN, INC., *Tokyo*

Contents

Preface   *ix*

Chapter 1   Probability Spaces and Random Variables   *1*

   1. Probability Spaces   *1*
   2. Random Variables and Stochastic Processes   *6*
   3. Conditional Probability   *14*
   4. Exercises   *17*

Chapter 2   Expectations and Independence   *21*

   1. Expected Value   *22*
   2. Conditional Expectations   *33*
   3. Exercises   *40*

Chapter 3   Bernoulli Processes and
             Sums of Independent Random Variables   *43*

   1. Bernoulli Process   *44*
   2. Numbers of Successes   *45*
   3. Times of Successes   *53*
   4. Sums of Independent Random Variables   *60*
   5. Exercises   *66*

Chapter 4 Poisson Processes *70*

1. Arrival Counting Process *71*
2. Times of Arrivals *79*
3. Forward Recurrence Times *85*
4. Superposition of Poisson Processes *87*
5. Decomposition of Poisson Processes *88*
6. Compound Poisson Processes *90*
7. Non-stationary Poisson Processes *94*
8. Exercises *101*

Chapter 5 Markov Chains *106*

1. Introduction *106*
2. Visits to a Fixed State *119*
3. Classification of States *125*
4. Exercises *138*

Chapter 6 Limiting Behavior and Applications of Markov Chains *144*

1. Computation of R and F *144*
2. Recurrent States and the Limiting Probabilities *152*
3. Periodic States *160*
4. Transient States *166*
5. Applications to Queueing Theory: M/G/1 Queue *169*
6. Queueing System G/M/1 *178*
7. Branching Processes *183*
8. Exercises *188*

Chapter 7 Potentials, Excessive Functions, and Optimal Stopping of Markov Chains *195*

1. Potentials *195*
2. Excessive Functions *204*
3. Optimal Stopping *208*
4. Games with Discounting and Fees *220*
5. Exercises *226*

Chapter 8 Markov Processes *232*

1. Markov Processes *233*
2. Sample Path Behavior *239*
3. Structure of a Markov Process *246*
4. Potentials and Generators *253*
5. Limit Theorems *261*
6. Birth and Death Processes *271*
7. Exercises *277*

Chapter 9   Renewal Theory   *283*

1. Renewal Processes   *283*
2. Regenerative Processes and Renewal Theory   *293*
3. Delayed and Stationary Processes   *302*
4. Exercises   *307*

Chapter 10   Markov Renewal Theory   *313*

1. Markov Renewal Processes   *313*
2. Markov Renewal Functions
   and Classification of States   *318*
3. Markov Renewal Equations   *323*
4. Limit Theorems   *328*
5. Semi-Markov Processes   *337*
6. Semi-Regenerative Processes   *343*
7. Applications to Queueing Theory   *350*
8. Exercises   *357*

Afterword   *363*

Appendix.   Non-Negative Matrices   *364*

1. Eigenvalues and Eigenvectors   *364*
2. Spectral Representations   *367*
3. Positive Matrices   *371*
4. Non-Negative Matrices   *373*
5. Limits and Rates of Convergence   *378*

References   *383*

Answers to Selected Exercises   *387*

Index of Notations   *393*

Subject Index   *395*

Preface

A man wanted to dock the tail of his horse. He consulted a wise man on how short to make it. "Make it as short as it pleases you," said the wise man, "for, no matter what you do, some will say it is too long, some too short, and your opinion itself will change from time to time."

This book is an introduction to stochastic processes. It covers most of the basic processes of interest except for two glaring omissions. Topics covered are developed in some depth, some fairly recent results are included, and references are given for further reading. These features should make the book useful to scientists and engineers looking for models and results to use in their work. However, this is primarily a textbook; a large number of numerical examples are worked out in detail, and the exercises at the end of each chapter are there to illustrate the theory rather than to extend it.

When theorems are proved, this is done in some detail; even the ill-prepared student should be able to follow them. On the other hand, not all theorems are proved. If a result is of sufficient intrinsic interest, and if it can be explained and understood, then it is listed as a theorem even though it could not be proved with the elementary tools available in this book. This freedom made it possible to include two characterization theorems on Poisson processes, several limit theorems on the ratios of additive functionals of Markov processes, several results on the sample path behavior of (continuous parameter) Markov processes, and a large number of results dealing with stopping times and the strong Markov property.

The book assumes a background in calculus but no measure theory; thus, the treatment is elementary. At the same time, it is from a modern viewpoint. In the modern approach to stochastic processes, the primary object is the behavior of the sample paths. This is especially so for the applied scientist and the engineer, since it is the sample path which he observes and tries to control. Our approach capitalizes on this happy harmony between the methods of the better mathematician and the intuition of the honest engineer.

We have also followed the modern trends in preferring matrix algebra and recursive methods to transform methods. The early probability theorist's desire to cast his problem into one concerning distribution functions and Fourier transforms is understandable in view of his background in classical analysis and the notion of an acceptable solution prevailing then. The present generation, however, influenced especially by the availability of computers, prefers a characterization of the solution coupled with a recursive method of obtaining it to an explicit closed form expression in terms of the generating functions of the Laplace transforms of the quantities of actual interest.

The first part of the book is based on a set of lecture notes which were used by myself and several colleagues in a variety of "applied" courses on stochastic processes over the last five years. In that stage I was helped by P. A. Jacobs and C. G. Gilbert in collecting problems and abstracting papers from the applied literature. Most of the final version was written during my sabbatical stay at Stanford University; I should like to thank the department of Operations Research for their hospitality then. While there I have benefited from conversations with K. L. Chung and D. L. Iglehart; it is a pleasure to acknowledge my debt to them. I am especially indebted to A. F. Karr for his help throughout this project; he eliminated many inaccuracies and obscurities. I had the good fortune to have G. Lemmond to type the manuscript, and finally, the National Science Foundation to support my work.

E. ÇINLAR

Introduction
To
Stochastic
Processes

CHAPTER 1

Probability Spaces
and Random Variables

The theory of probability now constitutes a formidable body of knowledge of great mathematical interest and of great practical importance. It has applications in every area of natural science: in minimizing the unavoidable errors of observations, in detecting the presence of assignable causes in observed phenomena, and in discovering the basic laws obscured by chance variations. It is also an indispensable tool in engineering and business: in deducing the true lessons from statistics, in forecasting the future, and in deciding which course to pursue.

In this chapter the basic vocabulary of probability theory will be introduced. Most of these terms have cognates in ordinary language, and the reader should do well not to fall into a false sense of security because of his previous familiarity with them. Instead he should try to refine his own use of these terms and watch for the natural context within which each term appears.

1. Probability Spaces

The basic notion in probability theory is that of a *random experiment:* an experiment whose outcome cannot be determined in advance. The set of all possible outcomes of an experiment is called the *sample space* of that experiment.

An *event* is a subset of a sample space. An event A is said to occur if and only if the observed outcome ω of the experiment is an element of the set A.

(1.1) EXAMPLE. Consider an experiment that consists of counting the number of traffic accidents at a given intersection during a specified time interval. The sample space is the set $\{0, 1, 2, 3, \ldots\}$. The statement "the number of accidents is less than or equal to seven" describes the event $\{0, 1, \ldots, 7\}$. The event $A = \{5, 6, 7, \ldots\}$ occurs if and only if the number of accidents is 5 or 6 or 7 or $\ldots$. ☐

Given a sample space Ω and an event A, the *complement* A^c of A is defined to be the event which occurs if and only if A does not occur, that is,

$$(1.2) \qquad\qquad A^c = \{\omega \in \Omega : \omega \notin A\}.$$

Given two events A and B, their *union* is the event which occurs if and only if either A or B (or both) occurs, that is,

$$(1.3) \qquad\qquad A \cup B = \{\omega \in \Omega : \omega \in A \text{ or } \omega \in B\}.$$

The *intersection* of A and B is the event which occurs if and only if both A and B occur, that is,

$$(1.4) \qquad\qquad A \cap B = \{\omega \in \Omega : \omega \in A \text{ and } \omega \in B\}.$$

The operations of taking unions, intersections, and complements may be combined to obtain new events. In particular, the following identities are of value:

$$(1.5) \qquad (A \cup B)^c = A^c \cap B^c, \qquad (A \cap B)^c = A^c \cup B^c.$$

The set Ω is also called the *certain event*. The set containing no elements is called the *empty event* and is denoted by $\varnothing$. Note that $\varnothing = \Omega^c$ and $\Omega = \varnothing^c$. Two events are said to be *disjoint* if they have no elements in common, that is, A and B are disjoint if

$$A \cap B = \varnothing.$$

If two events are disjoint, the occurrence of one implies that the other has not occurred. A family of events is called disjoint if every pair of them are disjoint.

Event A is said to *imply* the event B, written $A \subset B$, if every ω in A belongs also to B. To show that two events A and B are the same, then, it is sufficient to show that A implies B and B implies A.

If $A_1, A_2, \ldots$ are events, then their union

$$\bigcup_{i=1}^{\infty} A_i$$

is the event which occurs if and only if at least one of them occurs. Their intersection

$$\bigcap_{i=1}^{\infty} A_i$$

is the event which occurs if and only if *all* of them occur.

Next, corresponding to our intuitive notion of the chances of an event occurring, we introduce a function defined on a collection of events.

(1.6) DEFINITION. Let Ω be a sample space and P a function which associates a number with each event. Then P is called a *probability measure* provided that
    (a)   for any event A, $0 \le P(A) \le 1$;
    (b) $P(\Omega) = 1$;
    (c)   for any sequence $A_1, A_2, \ldots$ of disjoint events,

$$P(\bigcup_{i=1}^{\infty} A_i) = \sum_{i=1}^{\infty} P(A_i). \qquad \square$$

By axiom (b), the probability assigned to Ω is 1. Usually, there will be other events $A \subset \Omega$ such that $P(A) = 1$. If a statement holds for all ω in such a set A with $P(A) = 1$, then it is customary to say that the statement is true *almost surely* or that the statement holds for *almost all* $\omega \in \Omega$.

Axiom (c) above is a severe condition on the manner in which probabilities are assigned to events. Indeed, it is usually impossible to assign a probability $P(A)$ to every subset A and still satisfy (c). Because of this, it is customary to define $P(A)$ only for certain subsets A. Throughout this book we will avoid the issue by using the term "event" only for those subsets A of Ω for which $P(A)$ is defined.

If the sample space Ω is $\{0, 1, 2, \ldots\}$ as in Example (1.1), then there are as many subsets of Ω as there are points on the real line. Therefore, it might be difficult to assign a probability to each event in an explicit fashion. Furthermore, almost any meaningful real-life problem requires considering much more complex sample spaces. Usually, in such situtations, the probabilities of only a few key events are specified and the remaining probabilities are left to be computed from the axioms (1.6a, b, c) by considering the various relationships which might exist between events. In fact, most of probability theory concerns itself with finding methods of doing just this. The following are the first few steps in this direction.

(1.7) PROPOSITION. *If $A_1, \ldots, A_n$ are disjoint events, then*

$$P(A_1 \cup \cdots \cup A_n) = P(A_1) + \cdots + P(A_n).$$

Proof. First, we establish that $P(\varnothing) = 0$. In (1.6c) we may take $A_1 = A_2 = \cdots = \varnothing$. Then $\bigcup_i A_i = \varnothing$ also, and (1.6c) implies that $P(\varnothing) = P(\varnothing) + P(\varnothing) + \cdots$. But this is possible only if $P(\varnothing) = 0$, since $0 \le P(\varnothing) \le 1$ by (1.6a).

Next, let $A_1, \ldots, A_n$ be disjoint events, and define $A_{n+1} = A_{n+2} = \cdots = \varnothing$. Then $A_1, A_2, \ldots$ are disjoint, and $\bigcup_{i=1}^{\infty} A_i = A_1 \cup \cdots \cup A_n$. By (1.6c) we have $P(A_1 \cup \cdots \cup A_n) = \sum_{i=1}^{\infty} P(A_i) = P(A_1) + \cdots + P(A_n)$, since $P(A_{n+1}) = P(A_{n+2}) = \cdots = P(\varnothing) = 0$.   ☐

(1.8) PROPOSITION. If $A \subset B$, then $P(A) \le P(B)$.

Proof. If $A \subset B$, we can write

$$B = A \cup (A^c \cap B).$$

Since A and $A^c \cap B$ are disjoint, by (1.7),

$$P(B) = P(A) + P(A^c \cap B).$$

By (1.6a), $P(A^c \cap B) \ge 0$; thus, $P(B) \ge P(A)$.   ☐

(1.9) PROPOSITION. For any event A, $P(A) + P(A^c) = 1$.

Proof. The events A and A^c are disjoint, and by (1.7), $P(A \cup A^c) = P(A) + P(A^c)$. On the other hand, $A \cup A^c = \Omega$, and $P(\Omega) = 1$ by (1.6b).   ☐

Proposition (1.8) expresses our intuitive feeling that, if the occurrence of A implies that of B, then the probability of B should be at least as great as the probability of A. Proposition (1.9), restated in words, states that the probability that A does not occur is one minus the probability that A does occur.

Next is a theorem which is used quite often. It is useful in evaluating the probability of a "complex" event by breaking it into components whose probabilities may be simpler to evaluate.

(1.10) THEOREM. If $B_1, B_2, \ldots$ are disjoint events with $\bigcup_{i=1}^{\infty} B_i = \Omega$, then for any event A,

$$P(A) = \sum_{i=1}^{\infty} P(A \cap B_i).$$

Proof. For any event A we can write

$$A = A \cap \Omega = A \cap \left(\bigcup_{i=1}^{\infty} B_i \right) = \bigcup_{i=1}^{\infty} (A \cap B_i).$$

Since $B_1, B_2, \ldots$ are disjoint, $A \cap B_1, A \cap B_2, \ldots$ are disjoint. By (1.6c) then,

$$P(A) = \sum_{i=1}^{\infty} P(A \cap B_i). \qquad \square$$

(1.11) PROPOSITION. Let $A_1, A_2, \ldots$ be a sequence of events such that $A_1 \subset A_2 \subset A_3 \subset \cdots$, and put $A = \bigcup_{i=1}^{\infty} A_i$. Then

$$P(A) = \lim_{n \to \infty} P(A_n).$$

Proof. Let $B_1 = A_1, B_2 = A_2 \cap A_1^c, B_3 = A_3 \cap A_2^c, \ldots$. Then $B_1, B_2, \ldots$ are disjoint, and

$$\bigcup_{i=1}^{n} B_i = A_n, \qquad \bigcup_{i=1}^{\infty} B_i = A.$$

Thus, by (1.7),

$$P(A_n) = \sum_{i=1}^{n} P(B_i).$$

Now taking limits,

$$\lim_{n \to \infty} P(A_n) = \lim_{n \to \infty} \sum_{i=1}^{n} P(B_i) = \sum_{i=1}^{\infty} P(B_i).$$

But by (1.6c),

$$\sum_{i=1}^{\infty} P(B_i) = P(\bigcup_{i=1}^{\infty} B_i) = P(A). \qquad \square$$

(1.12) COROLLARY. Let $A_1, A_2, \ldots$ be a sequence of events such that $A_1 \supset A_2 \supset A_3 \supset \ldots$, and put $A = \bigcap_{i=1}^{\infty} A_i$. Then

$$P(A) = \lim_{n \to \infty} P(A_n).$$

Proof. The sequence of events $A_1^c, A_2^c, \ldots$ satisfies $A_1^c \subset A_2^c \subset A_3^c \subset \cdots$, and by (1.5),

$$\bigcup_{i=1}^{\infty} A_i^c = (\bigcap_{i=1}^{\infty} A_i)^c = A^c.$$

Thus, by Proposition (1.11),

$$P(A^c) = \lim_{n \to \infty} P(A_n^c).$$

By Proposition (1.9), $P(A) = 1 - P(A^c)$, and $1 - P(A_n^c) = P(A_n)$ for any n. Thus,

$$P(A) = 1 - \lim_{n \to \infty} P(A_n^c)$$
$$= \lim_{n \to \infty} (1 - P(A_n^c)) = \lim_{n \to \infty} P(A_n). \qquad \square$$

Proposition (1.11) gives a continuity property of P: if the events A_1, $A_2, \ldots$ "increase to" A, then their probabilities $P(A_1), P(A_2), \ldots$ increase to $P(A)$. Corollary (1.12) is similar; if the events $A_1, A_2, \ldots$ "decrease to" A, then $P(A_1), P(A_2), \ldots$ decrease to $P(A)$.

2. Random Variables and Stochastic Processes

Suppose we are given a sample space Ω and a probability measure P. Most often, especially in applied problems, we are interested in functions of the outcomes rather than the outcomes themselves.

(2.1) DEFINITION.   A *random variable* X with values in the set E is a function which assigns a value $X(\omega)$ in E to each outcome ω in Ω.   □

The most usual examples of E are the set of non-negative integers $\mathbb{N} = \{0, 1, 2, \ldots\}$, the set of all integers $\{\ldots, -1, 0, 1, \ldots\}$, the set of all real numbers $\mathbb{R} = (-\infty, +\infty)$, and the set of all non-negative real numbers $\mathbb{R}_+ = [0, \infty)$. In the first two cases and, more generally, when E is finite or countably infinite, X is said to be a discrete random variable.

(2.2) EXAMPLE.   Consider the experiment of flipping a coin once. The two possible outcomes are "Heads" and "Tails," that is, $\Omega = \{H, T\}$. Suppose X is defined by putting $X(H) = 1$, $X(T) = -1$. Then X is a random variable taking values in the set $E = \{1, -1\}$. We may think of it as the earning of the player who receives or loses a dollar according as the outcome is heads or tails.   □

(2.3) EXAMPLE.   Let an experiment consist of measuring the lifetimes of twelve electric bulbs. The sample space Ω is the set of all 12-tuples $\omega = (\omega_1, \ldots, \omega_{12})$ where $\omega_i \geq 0$ for all i. Then

$$X(\omega) = \frac{1}{12}(\omega_1 + \omega_2 + \cdots + \omega_{12})$$

defines a random variable on this sample space Ω. It represents the average lifetime of the 12 bulbs.   □

(2.4) EXAMPLE.   Suppose an experiment consists of observing the acceleration of a vehicle during the first 60 seconds of a race. Then each possible outcome is a real-valued right continuous function ω defined for $0 \leq t \leq$

60, and the sample space Ω is the set of all such functions ω. For $t \in [0, 60]$, let

$$X_t(\omega) = \omega(t),$$

$$Y_t(\omega) = \int_0^t \omega(s)\, ds,$$

$$Z_t(\omega) = \int_0^t Y_u(\omega)\, du = \int_0^t \int_0^u \omega(s)\, ds\, du,$$

for each $\omega \in \Omega$. Then X_t, Y_t, and Z_t are random variables on Ω. For the outcome ω, $X_t(\omega)$ is the acceleration at time t, $Y_t(\omega)$ the velocity, and $Z_t(\omega)$ the position. $\qquad\Box$

Let X be a random variable taking values in a set E, and let f be a real-valued function defined on the set E. Then for each $\omega \in \Omega$, $X(\omega)$ is a point in E and f assigns the value $f(X(\omega))$ to that point. By $f(X)$ we mean the random variable whose value at $\omega \in \Omega$ is $f(X(\omega))$. A particular function of some use is the *indicator function* I_B of a subset B of E; $I_B(x)$ is 1 or 0 according as $x \in B$ or $x \notin B$. Then $I_B(X)$ is a random variable which is equal to 1 if the event $\{X \in B\}$ occurs and is equal to 0 otherwise. Quite often there will be a number of random variables $X_1, \ldots, X_n$ and we will be concerned with functions of them. If $X_1, \ldots, X_n$ take values in E, and if f is a real-valued function f defined on $E \times \cdots \times E = E^n$, then $f(X_1, \ldots, X_n)$ is a real-valued random variable whose value at $\omega \in \Omega$ is $f(X_1(\omega), \ldots, X_n(\omega))$.

A *stochastic process with state space E* is a collection $\{X_t; t \in T\}$ of random variables X_t defined on the same probability space and taking values in E. The set T is called its *parameter set*. If T is countable, especially if $T = \mathbb{N} = \{0, 1, \ldots\}$, the process is said to be a *discrete parameter* process. Otherwise, if T is not countable, the process is said to have a *continuous parameter*. In the latter case the usual examples are $T = \mathbb{R}_+ = [0, \infty)$ and $T = [a, b] \subset \mathbb{R} = (-\infty, \infty)$. It is customary to think of the index t as representing time, and then one thinks of X_t as the "state" or the "position" of the process at time t.

(2.5) EXAMPLE. In Example (2.4) Y_t is the velocity of the vehicle at time t, and the collection $\{Y_t; 0 \le t \le 60\}$ is a continuous time-parameter stochastic process with state space $E = \mathbb{R}_+ = [0, \infty)$. Similarly for $\{Z_t; 0 \le t \le 60\}$. $\qquad\Box$

(2.6) EXAMPLE. Consider the process of arrivals of customers at a store, and suppose the experiment is set up to measure the interarrival times. Then the sample space Ω is the set of all sequences $\omega = (\omega_1, \omega_2, \ldots)$ of non-negative real numbers ω_i. For each $\omega \in \Omega$ and $t \in \mathbb{R}_+ = [0, \infty)$, we put $N_t(\omega) = k$ if and only if the integer k is such that $\omega_1 + \cdots + \omega_k \le t < \omega_1 + \cdots + \omega_{k+1}$ ($N_t(\omega) = 0$ if $t < \omega_1$). Then for the outcome ω, $N_t(\omega)$ is the number of arrivals in the time interval $(0, t]$. For each $t \in \mathbb{R}_+$, N_t is a random variable taking values in the set $E = \{0, 1, \ldots \}$. Thus, $\{N_t; t \in \mathbb{R}_+\}$ is a continuous time-parameter stochastic process with state space $E = \{0, 1, \ldots \}$. Considered as a function in t, for a fixed ω, the function $N_t(\omega)$ is non-decreasing, right continuous, and increases by jumps only; see Figure 1.2.1.   □

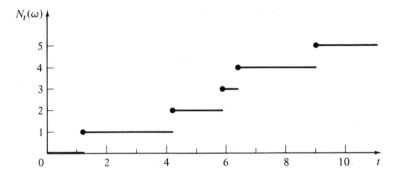

Figure 1.2.1 A possible realization of an arrival process. The picture is for the outcome $\omega = (1.2, 3.0, 1.7, 0.5, 2.6, \ldots)$.

Let X be a random variable taking values in $\mathbb{R} = (-\infty, \infty)$. If $b \in \mathbb{R}$, the set of all outcomes ω for which $X(\omega) \le b$ is an event, namely, the event $\{\omega: X(\omega) \le b\}$. We will write $\{X \le b\}$ for short, instead of $\{\omega: X(\omega) \le b\}$, and we will write $P\{X \le b\}$ for $P(\{X \le b\})$. If $a \le b$, then

$$(2.7) \qquad \{X \le a\} \subset \{X \le b\},$$

and Proposition (1.8) implies that

$$(2.8) \qquad P\{X \le a\} \le P\{X \le b\}.$$

Noting that

$$\{X \le a\} \cup \{a < X \le b\} = \{X \le b\}$$

and that the events $\{X \le a\}$ and $\{a < X \le b\}$ are disjoint, we get, by (1.7),

$$(2.9) \qquad P\{a < X \le b\} = P\{X \le b\} - P\{X \le a\}.$$

Next, note that $\Omega = \{X < +\infty\} = \bigcup_{n=1}^{\infty} \{X \le b_n\}$ for any sequence

$b_1, b_2, \ldots$ increasing to $+\infty$. Since $\{X \leq b_1\} \subset \{X \leq b_2\} \subset \cdots$ by (2.7), Proposition (1.11) applies, and we have

$$(2.10) \qquad \lim_{n \to \infty} P\{X \leq b_n\} = P\{X < \infty\} = 1.$$

Let $b_1, b_2, \ldots$ be a decreasing sequence with $\lim_n b_n = b$. Then $\{X \leq b_1\} \supset \{X \leq b_2\} \supset \cdots$ by (2.7), and $\bigcap_{n=1}^{\infty} \{X \leq b_n\} = \{X \leq b\}$. By Corollary (1.12), therefore,

$$(2.11) \qquad P\{X \leq b\} = \lim_{n \to \infty} P\{X \leq b_n\}.$$

In particular, if the b_n decrease to $-\infty$, the limit in (2.11) becomes zero.

The function φ defined by

$$(2.12) \qquad \varphi(b) = P\{X \leq b\}, \quad -\infty < b < +\infty,$$

is called the *distribution function* of the random variable X. If φ is a distribution function, then (also see Figure 1.2.2)

(2.13) (a) φ is non-decreasing by (2.8),
       (b) φ is right continuous by (2.11),
       (c) $\lim_{b \to \infty} \varphi(b) = 1$ by (2.10),
       (d) $\lim_{b \to -\infty} \varphi(b) = 0$ by (2.11) again.

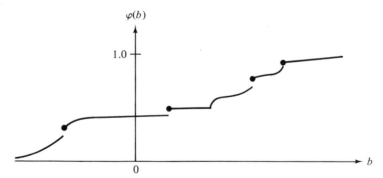

Figure 1.2.2 A distribution function is non-decreasing and right continuous and lies between 0 and 1.

In the opposite direction, if φ is any function defined on the real line such that (2.13a)–(2.13d) hold, then by taking

$$\Omega = (-\infty, +\infty)$$
$$P((-\infty, b]) = \varphi(b), \quad -\infty < b < +\infty,$$

and letting

$$X(\omega) = \omega, \quad \omega \in \Omega,$$

we see that X is a random variable with the distribution function φ. Hence, for any such function φ, there exists a random variable X which has φ as its distribution function.

Next, let X be a discrete random variable taking values in the (countable) set E. Then for any $a \in E$,

(2.14) $$\pi(a) = P\{X = a\}$$

is a non-negative number, and we must have

(2.15) $$\sum_{a \in E} \pi(a) = 1.$$

The collection $\{\pi(a): a \in E\}$ is called the *probability distribution* of X.

In the case of non-discrete X, it is sometimes possible to differentiate the distribution function. Then the derivative of the distribution function of X is called the *probability density function* of X.

(2.16) EXAMPLE. Consider Example (2.2). If the probability of "Heads" is 0.4, then

$$P(\varnothing) = 0, \quad P(\{H\}) = 0.4, \quad P(\{T\}) = 0.6, \quad P(\{H, T\}) = 1.$$

The random variable X defined there takes only two values: -1 and $+1$, and

$$P\{X = -1\} = 0.6 \qquad P\{X = 1\} = 0.4.$$

Then

$$\varphi(b) = P\{X \leq b\} = \begin{cases} 0 & b < -1, \\ 0.6 & -1 \leq b < 1, \\ 1 & 1 \leq b. \end{cases}$$

$\qquad\qquad\qquad\qquad\qquad\qquad\qquad\qquad\qquad\qquad\qquad\qquad\qquad\square$

(2.17) EXAMPLE. In simulation studies using computers, the following setup is utilized in "generating random variables from a given distribution φ."

A table of "random numbers" is a collection of numbers ω lying in the interval [0, 1] such that a number picked "at random" is in the interval $[a, b]$ with probability $b - a$. In our terminology, what this means is that we have a sample space $\Omega = [0, 1]$ and a probability measure P on Ω defined so that $P([a, b]) = b - a$ for all $0 \leq a \leq b \leq 1$. Then the event "the picked number ω is less than or equal to b," that is, the event $[0, b]$, has probability b.

Suppose the given distribution function φ is continuous and strictly increasing. Then for any $\omega \in \Omega = [0, 1]$, there is one and only one $a \in \mathbb{R}$ satisfying $\varphi(a) = \omega$. Therefore, setting

$$X(\omega) = a \quad \text{if and only if} \quad \varphi(a) = \omega$$

defines a function X from Ω into $\mathbb{R}$ (see Figure 1.2.3).

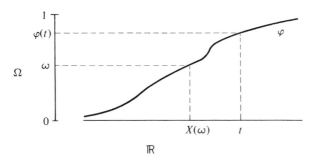

Figure 1.2.3 Defining a random variable X with a given distribution function φ.

For any $t \in \mathbb{R}$, $X(\omega) \leq t$ if and only if $\omega \leq \varphi(t)$, and consequently

$$P\{X \leq t\} = P(\{\omega : X(\omega) \leq t\}) = P([0, \varphi(t)]) = \varphi(t),$$

In other words, the random variable X we have defined has the given function φ as its distribution.

Hence, by picking a number ω at random from a table of "random numbers" and then computing $X(\omega)$ corresponding to ω from Figure 1.2.3, one obtains a possible value of a random variable X having the distribution function φ. □

(2.18) EXAMPLE. A mid-course trajectory correction calls for a velocity increase of 135 ft/sec. The spacecraft's engine provides a thrust which causes a constant acceleration of 15 ft/sec². Based on this, it is decided to fire the engine for 9 seconds. But the engine performance indicates that the actual length of burn time will be a random variable T with

$$P\{T \leq t\} = \begin{cases} 0 & t < 9, \\ 1 - e^{-5(t-9)} & t \geq 9. \end{cases}$$

What is the increase in velocity due to this burn?

Let Ω be the set of all possible burn times, i.e., $\Omega = [0, \infty)$, and define $T(\omega) = \omega$. Then, for the outcome ω, the velocity increase will be

$$X(\omega) = 15T(\omega) = 15\omega.$$

Thus,

$$P\{X \le b\} = P\{15T \le b\} = P\{T \le \tfrac{1}{15}b\}$$

$$= \begin{cases} 0 & \text{if } b < 135, \\ 1 - e^{-(b-135)/3} & \text{if } b \ge 135. \end{cases} \qquad \square$$

Suppose we have, defined on the sample space Ω, a number of random variables $X_1, \dots, X_n$ taking values in the countable set E. Then the probabilities of any events associated with $X_1, \dots, X_n$ can be computed (by using the results of Section 1) once their *joint distribution* is specified by giving

$$P\{X_1 = a_1, \dots, X_n = a_n\}$$

for all n-tuples $(a_1, \dots, a_n)$ with $a_i \in E$. In the case of random variables $X_1, \dots, X_n$ taking values in $\mathbb{R}$, the joint distribution is specified by giving

$$P\{X_1 \le b_1, \cdots, X_n \le b_n\}$$

for all numbers $b_1, \dots, b_n \in \mathbb{R}$. Specification of these probabilities themselves can be difficult at times. The concept we introduce next simplifies such tasks (when used properly).

(2.19) DEFINITION. The discrete random variables $X_1, \dots, X_n$ are said to be *independent* if

$$(2.20) \qquad P\{X_1 = a_1, \dots, X_n = a_n\} = P\{X_1 = a_1\} \cdots P\{X_n = a_n\}$$

for all $a_1, \dots, a_n \in E$. If the X_i take values in $\mathbb{R}$, they are said to be *independent* if

$$(2.21) \qquad P\{X_1 \le b_1, \dots, X_n \le b_n\} = P\{X_1 \le b_1\} \cdots P\{X_n \le b_n\}$$

for all $b_1, \dots, b_n \in \mathbb{R}$. An infinite collection $\{X_t; t \in T\}$ of random variables is called independent if any finite number of them are independent. $\qquad \square$

In particular, (2.21) implies, and is implied by, the condition that

$$(2.21') \qquad P\{X_1 \in B_1, \dots, X_n \in B_n\} = P\{X_1 \in B_1\} \cdots P\{X_n \in B_n\}$$

for all intervals $B_1, \dots, B_n \subset \mathbb{R}$.

We close this section by illustrating the concept through examples.

(2.22) EXAMPLE. Let X and Y be two discrete random variables taking values in $\{1, 2, 3, \dots\}$. Suppose

$$P\{X = m, Y = n\} = (0.64)(0.2)^{n+m-2}, \quad n, m = 1, 2, \dots.$$

For any $m = 1, 2, \ldots$, using Theorem (1.10) with $A = \{X = m\}$ and $B_i = \{Y = i\}$, we get†

$$P\{X = m\} = \sum_{i=1}^{\infty} P\{X = m, Y = i\}$$

$$= (0.64)(0.2)^{m-1} \sum_{i=1}^{\infty} (0.2)^{i-1}$$

$$= (0.8)(0.2)^{m-1}.$$

Similarly, for any $n \in \{1, 2, \ldots\}$, using (1.10) with $A = \{Y = n\}$ and $B_i = \{X = i\}$ now, we get

$$P\{Y = n\} = \sum_{i=1}^{\infty} P\{X = i, Y = n\} = (0.8)(0.2)^{n-1}.$$

Since

$$P\{X = m, Y = n\} = P\{X = m\}P\{Y = n\}$$

for all m and n, X and Y are independent.   □

(2.23) EXAMPLE. Let X and Y be two discrete random variables taking values in $\mathbb{N} = \{0, 1, 2, \ldots\}$ and having the joint distribution

$$P\{X = m, Y = n\} = \begin{cases} \dfrac{e^{-7}4^m 3^{n-m}}{m!(n-m)!} & \text{if } m = 0, 1, \ldots, n; n \in \mathbb{N}; \\ 0 & \text{otherwise.} \end{cases}$$

Then, using Theorem (1.10) with $A = \{X = m\}$ and $B_i = \{Y = i\}$, for any $m \in \mathbb{N}$, we have (see the footnote below)

(2.24) $$P\{X = m\} = \sum_{i=0}^{\infty} P\{X = m, Y = i\}$$

$$= \sum_{i=m}^{\infty} \frac{e^{-7}4^m 3^{i-m}}{m!(i-m)!} = \frac{e^{-4}4^m}{m!}.$$

Similarly, using Theorem (1.10) again, for any $n \in \mathbb{N}$,

$$P\{Y = n\} = \sum_{i=0}^{\infty} P\{X = i, Y = n\}$$

$$= \sum_{i=0}^{n} \frac{e^{-7}4^i 3^{n-i}}{i!(n-i)!}$$

$$= \frac{e^{-7}}{n!} \sum_{i=0}^{n} \frac{n!}{i!(n-i)!} 4^i 3^{n-i} = \frac{e^{-7}7^n}{n!}.$$

†As a reminder, $1 + x + x^2 + \cdots = 1/(1-x)$ for $x \in [0, 1)$. Also, in Example (2.23), $1 + x + x^2/2! + x^3/3! + \cdots = e^x$ for any $x \in \mathbb{R}$.

Since $P\{X = m, Y = n\} \neq P\{X = m\}P\{Y = n\}$, X and Y are *not* independent. $\qquad\square$

(2.25) EXAMPLE. Let X and Y be as in the preceding example. Then X and $Y - X$ are independent. To show this, we note that

$$P\{X = m, Y - X = k\} = P\{X = m, Y = m + k\} = \frac{e^{-7}4^m 3^k}{m!k!}$$

for all $m, k \in \mathbb{N}$ and that

$$P\{Y - X = k\} = \sum_{i=0}^{\infty} P\{X = i, Y - X = k\} = \frac{e^{-3}3^k}{k!}$$

for any $k \in \mathbb{N}$. Using (2.24) with these results, we see that

$$P\{X = m, Y - X = k\} = P\{X = m\}P\{Y - X = k\}$$

for all $m, k \in \mathbb{N}$, as claimed. $\qquad\square$

3. Conditional Probability

Let Ω be a sample space and P a probability measure on it.

(3.1) DEFINITION. Let A and B be two events. The *conditional probability of A given B*, written $P(A \mid B)$, is a number satisfying

(a) $0 \leq P(A \mid B) \leq 1$,
(b) $P(A \cap B) = P(A \mid B)P(B)$. $\qquad\square$

If $P(B) > 0$, then $P(A \mid B)$ is uniquely defined by (3.1b). Otherwise, if $P(B) = 0$, $P(A \mid B)$ can be taken to be any number in $[0, 1]$.

For fixed B with $P(B) > 0$, considered as a function of A, $P(A \mid B)$ satisfies the conditions (1.6) for a probability measure. That is,

(3.2) (a) $0 \leq P(A \mid B) \leq 1$,
       (b) $P(\Omega \mid B) = 1$,
       (c) $P(\bigcup_{i=1}^{\infty} A_i \mid B) = \sum_{i=1}^{\infty} P(A_i \mid B)$ provided the events $A_1, A_2, \ldots$
be disjoint; and thus Propositions (1.8)–(1.12) hold.

Heuristically, we think as follows. Suppose the outcome ω of the experiment is known to be in B, that is, B has occurred. Then event A can occur if and only if $\omega \in A \cap B$. And our estimate of the chances of A occurring given that B has occurred becomes the relative measure of $A \cap B$ with respect to B.

However, in practice, we usually have the various basic conditional probabilities specified, and our task then becomes the computation of other probabilities and conditional probabilities. The following proposition provides a simple tool for computing the probability of an event by conditioning on other events. It is sometimes referred to as the *theorem of total probability*.

(3.3) THEOREM. If $B_1, B_2, \ldots$ are disjoint events with $\bigcup_{i=1}^{\infty} B_i = \Omega$, then for any event A,

$$P(A) = \sum_{i=1}^{\infty} P(A \mid B_i) P(B_i).$$

Proof. By Theorem (1.10),

$$P(A) = \sum_{i=1}^{\infty} P(A \cap B_i);$$

and by (3.1b), $P(A \cap B_i) = P(A \mid B_i) P(B_i)$ for each i.   □

A simple consequence of this theorem is known as *Bayes' formula*.

(3.4) COROLLARY. If $B_1, B_2, \ldots$ are disjoint events with union Ω, then for any event A with $P(A) > 0$, and any j,

$$P(B_j \mid A) = \frac{P(A \mid B_j) P(B_j)}{\sum_{i=1}^{\infty} P(A \mid B_i) P(B_i)}.$$

Proof. Using (3.1b),

$$P(B_j \mid A) = \frac{P(A \cap B_j)}{P(A)}.$$

Writing $P(A)$ as the sum in Theorem (3.3) and using (3.1b) to write $P(A \cap B_j) = P(A \mid B_j) P(B_j)$ completes the proof.   □

(3.5) EXAMPLE. A coin is flipped until heads occur twice. Let X and Y denote, respectively, the trial numbers at which the first and the second heads are observed. If p is the probability of heads occurring at any one trial, then

$$P\{X = m, Y = n\} = \begin{cases} p^2 q^{n-2} & \text{if } m = 1, \ldots, n-1; n = 2, 3, \ldots; \\ 0 & \text{otherwise}; \end{cases}$$

where $q = 1 - p$, $0 < p < 1$. By Theorem (1.10), for any $m = 1, 2, \ldots$,

$$P\{X = m\} = \sum_i P\{X = m, Y = i\}$$

$$= \sum_{i=m+1}^{\infty} p^2 q^{i-2} = pq^{m-1}.$$

Similarly, for any $n = 2, 3, \ldots,$

$$P\{Y = n\} = \sum_i P\{X = i, Y = n\}$$

$$= \sum_{i=1}^{n-1} p^2 q^{n-2} = (n-1)p^2 q^{n-2}.$$

Thus, using Definition (3.1b),

$$(3.6) \qquad P\{X = m \mid Y = n\} = \frac{p^2 q^{n-2}}{(n-1)p^2 q^{n-2}} = \frac{1}{n-1}$$

for any $n = 2, 3, \ldots$ and $m = 1, 2, \ldots, n-1$. That is, if it is known that the second heads occurred at the nth trial, the first heads must have occurred during the first $n-1$ trials, and all these $n-1$ possibilities are equally likely.

We have, for any $m = 1, 2, \ldots, n-1$ and $n = 2, 3, \ldots,$

$$P\{Y = n \mid X = m\} = \frac{p^2 q^{n-2}}{pq^{m-1}} = pq^{(n-m)-1},$$

but a more instructive computation is the following. For $m = 1, 2, \ldots$ and $k = 1, 2, \ldots,$

$$(3.7) \qquad P\{Y - X = k \mid X = m\} = \frac{P\{X = m, Y - X = k\}}{P\{X = m\}}$$

$$= \frac{P\{X = m, Y = k + m\}}{P\{X = m\}}$$

$$= \frac{p^2 q^{k+m-2}}{pq^{m-1}} = pq^{k-1}.$$

There are two conclusions to be drawn from (3.7): $Y - X$ is independent of X, and $Y - X$ has the same distribution as X. $\qquad \square$

(3.8) EXAMPLE. Suppose the length of a telephone conversation between two ladies is a random variable X with distribution function

$$P\{X \le t\} = 1 - e^{-0.03t}, \quad t \ge 0,$$

where the time is measured in minutes. Given that the conversation has been going on for 30 minutes, let us compute the probability that it continues for at least another 20 minutes. That is, we want to compute $P\{X > 50 \mid X > 30\}$. Since the event $\{X > 50, X > 30\}$ is the same as the event $\{X > 50\}$, we have, by (3.1b),

$$P\{X > 50 \mid X > 30\} = \frac{P\{X > 50, X > 30\}}{P\{X > 30\}} = \frac{P\{X > 50\}}{P\{X > 30\}}.$$

But by (1.9), $P\{X > t\} = 1 - P\{X \leq t\} = e^{-0.03t}$ for any $t \geq 0$. So

$$P\{X > 50 \mid X > 30\} = \frac{e^{-1.5}}{e^{-0.9}} = e^{-0.6}.$$

Noting that $e^{-0.6} = P\{X > 20\}$, we have this interesting result: the probability that the conversation goes on another 20 minutes is independent of the fact that it has already lasted 30 minutes. Indeed, for any $t, s \geq 0$,

$$P\{X > t + s \mid X > t\} = \frac{P\{X > t + s, X > t\}}{P\{X > t\}} = \frac{P\{X > t + s\}}{P\{X > t\}}$$

$$= \frac{e^{-0.03(t+s)}}{e^{-0.03t}} = e^{-0.03s} = P\{X > s\}.$$

That is, the probability that the conversation continues another s units of time is independent of how long it has been going on. Or, at every instant t, the ladies' conversation starts afresh!   □

4. Exercises

(4.1) An experiment consists of drawing three flash bulbs from a lot and classifying each as defective (D) or non-defective (N). A drawing, then, can be described by a triplet; for example, (N, N, D) represents the outcome where the first and the second bulbs were found non-defective and the third defective. Let A denote the event "the first bulb drawn was defective," B the event "the second bulb drawn was defective," and C the event "the third bulb drawn was defective."
    (a) Describe the sample space by listing all possible outcomes.
    (b) List all outcomes in $A, B, B \cup C, A \cup C, A \cup B \cup C, A \cap B,$ $A^c \cap B^c \cap C, A \cap B^c \cap C, (A \cup B^c) \cap C, (A \cap C) \cup (B^c \cap C)$.

(4.2) Describe in detail the sample spaces for the following experiments:
    (a) Three tosses of a coin.
    (b) An infinite number of coin tosses.
    (c) Measurement of the speeds of cars passing a given point.
    (d) Scores of a class of 20 on an examination.
    (e) Measurement of noontime temperatures at a certain locality.
    (f) Observation of arrivals at a store.

(4.3) An experiment consists of firing a projectile at a target and observing the position of the point of impact. (Suppose the origin of the coordinate system is placed at the target.) Then, an outcome is a pair $\omega = (\omega_1, \omega_2)$, where ω_1 is the abscissa and ω_2 the ordinate of the point of impact. The sample space Ω consists of all such pairs ω. For each $\omega \in \Omega$, let

$$X(\omega) = |\omega_1|, \quad Y(\omega) = |\omega_2|, \quad Z(\omega) = \sqrt{\omega_1^2 + \omega_2^2}.$$

(a)   What do X, Y, Z stand for?
(b)   Suppose the probability measure P is such that

$$P(\{\omega : \omega_1 \leq a, \omega_2 \leq b\}) = \int_{-\infty}^{a} \int_{-\infty}^{b} \frac{1}{2\pi} \exp\left[-\frac{1}{2}(x^2 + y^2)\right] dx\, dy.$$

Then show that X and Y are independent random variables with the same distribution function

$$\varphi(t) = \int_{-t}^{t} \frac{1}{\sqrt{2\pi}} e^{-x^2/2} dx, \quad 0 \leq t < \infty.$$

(4.4)   Let X, Y be as defined in (2.22), and put $Z = X + Y$. Compute
(a) $P\{X = m, Z = k\}$ for $m = 5, k = 7$ first, and in general after that;
(b) $P\{X = 5 | Z = 7\}$;
(c) $P\{Z = 7 | X = 5\}$;
(d) $P\{Y = 3 | Z = 14\}$;
(e) $P\{X = 5, Y = 3 | Z = 8\}$;
(f) $P\{Z = 8 | X = 6, Y = 2\}$.

(4.5)   Let X be a discrete random variable with

$$P\{X = m\} = pq^{m-1}, \quad m = 1, 2, \ldots$$

where $p > 0$, $p + q = 1$. Show that for any $m, n \in \{1, 2, \ldots\}$,

$$P\{X > n + m | X > n\} = P\{X > m\}.$$

(4.6)   Let X and Y denote, respectively, the number of babies born on a certain day in a hospital and the number of them which are boys. Suppose their joint distribution is

$$P\{X = n, Y = m\} = \begin{cases} \dfrac{e^{-14}(7.14)^m(6.86)^{n-m}}{m!(n-m)!} & \text{if } m = 0, 1, \ldots, n; n = 0, 1, \ldots; \\ 0 & \text{otherwise.} \end{cases}$$

Find $P\{X = n\}$ and $P\{Y = m\}$ for all $m, n \in \{0, 1, \ldots\}$. Compute $P\{X = n | Y = m\}$, $P\{Y = m | X = n\}$, $P\{X - Y = k | X = n\}$, $P\{X - Y = k | Y = m\}$. Interpret the results.

(4.7)   Let A, B, and C be independent random variables with distributions indicated below:

$$P\{A = 1\} = 0.4, \quad P\{A = 2\} = 0.6,$$

$$P\{B = -3\} = 0.25, \quad P\{B = -2\} = 0.25, \quad P\{B = -1\} = 0.25, \quad P\{B = 1\} = 0.25,$$

$$P\{C = 1\} = 0.5, \quad P\{C = 2\} = 0.4, \quad P\{C = 3\} = 0.1.$$

What is the probability that

$$Ax^2 + Bx + C$$

has real roots?

(4.8) Reliability is the probability of a device performing its purpose adequately for the period of time intended under the operating conditions encountered.

A piece of equipment consists of three components in series: for the equipment to function, all three components must be functioning. Let X_1, X_2, and X_3 be the respective lifetimes of the components 1, 2, and 3 measured in hours. Suppose

$$P\{X_1 \le t\} = 1 - e^{-10^{-4}t}, \quad t \ge 0,$$
$$P\{X_2 \le t\} = 1 - e^{-2 \times 10^{-4}t}, \quad t \ge 0,$$
$$P\{X_3 \le t\} = 1 - e^{-3 \times 10^{-5}t}, \quad t \ge 0.$$

If the lifetimes of the components are independent, what is the reliability of the equipment in a mission requiring 4,000 hours?

(4.9) Let X be the lifetime of a component measured in hours, and let its distribution function be

$$P\{X \le t\} = 1 - e^{-0.2t} - 0.2te^{-0.2t}, \quad t \ge 0.$$

Given that the item has worked successfully for the first 3 hours, what is the probability that it does not fail within the next 1.5 hours?

In general, compute the conditional probability that $X > t + s$ given that $X > t$.

(4.10) A printing machine capable of printing any of n characters $a_1, \ldots, a_n$ is operated by electrical impulses, each character, in theory, being produced by a different impulse. Suppose the machine has probability p $(0 < p < 1)$ of producing the character corresponding to the impulse received, independent of past behavior. If it prints the wrong character, the probabilities that any of the $(n - 1)$ other characters will appear are equal.

     (a) Suppose that one of the n impulses is chosen at random and fed into the machine twice, and that the character a_i is printed both times. What is the probability that the impulse chosen is the one designed to produce a_i?

     (b) Suppose now that a_i was printed on the first trial and a_j $(j \ne i)$ on the second. What is the probability that the impulse was designed to produce a_i?

     (c) In (b), what is the probability that the impulse was designed to produce a_j?

     (d) In (b), suppose that a_i is printed on the first trial and that it is known only that some other character (not a_i) appeared on the second. Does this change the answer to (b)?

*   *   *

Then said they unto him, Say now Shibboleth: *and he said* Sibboleth: *for he could not frame to pronounce it right. Then they took him, and slew him.*

THE BIBLE, THE BOOK OF JUDGES 12:6

Failure to distinguish between an outcome and an event is unlikely to have as severe a consequence as the fate which befell the warrior challenged by the ancient man of Gilead. Yet, probabilistic reasoning is a part of our culture today, and an appreciation of it cannot be acquired without developing a precise feeling for its essential ingredients.

It is important to remember always that an event is a collection of outcomes, while a random variable is a function. A random variable assigns a value to each outcome; a probability measure assigns a value to each event. One talks of the probability of an event, never of the probability of an outcome. A certain amount of confusion is caused by the historical mistakes made while the subject was developing, and the insistence of certain teachers on repeating them.

The present axiomatic foundations of the theory were laid by KOLMOGOROV [1] in 1933. Since then, the progress in probability theory has been very rapid. This progress was especially aided by the discovery of unsuspected applications to pure mathematics on the one hand, and by an ever increasing demand from other scientists and engineers on the other.

CHAPTER 2

Expectations
and Independence

Consider a large piece of land whose area we take to be one unit. The land is divided into n lots which, for purposes of buying and selling, are indivisible. Let Ω denote the whole land and $P(A)$ the area of the region A. Let $X(\omega)$ denote the price per unit area of the lot containing the point ω. Then X takes only n values, say $b_1, b_2, \ldots, b_n$. The region on which X is equal to b_k has area $P\{X = b_k\}$, and hence, the value of the total land is

$$E[X] = \sum_{k=1}^{n} b_k P\{X = b_k\}.$$

Note that $E[X]$ is, in a sense, the *integral* of the function X over the set Ω, and since the total area of Ω is $P(\Omega) = 1$, $E[X]$ is also the average unit price.

If we think of Ω as a sample space and P as a probability measure (as we may), then X becomes a random variable. The integral of X that we obtain is then called the *expected value* of X. The justification for the term "expected value" lies in our interpretation of $E[X]$ as the average of X over Ω. This concept of integrating a random variable X over a sample space Ω with respect to a probability measure P extends also to arbitrary probability spaces. The present chapter is devoted to this and related concepts.

In Section 1 we give an account of expectation taking. Then in Section 2 we introduce conditional expectations and list many of their properties. The reader should study Section 1, and read Section 2 once or twice. He is urged not to dwell too long on Section 2 but to pass on to Chapters 3, 4, and 5 instead. In these later chapters there will be many opportunities to observe the workings of conditional expectations; by referring back to the cited theorems of this chapter, the reader will learn and appreciate them. Theorems on expectations and conditional expectations form the grammar of the language of probability, and are indispensable to anyone who wants to get acquainted with that language. But one does not start learning a language by memorizing the rules of grammar.

1. Expected Value

Let Ω be a sample space, P a probability measure, and X a discrete random variable defined on Ω. Let the values X takes be $b_0, b_1, b_2, \ldots \in \mathbb{R}_+ = [0, \infty)$, and put $B_n = \{\omega : X(\omega) = b_n\}$. Then $B_0, B_1, \ldots$ are disjoint and their union is Ω. The function X is equal to b_n on the set B_n whose measure is $P(B_n)$. So the integral of the function X with respect to the measure P is

$$(1.1) \qquad E[X] = b_0 P(B_0) + b_1 P(B_1) + b_2 P(B_2) + \cdots$$

(we allow it to be $+\infty$). (See Figure 2.1.1.) Note that the right hand side divided by $1 = P(\Omega)$ can also be looked upon as the weighted average of the

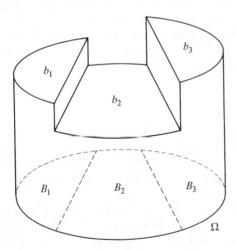

Figure 2.1.1 Expected value of a discrete random variable is the sum of its values weighted by the corresponding probabilities.

function X with respect to the weight distribution given by P. Replacing $P(B_n)$ by $P\{X = b_n\}$ in (1.1), we now make the following

(1.2) DEFINITION. The *expected value* of a discrete random variable X taking values in the set $E \subset \mathbb{R}_+$ is

$$E[X] = \sum_{a \in E} a P\{X = a\}. \qquad \square$$

The preceding defines the expected value of X when it is a discrete non-negative random variable. We extend this first to arbitrary non-negative random variables and then to arbitrary random variables.

Suppose X is a non-negative real-valued random variable. Then it is possible to find *discrete* random variables $X_1, X_2, \ldots$ such that

(1.3) $$X_1(\omega) \leq X_2(\omega) \leq \cdots$$

and

(1.4) $$\lim_{n \to \infty} X_n(\omega) = X(\omega)$$

for all ω except possibly for some $\omega \in \Omega_0$ with $P(\Omega_0) = 0$. Since each X_n is discrete, its expected value $E[X_n]$ is well defined by (1.2). By our interpretation of $E[X_n]$ as an integral it is easy to see that

$$E[X_1] \leq E[X_2] \leq \cdots,$$

and it seems reasonable to make the following

(1.5) DEFINITION. Let X be a non-negative random variable, and let $X_1, X_2, \ldots$ be discrete random variables satisfying (1.3) and (1.4). Then we define the *expected value* of X to be

$$E[X] = \lim_{n \to \infty} E[X_n] \leq +\infty. \qquad \square$$

Finally, if X is an arbitrary real-valued random variable (not necessarily non-negative), and if we define

(1.6) $$Y(\omega) = \begin{cases} 0 & \text{if } X(\omega) < 0, \\ X(\omega) & \text{if } X(\omega) \geq 0, \end{cases}$$

and

(1.7) $$Z(\omega) = \begin{cases} -X(\omega) & \text{if } X(\omega) < 0, \\ 0 & \text{if } X(\omega) \geq 0 \end{cases}$$

for all $\omega \in \Omega$, then both Y and Z are non-negative random variables, and

$$X = Y - Z.$$

Definition (1.5) provides the meanings for the expected values of Y and Z, and we now make the following

(1.8) DEFINITION. Let X be an arbitrary random variable with values in $\mathbb{R}$, and let Y and Z be defined by (1.6) and (1.7). Then

$$E[X] = E[Y] - E[Z]$$

provided that at least one of the numbers $E[Y]$ and $E[Z]$ is finite. If $E[Y] = E[Z] = +\infty$, then X is said to have *no expected value*.   □

Definitions (1.2) and (1.8) are quite workable, but (1.5) is not. In fact, we have not even settled the matter of nonambiguity. If $\{X_n\}$ is a sequence of discrete random variables increasing to X, and if $\{Y_n\}$ is another sequence of discrete random variables also increasing to X, then Definition (1.5) would put $E[X] = \lim_n E[X_n]$ and $E[X] = \lim_n E[Y_n]$. How do we know that these two numbers are the same? Indeed, they are the same, as the proof of the next theorem shows. As a by-product we obtain a nice computational formula.

(1.9) THEOREM.   For any non-negative random variable X,

(1.10) $$E[X] = \int_0^\infty P\{X > t\}\, dt.$$

Proof. First, suppose X is discrete with values in E. Then using Definition (1.2) and changing the order of summation and integration, we get (see Figure 2.1.2)

(1.11) $$E[X] = \sum_{a \in E} a P\{X = a\}$$

$$= \sum_{a \in E} \int_0^a dt\, P\{X = a\}$$

$$= \int_0^\infty dt \sum_{a > t} P\{X = a\} = \int_0^\infty P\{X > t\}\, dt.$$

This establishes (1.10) for X discrete.

Let X be an arbitrary non-negative random variable, and let $X_1, X_2, \ldots$ be discrete random variables increasing to X. Then (1.11) applies to each

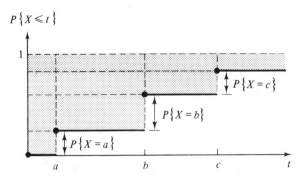

Figure 2.1.2   For non-negative discrete X, $E[X]$ is the shaded area no matter how that is sliced.

X_n, and we have

(1.12) $$E[X_n] = \int_0^\infty P\{X_n > t\}\, dt.$$

On the other hand, since the X_n increase to X, for any $t \geq 0$,

$$\{X_1 > t\} \subset \{X_2 > t\} \subset \cdots; \qquad \bigcup_{n=1}^\infty \{X_n > t\} = \{X > t\}.$$

Thus, Proposition (1.1.11) applies to give

(1.13) $$\lim_{n \to \infty} P\{X_n > t\} = P\{X > t\}.$$

It follows from (1.12) and (1.13) and the monotonicity of the convergence that, by Definition (1.5), we have

$$E[X] = \lim_{n \to \infty} E[X_n] = \lim_{n \to \infty} \int_0^\infty P\{X_n > t\}\, dt = \int_0^\infty P\{X > t\}\, dt.$$

This completes the proof. □

We note that, in Definition (1.5), the sequence chosen to approximate X has nothing to do with the value $E[X]$. Formula (1.10) is in general easy to use if the distribution of X is known. (See Figure 2.1.3). In the case of discrete random variables taking integer values $0, 1, 2, \ldots$, it reduces further to a simpler sum:

(1.14) COROLLARY. If X is a random variable taking values in $\mathbb{N} = \{0, 1, 2, \ldots\}$, then

$$E[X] = \sum_{n=0}^\infty P\{X > n\}.$$

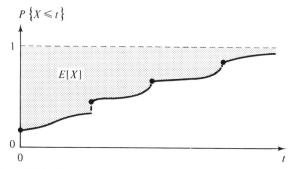

Figure 2.1.3 Expected value of a non-negative random variable is the shaded area lying above its distribution function.

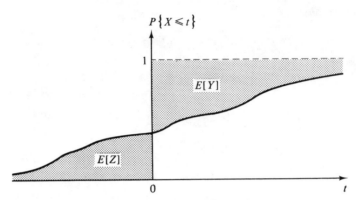

Figure 2.1.4 Expected value of a random variable X is the difference $E[Y] - E[X]$ of the two shaded areas.

In the case of arbitrary random variables, using Theorem (1.9) to compute $E[Y]$ and $E[Z]$ in Definition (1.8), we obtain (see Figure 2.1.4)

(1.15) COROLLARY.   For any real-valued random variable X,

$$E[X] = \int_0^\infty P\{X > t\} \, dt - \int_{-\infty}^0 P\{X \le t\} \, dt,$$

provided that at least one term on the right is finite.

In the formula given by (1.15), if we integrate by parts we obtain

(1.16) COROLLARY.   For any real-valued random variable X with distribution function φ,

$$E[X] = \int_{-\infty}^\infty t \, d\varphi(t),$$

provided the integral converges absolutely.

In computing a particular expectation, the choice of one formula over another is largely a matter of convenience. If there is a closed form expression for $P\{X > t\}$, in general, it is easier to use Theorem (1.9) and Corollary (1.15). Otherwise, it is easier to use Corollary (1.16) or its discrete equivalent, Definition (1.2).

(1.17) EXAMPLE.   The number of arrivals into a store during a specified time interval is a random variable X with

$$P\{X = n\} = \frac{e^{-8}8^n}{n!}, \quad n = 0, 1, \dots.$$

Then, using Definition (1.2),

$$E[X] = \sum_{n=0}^{\infty} n \frac{e^{-8}8^n}{n!}$$

$$= 8e^{-8} \sum_{n=1}^{\infty} \frac{8^{n-1}}{(n-1)!} = 8e^{-8}e^8 = 8. \qquad \square$$

(1.18) EXAMPLE. The lifetime X of an item has the distribution

$$P\{X \le t\} = 1 - e^{-0.02t}, \quad t \ge 0.$$

This is a non-negative random variable; it is easier to compute $E[X]$ by using Theorem (1.9). We obtain

$$E[X] = \int_0^{\infty} P\{X > t\}\, dt = \int_0^{\infty} e^{-0.02t}\, dt = \frac{1}{0.02} = 50. \qquad \square$$

(1.19) EXAMPLE. The intensity X of light falling on a certain surface has a distribution φ given by

$$d\varphi(t) = \frac{1}{\sqrt{2\pi\beta^2}} \exp\left[-\frac{1}{2\beta^2}(t-\alpha)^2\right] dt, \quad -\infty < t < \infty.$$

This distribution is called "the normal distribution with mean α and variance β^2." Using Corollary (1.16),

$$E[X] = \int_{-\infty}^{\infty} \frac{1}{\sqrt{2\pi\beta^2}}\, t \exp\left[-\frac{1}{2\beta^2}(t-\alpha)^2\right] dt = \alpha. \qquad \square$$

(1.20) EXAMPLE. A discrete random variable X has the distribution

$$P\{X = n\} = pq^{n-1}, \quad n = 1, 2, \ldots$$

where $p, q > 0, p + q = 1$. If we use Definition (1.2),†

$$E[X] = \sum_{n=1}^{\infty} npq^{n-1} = p \cdot \frac{1}{(1-q)^2} = \frac{1}{p}.$$

On the other hand, if we choose to use Corollary (1.14), we first compute

$$P\{X > n\} = \sum_{k=n+1}^{\infty} P\{X = k\} = \sum_{k=n+1}^{\infty} pq^{k-1} = q^n$$

†Note that, for $x \in (0, 1)$, $\sum_{n=0}^{\infty} x^n = 1/(1-x)$. If we differentiate both sides we get $\sum_{n=1}^{\infty} nx^{n-1} = 1/(1-x)^2$ and $\sum_{n=2}^{\infty} n(n-1)x^{n-2} = 2/(1-x)^3$.

for all $n \in \mathbb{N}$; then

$$E[X] = \sum_{n=0}^{\infty} q^n = \frac{1}{1-q} = \frac{1}{p}. \qquad \square$$

(1.21) EXAMPLE. A piece of equipment has two components whose life-times X and Y are independent random variables with distributions

$$P\{X \le t\} = 1 - e^{-2t}, \qquad P\{Y \le t\} = 1 - e^{-3t}, \quad t \ge 0.$$

The equipment fails if either one of the two components does, namely, the lifetime of the equipment is $Z = \min(X, Y)$. To compute $E[Z]$ we use Theorem (1.9). Now, $Z > t$ if and only if both $X > t$ and $Y > t$. So

$$P\{Z > t\} = P\{X > t, Y > t\}$$
$$= P\{X > t\}P\{Y > t\} = e^{-2t} \cdot e^{-3t} = e^{-5t},$$

where the second equality follows from the independence of X and Y (see Definition (1.2.21)). So

$$E[Z] = \int_0^{\infty} e^{-5t} \, dt = \frac{1}{5}. \qquad \square$$

If X is a random variable taking values in E, and if f is a function from E into $\mathbb{R}$, then $f(X)$ is a random variable with values in $\mathbb{R}$. Given the dis-tribution of X, one can obtain the distribution of $Y = f(X)$ and, using that, compute the expected value of Y by using the formulae of the preceding propositions. However, it is much easier to think of $E[Y]$ as the integral of the function Y with respect to P and obtain it as in Definition (1.2):

(**1.22**) PROPOSITION. Let X be a discrete random variable taking values in E, and let f be a function from E into $\mathbb{R}$. Then

$$E[f(X)] = \sum_{a \in E} f(a)P\{X = a\},$$

provided the sum is absolutely convergent.

Proof. The random variable $Y = f(X)$ takes the value $f(a)$ on the set $\{X = a\}$, whose measure is $P\{X = a\}$. The integral of Y therefore is $\sum f(a)$ $P\{X = a\}$ where the summation is over all $a \in E$. $\qquad \square$

In the case of arbitrary (instead of discrete) random variables, the same reasoning gives the following

(**1.23**) PROPOSITION. Let X be a random variable with values in E, and let

f be a function from E into $\mathbb{R}$. Then

$$E[f(X)] = \int_E f(t)\, d\varphi(t),$$

provided that the integral is absolutely convergent.

Proof is omitted. If instead we had a function of more than one random variable, the preceding two propositions become as follows. Again, we omit the proof.

(1.24) THEOREM. Let $X_1, \ldots, X_n$ be random variables taking values in E, and let f be a function from $E \times \cdots \times E = E^n$ into $\mathbb{R}_+$. Then

$$E[f(X_1, \ldots, X_n)] = \int_{E^n} f(t_1, \ldots, t_n)\, d\varphi(t_1, \ldots, t_n)$$

where φ is the joint distribution of $X_1, \ldots, X_n$. In case $X_1, \ldots X_n$ are discrete, this becomes

$$E[f(X_1, \ldots, X_n)] = \sum_{\mathbf{a}} f(a_1, \ldots, a_n) P\{X_1 = a_1, \ldots, X_n = a_n\}$$

where the summation is over all n-tuples $\mathbf{a} = (a_1, \ldots, a_n)$ with $a_i \in E$.

(1.25) COROLLARY.
 (a) $E[cX] = cE[X]$;
 (b) $E[X + Y] = E[X] + E[Y]$;
 (c) $E[c_1X_1 + \cdots + c_nX_n] = c_1E[X_1] + \cdots + c_nE[X_n]$.

Proof of (a) is immediate from Proposition (1.23), where we take $f(a) = ca$ and then use Corollary (1.16). Proof of (b) follows from (1.24) by taking $f(a, b) = a + b$, and (c) is immediate from (a) and (b). $\qquad \square$

We note that in the preceding corollary, we made no assumption of independence: whether or not the random variables are independent, the expected value of any linear combination of them is equal to the same linear combination of their expectations. The following is the analog for the case of multiplication.

(1.26) PROPOSITION. Let X and Y be two independent random variables taking values in E, and let g and h be two functions from E into $\mathbb{R}_+$. Then

$$E[g(X)h(Y)] = E[g(X)]E[h(Y)].$$

Proof for discrete X, Y. Put $f(a, b) = g(a)h(b)$ in Theorem (1.24). Then

$$E[g(X)h(Y)] = \sum_a \sum_b g(a)h(b)P\{X = a, Y = b\}.$$

But the independence of X and Y implies that

$$P\{X = a, Y = b\} = P\{X = a\}P\{Y = b\}$$

for any a and b. So

$$
\begin{aligned}
E[g(X)h(Y)] &= \sum_a \sum_b g(a)h(b)P\{X = a\}P\{Y = b\} \\
&= \sum_a g(a)P\{X = a\} \sum_b h(b)P\{Y = b\} \\
&= E[g(X)]E[h(Y)].
\end{aligned}
$$
$\square$

The assumption of independence in this proposition is crucial: if X and Y are *not* independent, then $E[g(X)h(Y)]$ might differ from $E[g(X)]E[h(Y)]$.

For certain special functions f, $E[f(X)]$ is given certain special names. In particular, if $f(b) = b^n$ then $E[f(X)] = E[X^n]$ is called the nth *moment* of X about the origin. If $f(b) = (b - \mu)^2$, where $\mu = E[X]$, then $E[f(X)] = E[(X - \mu)^2]$ is called the *variance* of X and is denoted by $\mathrm{Var}(X)$. If X is a non-negative integer-valued random variable, and if $f(b) = \alpha^b$ for some $\alpha \in [0, 1]$, then $E[f(X)] = E[\alpha^X]$ is a number between 0 and 1. Considered as a function of $\alpha \in [0, 1]$, $G(\alpha) = E[\alpha^X]$ is called the *generating function* of X. If X is a non-negative random variable, and if $f(b) = e^{-\alpha b}$ for some $\alpha \geq 0$, then $E[f(X)]$ is again a number between 0 and 1. Considered as a function of α, $F(\alpha) = E[e^{-\alpha X}]$ is called the *Laplace transform* of X.

The expected value of X is a rough guide to the value X is likely to be near. The variance of X measures the deviation of X from this likely value $E[X]$. If the variance is small, then X is more likely to be near $E[X]$. The following is an estimate that can be used when the distribution of X is not known. It is called *Chebyshev's inequality*.

(1.27) PROPOSITION. Let X be a random variable with expectation a and variance b^2. Then for any $\varepsilon > 0$,

$$P\{|X - a| > \varepsilon\} \leq \frac{b^2}{\varepsilon^2}.$$

Proof. Consider the expectation of the positive random variable $Y = (X - a)^2$; it is $E[Y] = b^2$. Now, $E[Y]$ is the integral of Y over all of Ω, and as such it is greater than the integral of Y on the set $\{Y > \varepsilon^2\}$. The measure of that set is $P\{Y > \varepsilon^2\}$, and $Y > \varepsilon^2$ on that set. So the integral must be greater than $\varepsilon^2 P\{Y > \varepsilon^2\}$. That is,

$$b^2 \geq \varepsilon^2 P\{Y > \varepsilon^2\},$$

from which the proposition follows.   □

In computing the variance it is usually worth noting that

(1.28) $$\begin{aligned} \text{Var}(X) &= E[(X - E[X])^2] \\ &= E[X^2 - 2XE[X] + (E[X])^2] \\ &= E[X^2] - (E[X])^2. \end{aligned}$$

Following are some examples of these computations.

(1.29) EXAMPLE. Consider the random variable X of Example (1.17). We had already computed $E[X] = 8$. Now, to obtain the variance we use the formula (1.28). To compute $E[X^2]$, note that it is easier to compute $E[X(X - 1)]$ first and then use $E[X^2] = E[X(X - 1)] + E[X]$. Now,

$$\begin{aligned} E[X(X - 1)] &= \sum_{n=0}^{\infty} n(n - 1)\frac{e^{-8}8^n}{n!} \\ &= e^{-8}8^2 \sum_{n=2}^{\infty} \frac{8^{n-2}}{(n - 2)!} = 8^2 e^{-8} e^8 = 64. \end{aligned}$$

Hence,

$$E[X^2] = 64 + 8,$$

and

$$\text{Var}(X) = E[X^2] - (E[X])^2 = 8.$$

Next we compute its generating function. We have

$$\begin{aligned} G(\alpha) &= E[\alpha^X] \\ &= \sum_{n=0}^{\infty} \alpha^n \frac{e^{-8}8^n}{n!} = e^{-8} \cdot e^{8\alpha} = e^{-8(1-\alpha)}, \end{aligned}$$

for any $\alpha \in [0, 1]$. Note that the derivative of G at $\alpha = 1$ is

$$G'(1) = 8e^{-8(1-\alpha)}|_{\alpha=1} = 8 = E[X],$$

whereas the second derivative of G at $\alpha = 1$ is

$$G''(1) = 64e^{-8(1-\alpha)}|_{\alpha=1} = 64 = E[X(X - 1)],$$

and the third derivative is

$$G'''(1) = 8^3 e^{-8(1-\alpha)}|_{\alpha=1} = 8^3 = E[X(X-1)(X-2)]. \qquad \square$$

(1.30)   EXAMPLE.   Consider the lifetime X of the item discussed in Example (1.18). Its expectation was $E[X] = 50$. Now,

$$E[X^2] = \int_0^\infty t^2 \, d\varphi(t)$$

$$= \int_0^\infty t^2 \, 0.02 e^{-0.02t} \, dt$$

$$= \int_0^\infty 2t \, e^{-0.02t} \, dt$$

$$= \frac{2}{0.02} \int_0^\infty t \, 0.02 e^{-0.02t} \, dt = 2(E[X])^2,$$

and hence

$$\mathrm{Var}(X) = E[X^2] - (E[X])^2 = (E[X])^2 = 2500.$$

Computing the Laplace transform of X, we find

$$F(\alpha) = E[e^{-\alpha X}] = \int_0^\infty e^{-\alpha t} \, 0.02 e^{-0.02t} \, dt$$

$$= \frac{0.02}{\alpha + 0.02} \int_0^\infty (\alpha + 0.02) e^{-(\alpha + 0.02)t} \, dt = \frac{0.02}{\alpha + 0.02}.$$

We note that the derivative of F at $\alpha = 0$ is

$$F'(0) = -\frac{0.02}{(\alpha + 0.02)^2}\bigg|_{\alpha=0} = -\frac{1}{0.02} = -E[X]$$

and that the second derivative at $\alpha = 0$ is

$$F''(0) = 2\frac{0.02}{(\alpha + 0.02)^3}\bigg|_{\alpha=0} = 2\frac{1}{0.02^2} = E[X^2]. \qquad \square$$

The results concerning the derivatives of the generating function in Example (1.29) hold in general: we have, for the generating function G of a non-negative integer-valued random variable X,

(1.31) $G^{(k)}(1) = E[X(X-1) \cdots (X - k + 1)]$

where $G^{(k)}$ is the kth derivative of G. Similarly, the results in Example (1.30) concerning the Laplace transform also generalize. For any non-negative random variable X with Laplace transform F,

(1.32) $$F^{(k)}(0) = (-1)^k E[X^k].$$

It is also worth mentioning that a generating function determines the probability distribution associated with it; this is true because

(1.33) $$G(\alpha) = \sum_{n=0}^{\infty} \alpha^n P\{X = n\}, \quad \alpha \in [0, 1],$$

which means that $P\{X = n\}$ is the coefficient of α^n in the power-series expansion of $G(\alpha)$. Similarly, the Laplace transform determines the associated distribution function.

We close this section with two theorems on the expected value of the limit of a sequence of random variables. The first is called the *monotone convergence theorem* and the second the *bounded convergence theorem*. The proof of the first is the same as that of (1.9), and we will not repeat it; we also omit the proof of the second.

(1.34) THEOREM. If $X_1, X_2, \ldots$ is a sequence of non-negative random variables increasing to the random variable X, then the expectations $E[X_1]$, $E[X_2], \ldots$ increase to $E[X]$.

(1.35) THEOREM. Let $X_1, X_2, \ldots$ be a sequence of random variables which are bounded in absolute value by a random variable Y such that $E[Y] < \infty$. If

$$\lim_{n \to \infty} X_n(\omega) = X(\omega)$$

for almost all $\omega \in \Omega$, then

$$\lim_{n \to \infty} E[X_n] = E[X].$$

2. Conditional Expectations

Let Y be a discrete random variable taking values in $\mathbb{R}_+$, and let A be an event with $P(A) > 0$. Then the conditional probability that $Y = b$ given the event A is (see (1.3.1))

(2.1) $$P\{Y = b \mid A\} = \frac{P(\{Y = b\} \cap A)}{P(A)}.$$

As b varies, this is called the *conditional distribution of Y given the event A*.

We define the *conditional expectation of Y given the event A* as

(2.2) $$E[Y|A] = \sum_b bP\{Y = b \,|\, A\}.$$

In particular, when $A = \{X = a\}$ for a discrete random variable X taking values in a set E,

(2.3) $$E[Y|X = a] = \sum_b bP\{Y = b \,|\, X = a\}$$

is called the *conditional expectation* of Y given that $X = a$. As a varies, (2.3) defines a function f on E by

(2.4) $$f(a) = E[Y|X = a].$$

By the *conditional expectation of Y given X*, written $E[Y|X]$, we mean the random variable $f(X)$; that is,

(2.5) $$E[Y|X] = f(X)$$

where f is as defined by (2.4). The following definition is the generalized version of this.

(2.6) DEFINITION. Let $X_1, \ldots, X_n$ be discrete random variables taking values in E, and let Y be a discrete random variable with values in $\mathbb{R}_+$. Then the conditional expectation of Y given $X_1, \ldots, X_n$ is

$$E[Y|X_1, \ldots, X_n] = f(X_1, \ldots, X_n)$$

where for any n-tuple $(a_1, \ldots, a_n)$ with $a_i \in E$,

$$f(a_1, \ldots, a_n) = \sum_b bP\{Y = b \,|\, X_1 = a_1, \ldots, X_n = a_n\}. \qquad \square$$

If Y is not discrete, then a similar definition is given in terms of its conditional distribution $P\{Y \le t \,|\, X_1 = a_1, \ldots, X_n = a_n\}$. For example, if Y is non-negative,

(2.7) $$E[Y|X_1, \ldots, X_n] = f(X_1, \ldots, X_n)$$

where

(2.8) $$f(a_1, \ldots, a_n) = \int_0^\infty P\{Y > t \,|\, X_1 = a_1, \ldots, X_n = a_n\} \, dt$$

for all $a_1, \ldots, a_n \in E$.

For any event A, its indicator function I_A (which is such that $I_A(\omega) = 1$ or 0 according as $\omega \in A$ or not) is a random variable. Then we define the *conditional probability of A given* $X_1, \ldots, X_n$ as

(2.9) $$P\{A|X_1, \ldots, X_n\} = E[I_A|X_1, \ldots, X_n].$$

The following are some easy properties of conditional expectations. These are analogous to Propositions (1.22), (1.23), (1.24), and (1.25). We omit the proofs.

(2.10) PROPOSITION. Let Y be a discrete random variable with values in E and g a function from E into $\mathbb{R}_+$. Then

$$E[g(Y)\,|\,X_1, \ldots, X_n] = \sum_{b \in E} g(b)P\{Y = b\,|\,X_1, \ldots, X_n\}.$$

(2.11) PROPOSITION. Let $Y_1, \ldots, Y_m$ be discrete random variables with values in E, and let g be a function from E^m into $\mathbb{R}_+$. Then

$$E[g(Y_1, \ldots, Y_m)\,|\,X_1, \ldots, X_n]$$
$$= \sum_{\mathbf{b}} g(b_1, \ldots, b_m)P\{Y_1 = b_1, \ldots, Y_m = b_m\,|\,X_1 \ldots, X_n\}.$$

(2.12) COROLLARY. If $Y_1, \ldots, Y_m$ take values in $\mathbb{R}_+$ and $c_1, \ldots, c_m$ are constants, then

$$E[c_1 Y_1 + \cdots + c_m Y_m\,|\,X_1, \ldots, X_n]$$
$$= c_1 E[Y_1\,|\,X_1, \ldots, X_n] + \cdots + c_m E[Y_m\,|\,X_1, \ldots, X_n].$$

(2.13) EXAMPLE. Let X and Y be two random variables with

$$P\{Y = 2\,|\,X = 1\} = 0.4 \qquad P\{Y = 3\,|\,X = 1\} = 0.6$$
$$P\{Y = 4\,|\,X = 2\} = 0.4 \qquad P\{Y = 9\,|\,X = 2\} = 0.6.$$

Let $f(b) = E[Y\,|\,X = b]$, $b = 1, 2$. Then

$$f(1) = 2(0.4) + 3(0.6) = 2^1(0.4) + 3^1(0.6) = 2.6,$$
$$f(2) = 4(0.4) + 9(0.6) = 2^2(0.4) + 3^2(0.6) = 7.$$

Thus,

$$E[Y\,|\,X] = f(X) = (0.4)2^X + (0.6)3^X. \qquad \square$$

(2.14) EXAMPLE. Consider three random variables X, Y, and Z with joint distribution

$$P\{X = k, Y = m, Z = n\} = p^3 q^{n-3}$$

for $k = 1, \ldots, m - 1$; $m = 2, \ldots, n - 1$; $n = 3, 4, \ldots$, where $0 < p < 1$, $p + q = 1$.

Then for $k = 1, \ldots, m - 1; m = 2, 3, \ldots$;

$$P\{X = k, Y = m\} = \sum_{i=m+1}^{\infty} p^3 q^{i-3} = p^2 q^{m-2}.$$

Thus, for $k = 1, \ldots, m - 1$ and $m = 2, \ldots, n - 1$, we have

$$P\{Z = n \,|\, X = k, Y = m\} = pq^{n-m-1}.$$

Hence, for $k = 1, \ldots, m - 1$ and $m = 2, 3, \ldots$,

$$E[Z \,|\, X = k, Y = m] = \sum_{n=m+1}^{\infty} npq^{n-m-1}$$

$$= \sum_{j=1}^{\infty} (j + m)pq^{j-1}$$

$$= m + \sum_{j=1}^{\infty} jpq^{j-1} = m + \frac{1}{p}.$$

Thus,

$$E[Z \,|\, X, Y] = Y + \frac{1}{p}.$$

We note that for any bounded function g,

$$E[g(Z) \,|\, X = k, Y = m] = \sum_{j=1}^{\infty} g(j + m)pq^{j-1},$$

so that

$$E[g(Z) \,|\, X, Y] = \sum_{j=1}^{\infty} g(Y + j)pq^{j-1}.$$

In particular, if $g(b) = \alpha^b$ for some $\alpha \in [0, 1]$, then

$$E[g(Z) \,|\, X, Y] = \frac{p\alpha}{1 - q\alpha} \alpha^Y. \qquad\qquad \square$$

The next proposition states that if the knowledge of $X_1, \ldots, X_n$ determines Y completely, then the conditional expectation of Y given $X_1, \ldots, X_n$ is equal to Y itself. The proof is very easy and we omit it.

(2.15) PROPOSITION. If Y can be written as

$$Y = f(X_1, \ldots, X_n)$$

for some function f, then

$$E[Y \,|\, X_1, \ldots, X_n] = Y.$$

Next is a very useful result used in situations where $E[Y|X_1, \ldots, X_n]$ is easy to obtain or known somehow. Since $E[Y|X_1, \ldots, X_n]$ is a random variable taking real values, we can talk about *its* expected value. That expected value is the same as the expectation of Y. In words, the expected value of any conditional expectation of Y is equal to the expected value of Y.

(2.16) PROPOSITION. $E[E[Y|X_1, \ldots, X_n]] = E[Y]$.

Proof for discrete $Y, X_1, \ldots, X_n$. Let

$$E[Y|X_1, \ldots, X_n] = f(X_1, \ldots, X_n);$$

then

$$(2.17) \quad E[E[Y|X_1, \ldots, X_n]] = E[f(X_1, \ldots, X_n)]$$
$$= \sum_{\mathbf{a}} f(a_1, \ldots, a_n) P\{X_1 = a_1, \ldots, X_n = a_n\}.$$

On the other hand,

$$(2.18) \quad f(a_1, \ldots, a_n) = \sum_b b P\{Y = b | X_1 = a_1, \ldots, X_n = a_n\}.$$

Putting (2.18) into (2.17) and changing the order of summation, noting Definition (1.3.1) of conditional probabilities, we obtain

$$E[E[Y|X_1, \ldots, X_n]]$$
$$= \sum_b b \sum_{\mathbf{a}} P\{Y = b | X_1 = a_1, \ldots, X_n = a_n\} P\{X_1 = a_1, \ldots, X_n = a_n\}$$
$$= \sum_b b \sum_{\mathbf{a}} P\{Y = b, X_1 = a_1, \ldots, X_n = a_n\}$$
$$= \sum_b b P\{Y = b\} = E[Y]. \qquad \square$$

The next result is very important in the theory of stochastic processes. It shows how to obtain the conditional expectation of Y given $X_1, \ldots, X_n$ when it is easy to obtain the same given $X_1, \ldots, X_n$ plus some extra information contained in $X_{n+1}, \ldots, X_{n+m}$.

(2.19) THEOREM. For any $n, m \geq 1$

$$E[E[Y|X_1, \ldots, X_{n+m}] | X_1, \ldots, X_n] = E[Y|X_1, \ldots, X_n].$$

Proof for $n = 2, m = 1, X_1, X_2, X_3, Y$ discrete. Let $Z = f(X_1, X_2, X_3) = E[Y|X_1, X_2, X_3]$. We need to show that

$$(2.20) \quad E[Z|X_1, X_2] = E[Y|X_1, X_2].$$

We have

$$E[Z|X_1 = a_1, X_2 = a_2] = E[f(X_1, X_2, X_3)|X_1 = a_1, X_2 = a_2]$$
$$= \sum_{a_3} f(a_1, a_2, a_3)P\{X_3 = a_3|X_1 = a_1, X_2 = a_2\}$$

and

$$f(a_1, a_2, a_3) = \sum_{b} bP\{Y = b|X_1 = a_1, X_2 = a_2, X_3 = a_3\}.$$

Putting the two computations together, we get

$$(2.21) \quad E[Z|X_1 = a_1, X_2 = a_2] = \sum_{b} bP\{Y = b|X_1 = a_1, X_2 = a_2\}$$
$$= E[Y|X_1 = a_1, X_2 = a_2],$$

since

$$\sum_{a_3} P\{Y = b|X_1 = a_1, X_2 = a_2, X_3 = a_3\}P\{X_3 = a_3|X_1 = a_1, X_2 = a_2\}$$
$$= \sum_{a_3} P\{Y = b, X_3 = a_3|X_1 = a_1, X_2 = a_2\}$$
$$= P\{Y = b|X_1 = a_1, X_2 = a_3\}$$

by (1.3.1) and (1.3.2). Noting that (2.21) is the same as (2.20) completes the proof.

(2.22) COROLLARY. If

$$E[Y|X_1, \ldots, X_n, \ldots, X_{n+m}] = g(X_1, \ldots, X_n),$$

then

$$E[Y|X_1, \ldots, X_n] = g(X_1, \ldots, X_n).$$

Proof. By Theorem (2.19) and Proposition (2.15),

$$E[Y|X_1, \ldots, X_n] = E[E[Y|X_1, \ldots, X_n, \ldots, X_{n+m}]|X_1, \ldots, X_n]$$
$$= E[g(X_1, \ldots, X_n)|X_1, \ldots, X_n]$$
$$= g(X_1, \ldots, X_n). \qquad \square$$

The preceding corollary has the following interpretation. Suppose, given $X_1, \ldots, X_{n+m}$, that the conditional expectation of Y depends only on $X_1, \ldots, X_n$. This means that, as far as predicting the value of Y is concerned, knowledge of $X_1, \ldots, X_n$ makes further knowledge concerning $X_{n+1}, \ldots, X_{n+m}$ irrelevant. Therefore, the conditional expectation of Y given $X_1, \ldots, X_n$ is the same as that of Y given $X_1, \ldots, X_{n+m}$.

A particular case of this happens to come up fairly often. Suppose we have $E[Y|X_1, \ldots, X_n]$ computed; and suppose $Y_1, \ldots, Y_m$ are functions of $X_1, \ldots, X_n$, that is, $Y_1 = g_1(X_1, \ldots, X_n), \ldots, Y_m = g_m(X_1, \ldots, X_n)$. Then $E[Y|X_1, \ldots, X_n, Y_1, \ldots Y_m] = E[Y|X_1, \ldots, X_n]$.

Another important concept is contained in the following theorem.

Suppose $\{Y_1, \ldots, Y_m\}$ and $\{X_1, \ldots, X_n\}$ are such that knowing the values of one set determines the values of the other. This is especially the case when $Y_1 = g_1(X_1, \ldots, X_n), \ldots, Y_m = g_m(X_1, \ldots, X_n)$ and conversely $X_1 = f_1(Y_1, \ldots, Y_m), \ldots, X_n = f_n(Y_1, \ldots, Y_m)$. Then for any random variable Y, the conditional expectation of Y given $X_1, \ldots, X_n$ is the same as the conditional expectation of Y given $Y_1, \ldots, Y_m$. This is so since $\{X_1, \ldots, X_n\}$ carries the same information as $\{Y_1, \ldots, Y_m\}$. The proof is easy and will be omitted.

(2.23) THEOREM. Suppose the collections $\{X_1, \ldots, X_n\}$ and $\{Y_1, \ldots, Y_m\}$ are such that the knowledge of the random variables in one collection determines the values of the random variables in the other. Then for any Y,

$$E[Y \mid X_1, \ldots, X_n] = E[Y \mid Y_1, \ldots, Y_m].$$

We close this section by giving an extension of the concept of independence.

(2.24) DEFINITION. The set of random variables $\{Y_1, \ldots, Y_m\}$ is said to be *independent* of $\{X_1, \ldots, X_n\}$ if

$$E[g(Y_1, \ldots, Y_m) \mid X_1, \ldots, X_n] = E[g(Y_1, \ldots, Y_m)]$$

for all non-negative functions g. Two stochastic processes $\{Y_t; t \in T_1\}$ and $\{X_t; t \in T_2\}$ are said to be *independent* of each other if any finite collection $\{Y_{t_1}, \ldots, Y_{t_m}\}$ from the first is independent of any finite collection $\{X_{s_1}, \ldots, X_{s_n}\}$ from the second.   □

We note that in a collection of random variables, independence in the sense of (1.2.19) is equivalent to the independence, in the sense of (2.24), of any two subcollections. As such, we will not distinguish between the two.

Next is a new concept, that of conditional independence.

(2.25) DEFINITION. $\{Y_1, \ldots, Y_m\}$ is said to be *conditionally independent of* $\{Z_1, \ldots, Z_k\}$ given $\{X_1, \ldots, X_n\}$ provided that

$$E[g(Y_1, \ldots, Y_m) \mid X_1, \ldots, X_n; Z_1, \ldots, Z_k] = E[g(Y_1, \ldots, Y_m) \mid X_1, \ldots, X_n]$$

for all non-negative functions g. The collection $\{Y_t; t \in T_1\}$ is said to be conditionally independent of the collection $\{Z_t; t \in T_2\}$ given the collection $\{X_t; t \in T_3\}$ provided that for any finite collection $\{Y_{t_1}, \ldots, Y_{t_m}\}$ from the first and any finite collection $\{Z_{s_1}, \ldots, Z_{s_n}\}$ from the second,

$$E[g(Y_{t_1}, \ldots, Y_{t_m}) \mid Z_{s_1}, \ldots, Z_{s_n}; X_t, t \in T_3]$$
$$= E[g(Y_{t_1}, \ldots, Y_{t_m}) \mid X_t, t \in T_3]$$

for all non-negative functions g.   □

In words, $\{Y_1, \ldots, Y_m\}$ is conditionally independent of $\{Z_1, \ldots, Z_k\}$ given $\{X_1, \ldots, X_n\}$ provided that, as far as predicting the value of any function of $Y_1, \ldots, Y_m$ is concerned, the extra knowledge provided by $Z_1, \ldots,$ Z_k loses all its significance once the values of $X_1, \ldots, X_n$ are known.

(2.26) EXAMPLE. Consider the random variables X, Y, Z of Example (2.14). We had shown that

$$E[g(Z) \mid X, Y] = \sum_{j=1}^{\infty} g(Y + j)pq^{j-1}.$$

The right hand side being independent of X, we see that Z is *conditionally independent of X given Y*. We also note that Z is *not independent* of X. $\quad\square$

(2.27) EXAMPLE. Let $X_1, X_2, \ldots$ be a sequence of random variables with $E[X_i] = \mu$ for all i. Let N be a non-negative integer-valued random variable independent of $X_1, X_2, \ldots$ with $E[N] = \lambda$. For each $\omega \in \Omega$, let

$$Y(\omega) = \begin{cases} 0 & \text{if } N(\omega) = 0 \\ X_1(\omega) + \cdots + X_n(\omega) & \text{if } N(\omega) = n. \end{cases}$$

We would like to compute $E[Y]$. We may think of $X_1, X_2, \ldots$ as the amounts spent by customers $1, 2, \ldots$ and of N as the number of arrivals within the first hour. Then Y is the total revenue within that hour.

By Proposition (2.16),

$$(2.28) \qquad\qquad E[Y] = E[E[Y \mid N]].$$

On the other hand, since N is independent of $X_1, X_2, \ldots$, for $n \geq 1$,

$$\begin{aligned} E[Y \mid N = n] &= E[X_1 + \cdots + X_n \mid N = n] \\ &= E[X_1 + \cdots + X_n] \\ &= E[X_1] + \cdots + E[X_n] = n\mu. \end{aligned}$$

Hence

$$E[Y \mid N] = N\mu,$$

and by (2.28)

$$E[Y] = E[\mu N] = \mu E[N] = \lambda\mu. \qquad\square$$

3.   Exercises

(3.1) Find the expected value of the random variable X taking the values $-5, 1,$ $4, 8, 10$ with probabilities $0.3, 0.2, 0.2, 0.1, 0.2$ respectively.

(3.2) Consider the random variable X taking the values $-2, 0, 2$ with probabilities $0.4, 0.3, 0.3$ respectively. Compute the expected values of X, X^2, $3X^2 + 5$.

(3.3) Compute the variance and the generating function of the random variable in Example (1.20).

(3.4) A random variable X is said to have the uniform distribution over $[a, b]$ if

$$P\{X \le t\} = \frac{t - a}{b - a}, \quad a \le t \le b.$$

    (a) Compute $E[X]$, $\text{Var}(X)$, $E[(X - a)/(b - a)]$.
    (b) Find the distribution of $Y = (X - a)/(b - a)$.

(3.5) Compute the variance and the Laplace transform of the lifetime in Example (1.21).

(3.6) Compute the variance of the intensity of light in Example (1.19).

(3.7) The headway X between two vehicles at a fixed instant is a random variable with

$$P\{X \le t\} = 1 - 0.6e^{-0.02t} - 0.4e^{-0.03t}, \quad t \ge 0.$$

Find the expected value and the variance of the headway.

(3.8) Show that for any constants a and b,

$$\text{Var}(aX + b) = a^2 \text{Var}(X)$$

for any random variable X.

(3.9) Show that for any two independent random variables X and Y,

$$\text{Var}(X + Y) = \text{Var}(X) + \text{Var}(Y).$$

(3.10) The lifetime X of a device has the distribution

$$P\{X \le t\} = 1 - e^{-ct}, \quad t \ge 0.$$

    (a) Show that $E[X] = 1/c$.
    (b) Show that (see also Example (1.3.8))

$$E[X \mid X > t] = t + \frac{1}{c}.$$

(3.11) Let X, Y be as defined in Example (1.2.23).
    (a) Compute $E[X]$, $E[Y - X]$, $E[Y]$.
    (b) Find $E[Y - X \mid X]$, $E[Y \mid X]$.
    (c) Show by direct computation that $E[E[Y \mid X]] = E[Y]$.

(3.12) Suppose $X_1, X_2, \ldots$ are independent and identically distributed random variables with

$$E[X_n] = a, \quad \text{Var}(X_n) = b^2.$$

Let N be a non-negative integer-valued random variable which is independent of $\{X_1, X_2, \ldots\}$, and let

$$E[N] = c, \qquad \text{Var}(N) = d^2.$$

Let $S_0 = 0$, $S_1 = X_1$, $S_2 = X_1 + X_2, \ldots$, and let $Y = S_N$.

    (a)   Compute $E[Y|N]$, $E[Y^2|N]$.

    (b)   Compute $E[Y]$, $E[Y^2]$, $\text{Var}(Y)$.

    (c)   Show that for any $\alpha \in \mathbb{R}_+$,

$$E[e^{-\alpha Y}] = G(F(\alpha))$$

where

$$F(\alpha) = E[e^{-\alpha X_n}], \qquad G(\beta) = E[\beta^N]; \quad \alpha \in \mathbb{R}_+, \beta \in [0, 1].$$

*   *   *

The chapter just finished completes our account of the preliminaries necessary for studying stochastic processes. Especially in view of the monstrous looks of the last section, it seems all too advisable to inquire how the reader's patience is holding out and to assure him that he will in time come to appreciate the true friendliness of these concepts. For a deeper treatment and for the proofs which we have omitted, we refer the reader to CHUNG *[2].*

CHAPTER 3

Bernoulli Processes
and Sums of Independent
Random Variables

Consider an experiment consisting of an infinite sequence of trials. Suppose that the trials are independent of each other and that each one has only two possible outcomes: "success" and "failure." A possible outcome of such an experiment is $(S, F, F, S, F, S, \ldots)$, which stands for the outcome where the first trial resulted in success, the second and third in failure, fourth in success, fifth in failure, sixth in success, etc. Thus, the sample space Ω of the experiment consists of all sequences of two letters S and F, that is,

$$\Omega = \{\omega : \omega = (\omega_1, \omega_2, \ldots), \omega_i \text{ is either } S \text{ or } F\}.$$

We next describe a probability measure P on all subsets of Ω. Let $0 \leq p \leq 1$, and define $q = 1 - p$. We think of p and q as the probabilities of success and failure at any one trial. The probability of the event $\{\omega : \omega_1 = S, \omega_2 = S, \omega_3 = F\}$ should then be $p \cdot p \cdot q = p^2 q$, and the probability of the event $\{\omega : \omega_1 = S, \omega_2 = F, \omega_3 = F, \omega_4 = F, \omega_5 = S\}$ should be $p \cdot q \cdot q \cdot q \cdot p = p^2 q^3$. For each n, we consider all events which are specified by the first n trials and define their probabilities in this manner. This, in addition to the conditions in Definition (1.1.6), completely specifies the probability P.

For each $\omega \in \Omega$ and $n \in \{1, 2, \ldots\}$, define $X_n(\omega) = 1$ or 0 according as $\omega_n = S$ or $\omega_n = F$. Then for each n, we have a random variable X_n whose only values are 1 and 0. It follows from the description of P that $X_1, X_2, \ldots$ are independent and identically distributed with $P\{X_n = 1\} = p$, $P\{X_n = 0\} = q$. In the first three sections of this chapter we will be interested in the properties of stochastic processes such as $\{X_n; n = 1, 2, \ldots\}$ and other processes defined in terms of $\{X_n\}$.

The process $\{X_n\}$ is very simple in nature, and the *answers* to most problems raised here are easy to obtain. Our object here is to demonstrate some of the *questions* that are raised in studying a stochastic process, and to provide some experience in using the tools of Chapters 1 and 2.

The processes of Sections 2 and 3 provide two examples of sums of independent and identically distributed random variables. We collect results of a more general nature about such processes in Section 4 along with certain classical limit theorems.

1. Bernoulli Processes

Let Ω be a sample space and P a probability measure on Ω. Let $\{X_n; n = 1, 2, \ldots\}$ be a sequence of random variables defined on Ω and taking only the two values 0 and 1.

(1.1) DEFINITION. The stochastic process $\{X_n; n = 1, 2, \ldots\}$ is called a *Bernoulli process* with success probability p provided that
   (a) $X_1, X_2, \ldots$ are independent, and
   (b) $P\{X_n = 1\} = p$, $P\{X_n = 0\} = q = 1 - p$ for all n.   □

(1.2) EXAMPLE. Finished products coming off an assembly line are given a routine inspection. If the nth item is "defective" we put $X_n = 1$, otherwise $X_n = 0$. If the production process is "under control," the successive items produced will show only random chance effects, and $X_1, X_2, \ldots$ will be independent. If, further, the defective rate p remains constant over time, then $\{X_n; n = 1, 2, \ldots\}$ will be a Bernoulli process.   □

(1.3) EXAMPLE. At a certain fork on a road, about 62 percent of the vehicles turn left. We define X_n to be 1 or 0 according as the nth vehicle turns left or right. The random variables $X_1, X_2, \ldots$ are independent of each other if the drivers act independently of each other in choosing their directions to turn. Then $\{X_n; n = 1, 2, \ldots\}$ is a Bernoulli process with probability 0.62 for success.   □

(1.4) EXAMPLE. Diameters of bearings coming off a production line are measured, and those that do not meet the specifications are rejected. Let Y_1, $Y_2, \ldots$ be the diameters in inches of the first, second, $\ldots$ bearings. Let $a = 2.994$ and $b = 3.006$ be the lower and the upper tolerance limits. Thus, the nth bearing is not rejected if and only if $2.994 \le Y_n \le 3.006$.

Suppose $Y_1, Y_2, \ldots$ are independent and identically distributed with the common distribution function φ given by

$$d\varphi(t) = \frac{1}{\sqrt{8\pi \times 10^{-6}}} \exp\left[-\frac{1}{8 \times 10^{-6}}(t - 3)^2\right] dt, \quad -\infty < t < \infty$$

(this is the normal distribution with mean 3 and variance $(0.002)^2$). For each n let

$$X_n = I_B(Y_n)$$

where $B = [2.994, 3.006]$. Then X_n is 1 if the nth bearing meets the specifications and is 0 otherwise. Since $Y_1, Y_2, \ldots$ are independent, $X_1, X_2, \ldots$ are also independent; and since the Y_n have a common distribution, the X_n also have a common distribution. Hence, $\{X_n; n = 1, 2, \ldots\}$ is a Bernoulli process. The probability $p = P\{X_n = 1\}$ is now computed as

$$
\begin{aligned}
p &= P\{I_B(Y_n) = 1\} \\
&= P\{2.994 \le Y_n \le 3.006\} \\
&= \int_{2.994}^{3.006} \frac{1}{\sqrt{8\pi \times 10^{-6}}} \exp\left[-\frac{1}{8 \times 10^{-6}}(t - 3)^2\right] dt \\
&= \int_{-3}^{3} \frac{1}{\sqrt{2\pi}} \exp\left[-\frac{1}{2}x^2\right] dx = 0.9974,
\end{aligned}
$$

where the table for the standard normal distribution is used to obtain the last number (see Figure 3.4.3 on page 65).   □

Let $\{X_n; n = 1, 2, \ldots\}$ be a Bernoulli process with success probability p. Then for any n,

$$(1.5) \qquad E[X_n] = E[X_n^2] = E[X_n^3] = \cdots = p,$$

$$(1.6) \qquad \mathrm{Var}(X_n) = E[X_n^2] - E[X_n]^2 = p - p^2 = pq,$$

$$(1.7) \qquad E[\alpha^{X_n}] = \alpha^0 P\{X_n = 0\} + \alpha P\{X_n = 1\} = q + \alpha p$$

for any $\alpha \ge 0$.

2. Numbers of Successes

Let $\{X_n; n = 1, 2, \ldots\}$ be a Bernoulli process with probability of success p. We think of X_n as the number of successes at the nth trial. Define

$$(2.1) \qquad N_n(\omega) = \begin{cases} 0 & \text{if } n = 0, \\ X_1(\omega) + \cdots + X_n(\omega) & \text{if } n = 1, 2, \ldots \end{cases}$$

for each $\omega \in \Omega$. Then N_n is the number of successes in the first n trials, and $N_{n+m} - N_n$ is the number of successes in the trials numbered $n + 1, n + 2, \ldots, n + m$. In this section we are interested in the stochastic process

$\{N_n; n \in \mathbb{N}\}$ whose time parameter is discrete $\mathbb{N} = \{0, 1, \ldots\}$ and whose state space is discrete and again is $\{0, 1, \ldots\}$.

We start with simple descriptive quantities. Using the definition of N_n, Corollary (2.1.25), and the result (1.5) above, we obtain

$$(2.2) \qquad E[N_n] = E[X_1 + \cdots + X_n]$$
$$= E[X_1] + \cdots + E[X_n] = np$$

for all $n \geq 1$, and clearly, $E[N_0] = 0$, so $E[N_n] = np$ for all $n \geq 0$. Similarly, using the result in Exercise (2.3.9) and the independence of $X_1, \ldots, X_n$,

$$(2.3) \qquad \mathrm{Var}(N_n) = \mathrm{Var}(X_1) + \cdots + \mathrm{Var}(X_n) = npq$$

by (1.6). From (2.2) and (2.3) we can get, by using the formula (2.1.28) for the variance,

$$(2.4) \qquad E[N_n^2] = \mathrm{Var}(N_n) + (E[N_n])^2 = npq + n^2p^2.$$

We could have computed $E[N_n^2]$ directly by noting that

$$N_n^2 = \sum_{i=1}^{n} X_i^2 + \sum_{i=1}^{n} \sum_{j \neq i} X_i X_j$$

so that

$$E[N_n^2] = \sum_{i=1}^{n} E[X_i^2] + \sum_{i=1}^{n} \sum_{j \neq i} E[X_i X_j]$$
$$= np + n(n-1)p^2 = n^2p^2 + npq.$$

This last computation indicates how to go about obtaining the higher moments of N_n.

Note that we have computed the expected value and variance without knowing the distribution of N_n. However, if one needs to compute $E[g(N_n)]$ for arbitrary g, then we need the distribution of N_n. This we do next. The use of conditional probabilities in the proof below illustrates one of the main techniques found useful in studying stochastic processes.

(2.5) LEMMA.   For any $n, k \in \mathbb{N}$

$$P\{N_{n+1} = k\} = pP\{N_n = k - 1\} + qP\{N_n = k\}.$$

Proof. Fix n and k. Using Theorem (1.3.3) with $A = \{N_{n+1} = k\}$ and $B_j = \{N_n = j\}$, we get

$$(2.6) \qquad P\{N_{n+1} = k\} = \sum_j P\{N_{n+1} = k \mid N_n = j\} P\{N_n = j\}.$$

Since $\{X_1, \ldots, X_n\}$ is independent of X_{n+1}, by Definition (2.2.24), $N_n = X_1 + \cdots + X_n$ is independent of X_{n+1}. Using this fact along with $N_{n+1} = N_n + X_{n+1}$, we obtain

$$P\{N_{n+1} = k \mid N_n = j\} = P\{X_{n+1} = k - j\} = \begin{cases} p & \text{if } j = k - 1, \\ q & \text{if } j = k, \\ 0 & \text{otherwise.} \end{cases}$$

Putting this in (2.6), we obtain the desired result. $\qquad\qquad\qquad\square$

For $n = 0$, $N_0 = 0$ and therefore $P\{N_0 = 0\} = 1$, and $P\{N_0 = k\} = 0$ for $k \neq 0$. These, along with Lemma (2.5), specify the values of $P\{N_1 = k\}$. The values of $P\{N_1 = k\}$ found, along with Lemma (2.5), give the values of $P\{N_2 = k\}$, etc. Doing this a few steps, we obtain the table shown in Figure 3.2.1. From the table we recognize the nth row as the terms in the binomial expansion of $(q + p)^n$. Then we prove this conjecture.

	$k = 0$	$k = 1$	$k = 2$	$k = 3$	$k = 4$	...
$n = 0$:	1	0	0	0	0	
$n = 1$:	q	p	0	0	0	
$n = 2$:	q^2	$2qp$	p^2	0	0	
$n = 3$:	q^3	$3q^2 p$	$3qp^2$	p^3	0	
$n = 4$:	q^4	$4q^3 p$	$6q^2 p^2$	$4qp^3$	p^4	

Figure 3.2.1   Using the probabilities $P\{N_n = k\}$ in one row, one obtains the values $P\{N_{n+1} = k\}$ in the next row.

(2.7) THEOREM.   For any $n \in \mathbb{N}$

$$P\{N_n = k\} = \frac{n!}{k!\,(n - k)!}p^k q^{n-k}, \quad k = 0, \ldots, n.$$

Proof. For $n = 0$, the claim is true, since $N_0 = 0$. We now make the induction hypothesis that the formula is true for $n = m$ and all k. To complete the proof we need to show that, then, the claim is also true for $n = m + 1$ and all k. For $n = m + 1$, $k = 0$, Lemma (2.5) and the induction hypothesis give

$$P\{N_{m+1} = 0\} = p \cdot 0 + q \cdot q^m = q^{m+1}$$

as claimed. For $n = m + 1$ and $0 < k \leq m + 1$, again by Lemma (2.5) and the induction hypthesis, we have

$$P\{N_{m+1} = k\} = p \cdot \frac{m!}{(k-1)!\,(m-k+1)!} p^{k-1} q^{m-k+1}$$

$$+ q \cdot \frac{m!}{k!\,(m-k)!} p^k q^{m-k}$$

$$= \frac{m!}{(k-1)!\,(m-k)!} p^k q^{m-k+1} \left[\frac{1}{m-k+1} + \frac{1}{k} \right]$$

$$= \frac{(m+1)!}{k!\,(m+1-k)!} p^k q^{m+1-k};$$

so the claim is again true. □

We have now specified the distribution of the number of successes in the first n trials. Noting that $N_{m+n} - N_m = X_{m+1} + \cdots + X_{m+n}$ also is the sum of n independent and identically distributed Bernoulli variables, we deduce that the distribution of $N_{m+n} - N_m$ is the same as that of N_n. This we state as a

(2.8) COROLLARY. For any $m, n \in \mathbb{N}$

$$P\{N_{m+n} - N_m = k\} = \frac{n!}{k!\,(n-k)!} p^k q^{n-k}, \quad k = 0, \ldots, n,$$

independent of m.

The distribution figuring above is called the *binomial distribution*. As we had noted before, $P\{N_n = k\}$ is the kth term in the binomial expansion of $(q + p)^n$. The coefficient of the kth term is usually denoted by $\binom{n}{k}$ (read n above k), that is,

(2.9)
$$\binom{n}{k} = \frac{n!}{k!\,(n-k)!}.$$

(2.10) EXAMPLE. Using Corollary (2.8), we obtain

$$P\{N_5 = 4\} = \binom{5}{4} p^4 q = 5 p^4 q,$$

$$P\{N_{13} - N_7 = 3\} = \binom{6}{3} p^3 q^3 = 20 p^3 q^3.$$
□

So the corollary above enables us to compute the distribution of $N_{m+n} - N_m$ for all n, m. But we still are not able to compute compound probabilities such as $P\{N_5 = 4, N_7 = 5, N_{13} = 8\}$. The means for doing that is furnished by the following

(2.11) THEOREM. For any $m, n \in \mathbb{N}$

$$P\{N_{m+n} - N_m = k \mid N_0, \ldots, N_m\} = P\{N_{m+n} - N_m = k\} = \binom{n}{k} p^k q^{n-k}$$

for $k = 0, \ldots, n$.

(2.12) EXAMPLE. To illustrate the use of this theorem, suppose we are interested in evaluating the joint probability

$$P\{N_5 = 4, N_7 = 5, N_{13} = 8\}.$$

The event whose probability is desired is equal to the event $\{N_5 = 4, N_7 - N_5 = 1, N_{13} - N_7 = 3\}$. By Theorem (2.11), the random variable $N_{13} - N_7$ is independent of $N_0, \ldots, N_7$. Thus,

$$P\{N_5 = 4, N_7 - N_5 = 1, N_{13} - N_7 = 3\}$$
$$- P\{N_{13} - N_7 = 3\} P\{N_5 = 4, N_7 - N_5 = 1\}.$$

Again by Theorem (2.11), $N_7 - N_5$ is independent of $N_0, \ldots, N_5$ and therefore of N_5 in particular. Hence,

$$P\{N_5 = 4, N_7 - N_5 = 1\} = P\{N_5 = 4\} \, P\{N_7 - N_5 = 1\}.$$

Putting all these together,

$$P\{N_5 = 4, N_7 = 5, N_{13} = 8\}$$
$$= P\{N_5 = 4\} P\{N_7 - N_5 = 1\} P\{N_{13} - N_7 = 3\}$$
$$= \binom{5}{4} p^4 q \binom{2}{1} pq \binom{6}{3} p^3 q^3 = 200 p^8 q^5,$$

where the last line followed from Corollary (2.8), which was also put in Theorem (2.11).

Finally, suppose we want to compute $E[N_5 N_8]$. Writing $N_8 = N_5 + (N_8 - N_5)$, we have

$$E[N_5 N_8] = E[N_5(N_5 + (N_8 - N_5))]$$
$$= E[N_5^2 + N_5(N_8 - N_5)]$$
$$= E[N_5^2] + E[N_5]E[N_8 - N_5],$$

where to get the last line we first used Corollary (2.1.25), and then used the independence of N_5 and $N_8 - N_5$ (which is implied by Theorem (2.11) above) to apply Proposition (2.1.26) on the expectation of the products of independent random variables. Now using (2.2) and (2.3), we have

$$E[N_5 N_8] = 25p^2 + 5pq + 5p \cdot 3p = 40p^2 + 5pq. \qquad \square$$

Proof of Theorem (2.11). The random variables $N_0, \ldots, N_m$ are completely determined by $X_1, \ldots, X_m$ through (2.1), and conversely, $N_0, \ldots, N_m$ completely determine $X_1, \ldots, X_m$ through $X_1 = N_1, X_2 = N_2 - N_1, \ldots,$ $X_m = N_m - N_{m-1}$. Hence, the knowledge of $\{N_0, \ldots, N_m\}$ determines $\{X_1, \ldots, X_m\}$ and vice versa. By Theorem (2.2.23), then,

$$P\{N_{m+n} - N_m = k \,|\, N_0, \ldots, N_m\} = P\{N_{m+n} - N_m = k \,|\, X_1, \ldots, X_m\}.$$

On the other hand, $N_{m+n} - N_m = X_{m+1} + \cdots + X_{m+n}$, and $\{X_{m+1}, \ldots,$ $X_{m+n}\}$ is independent of $\{X_1, \ldots, X_m\}$. Hence, Definition (2.2.24) gives (note that, here, $g(a, \ldots, b) = I_{\{k\}}(a + \cdots + b)$)

$$P\{N_{m+n} - N_m = k \,|\, X_1, \ldots, X_m\} = P\{N_{m+n} - N_m = k\}.$$

The last equality of the theorem follows now from Corollary (2.8). $\qquad \square$

We restate Theorem (2.11) in a slightly different wording:

(2.13) COROLLARY.   Let $n_0 = 0 < n_1 < n_2 < \cdots < n_j$ be integers. Then the random variables $N_{n_1} - N_{n_0}, N_{n_2} - N_{n_1}, \ldots, N_{n_j} - N_{n_{j-1}}$ are independent.

Theorem (2.11) states that the increase in the value of the function N during the trials numbered $m + 1, \ldots, m + n$ is independent of the past history until m (this history being given by $N_0, \ldots, N_m$ or, equivalently, by $X_1, \ldots, X_m$). Further, the distribution of that increase depends only on n (independent of m).

A stochastic process satisfying Theorem (2.11) or, equivalently, Corollary (2.13) is called a process with *independent increments*. Note that the independence of $N_{m+n} - N_m$ from $N_0, \ldots, N_m$ was proved without any appeal to the distribution of the X_n. Thus, Corollary (2.13) is true for any stochastic process $\{Z_n; n \in \mathbb{N}\}$ defined by

$$Z_n = \begin{cases} 0 & \text{if } n = 0, \\ Y_1 + \cdots + Y_n & \text{if } n \geq 1, \end{cases}$$

with $Y_1, Y_2, \ldots$ being independent random variables. If the Y_n are also identically distributed (in addition to being independent), then the distribution of the increment $Z_{m+n} - Z_m$ does not depend on m. Then $\{Z_n\}$ is said to have *stationary and independent increments*.

Going back to Bernoulli processes, suppose we are given the past history until the time m, namely, we are given $\{N_0, \ldots, N_m\}$ or, equivalently, we are given $\{X_1, \ldots, X_m\}$. Suppose we are to predict the value of a random variable Y depending on the future of the process $\{N_n\}$, namely, $Y = g(N_m, N_{m+1}, \ldots)$ for some function g. The next theorem states that, as far as predicting the value of Y is concerned, all past data concerning $N_0, \ldots, N_{m-1}$ become worthless provided that N_m be known.

(2.14) THEOREM. Let Y be a random variable depending on a finite number of the random variables $N_m, N_{m+1}, \ldots$; that is, Y can be written as

$$Y = g(N_m, N_{m+1}, \ldots, N_{m+n})$$

for some n and some function g. Then

$$E[Y \mid N_0, \ldots, N_m] = E[Y \mid N_m].$$

Proof. Since $N_{m+1} = N_m + X_{m+1}, \ldots, N_{m+n} = N_m + X_{m+1} + \cdots + X_{m+n}$, there is a unique function h such that

$$Y = g(N_m, N_{m+1}, \ldots, N_{m+n}) = h(N_m, X_{m+1}, \ldots, X_{m+n}).$$

Thus,

$$(2.15) \quad E[Y \mid N_0, \ldots, N_m]$$
$$= \sum_k \sum_{\mathbf{i}} h(k, i_1, \ldots, i_n) P\{N_m = k, X_{m+1} = i_1, \ldots, X_{m+n} = i_n \mid N_0, \ldots, N_m\}$$

where the second summation is over all n-tuples $\mathbf{i} = (i_1, \ldots, i_n)$ of zeros and ones.

Since $\{X_{m+1}, \ldots, X_{m+n}\}$ is independent of $\{X_1, \ldots, X_m\}$, by Theorem (2.2.23), $\{X_{m+1}, \ldots, X_{m+n}\}$ is also independent of the random variables $N_0, \ldots, N_m$ determined by $\{X_1, \ldots, X_m\}$. Hence,

$$P\{N_m = k, X_{m+1} = i_1, \ldots, X_{m+n} = i_n \mid N_0, \ldots, N_m\}$$
$$= P\{X_{m+1} = i_1, \ldots, X_{m+n} = i_n\} P\{N_m = k \mid N_0, \ldots, N_m\}$$
$$= \pi(i_1) \cdots \pi(i_n) P\{N_m = k \mid N_0, \ldots, N_m\},$$

where $\pi(i) = P\{X_n = i\} = p$ or q according as $i = 1$ or $i = 0$. On the other hand,

$$P\{N_m = k \mid N_0, \ldots, N_m\} = E[I_{\{k\}}(N_m) \mid N_0, \ldots, N_m]$$
$$= I_{\{k\}}(N_m) = \begin{cases} 1 & \text{if } k = N_m, \\ 0 & \text{otherwise,} \end{cases}$$

by Proposition (2.2.15). Putting the last two results in (2.15), we obtain

$$E[Y\,|\,N_0,\ldots,N_m] = \sum_i h(N_m, i_1,\ldots,i_n)\pi(i_1)\cdots\pi(i_n) = f(N_m)$$

independent of $N_0,\ldots,N_{m-1}$. Finally, by Corollary (2.2.22), this implies that

$$E[Y\,|\,N_0,\ldots,N_m] = f(N_m) = E[Y\,|\,N_m],$$

as was to be shown. □

Theorem (2.14) above is satisfied by many processes of much less specific structure. Such processes are called Markov chains, and we will later study them at some length. The particular Markov chain we have, namely $\{N_n; n \in \mathbb{N}\}$, has the further property mentioned in Corollary (2.13), and this makes our present job easier. The following is to illustrate the workings of Theorem (2.14).

(2.16) EXAMPLE. We want to compute the conditional expectation of N_{11} given N_5. We have

$$
\begin{aligned}
E[N_{11}\,|\,N_5] &= E[N_5 + (N_{11} - N_5)\,|\,N_5] \\
&= E[N_5\,|\,N_5] + E[N_{11} - N_5\,|\,N_5] \\
&= N_5 + E[N_{11} - N_5] = N_5 + 6p.
\end{aligned}
$$

The second equality made use of Corollary (2.2.12); to pass to the third equality we used the fact that $E[N_5\,|\,N_5] = N_5$ by Proposition (2.2.15), and the fact that $N_{11} - N_5$ is independent of N_5 by Corollary (2.13). □

(2.17) EXAMPLE. Let us compute $E[N_5 N_8]$ again (cf. Example (2.12) for the same) to show that conditional expectations, used properly, can reduce the work involved.

By Proposition (2.2.16), $E[N_5 N_8] = E[E[N_5 N_8\,|\,N_5]]$. On the other hand,

$$E[N_5 N_8\,|\,N_5] = N_5 E[N_8\,|\,N_5] = N_5(N_5 + 3p).$$

Hence,

$$
\begin{aligned}
E[N_5 N_8] &= E[N_5^2 + 3pN_5] \\
&= E[N_5^2] + 3pE[N_5] \\
&= 25p^2 + 5pq + 3p\cdot 5p = 40p^2 + 5pq.
\end{aligned}
$$
□

(2.18) EXAMPLE. We want to compute

$$Z = E[N_5 N_{11}\,|\,N_2, N_3].$$

By Theorems (2.2.19) and (2.14),

(2.19)
$$Z = E[E[N_5 N_{11} | N_0, \ldots, N_5] | N_2, N_3]$$
$$= E[E[N_5 N_{11} | N_5] | N_2, N_3].$$

Using the result of Example (2.16),

$$E[N_5 N_{11} | N_5] = N_5 E[N_{11} | N_5] = N_5(N_5 + 6p),$$

so that

$$Z = E[N_5^2 + 6pN_5 | N_2, N_3].$$

Again applying Theorems (2.2.19) and (2.14) and Proposition (2.2.15) in that order,

$$Z = E[E[N_5^2 + 6pN_5 | N_0, N_1, N_2, N_3] | N_2, N_3]$$
$$= E[E[N_5^2 + 6pN_5 | N_3] | N_2, N_3]$$
$$= E[N_5^2 + 6pN_5 | N_3]$$

(that is, given N_2 and N_3, we need only N_3 as far as predicting the future is concerned).

Writing $N_5 = N_3 + (N_5 - N_3)$, then using Corollary (2.13) to deduce the independence of $N_5 - N_3$ and N_3, we get

$$Z = E[N_3^2 + 2N_3(N_5 - N_3) + (N_5 - N_3)^2 + 6pN_5 | N_3]$$
$$= N_3^2 + 2N_3 E[N_5 - N_3 | N_3] + E[(N_5 - N_3)^2 | N_3] + 6pE[N_5 | N_3]$$
$$= N_3^2 + 2N_3 E[N_5 - N_3] + E[(N_5 - N_3)^2] + 6p(N_3 + 2p)$$
$$= N_3^2 + 2N_3 \cdot 2p + 4p^2 + 2pq + 6pN_3 + 12p^2$$
$$= N_3^2 + 10pN_3 + 16p^2 + 2pq. \qquad \square$$

We now have a number of theorems at our disposal to compute the probabilities or expectations related to the sequence $\{N_n\}$. In the next section we will obtain results of a similar nature about the sequence of times (random) at which the successes occur.

3. Times of Successes

Let $\{X_n; n \geq 1\}$ be a Bernoulli process with probability p of success. For fixed $\omega \in \Omega$, consider the realization $X_1(\omega), X_2(\omega), X_3(\omega), \ldots$ of this process. This is a sequence of ones and zeros. Denote by $T_1(\omega), T_2(\omega), \ldots$ the indices corresponding to the successive ones. For example, if $X_1(\omega) = 0$, $X_2(\omega) = 1, X_3(\omega) = 0, X_4(\omega) = 1, X_5(\omega) = 1, \ldots$, then $T_1(\omega) = 2, T_2(\omega)$

$= 4$, $T_3(\omega) = 5, \ldots$. Then T_k is the trial number at which the kth success occurs. In this section we are interested in the sequence of random variables $\{T_k; k \geq 1\}$.

(3.1) EXAMPLE. We have the following model as an approximation of the traffic flow through a fixed point on an uncongested highway. The time axis is divided into intervals of length 1 second each. If there is a vehicle crossing during the nth second, we put $X_n = 1$; otherwise put $X_n = 0$. The traffic flow is such that the numbers of vehicles passing the fixed point in disjoint time intervals are independent. If, furthermore, the traffic density remains constant over time, then the sequence $\{X_n; n \geq 1\}$ is a Bernoulli process. Now, N_n is the number of vehicles passing the fixed point in the interval $(0, n]$, and T_k is the second during which the kth vehicle crosses the given point. □

There is a fundamental relation between the times of successes T_k and the numbers of successes N_n. Fix $\omega \in \Omega$; suppose that for the realization ω, the kth success has occurred at or before the nth trial; that is, suppose $T_k(\omega) \leq n$. Then the number of successes in the first n trials must be at least k; that is, we must have $N_n(\omega) \geq k$. Thus, if $T_k(\omega) \leq n$, then $N_n(\omega) \geq k$. Conversely, if $N_n(\omega) \geq k$, then $T_k(\omega) \leq n$.

Another relation between the T_k and the N_n is obtained as follows. Fix $\omega \in \Omega$ again, and suppose $T_k(\omega) = n$. This implies that for that realization ω, there were exactly $k - 1$ successes in the first $n - 1$ trials and a success occurred at the nth trial, that is, $N_{n-1}(\omega) = k - 1$ and $X_n(\omega) = 1$. Conversely, if $N_{n-1}(\omega) = k - 1$ and $X_n(\omega) = 1$, then $T_k(\omega) = n$. We state these two relationships next as a lemma, and then use them to obtain the distribution of the T_k from the known distribution of the N_n.

(3.2) LEMMA. For any $\omega \in \Omega$, $k = 1, 2, \ldots$, and $n \geq k$, we have
   (a) $T_k(\omega) \leq n$ if and only if $N_n(\omega) \geq k$,
   (b) $T_k(\omega) = n$ if and only if $N_{n-1}(\omega) = k - 1$, $X_n(\omega) = 1$.

(3.3) THEOREM. For any $k \in \{1, 2, \ldots\}$,

(3.4) $$P\{T_k \leq n\} = \sum_{j=k}^{n} \binom{n}{j} p^j q^{n-j}, \quad n = k, k + 1, \ldots ;$$

(3.5) $$P\{T_k = n\} = \binom{n-1}{k-1} p^k q^{n-k}, \quad n = k, k + 1, \ldots .$$

Proof. Fix $k \in \{1, 2, \ldots\}$ and $n \geq k$. By Lemma (3.2a) the event $\{T_k \leq n\}$ is equal to the event $\{N_n \geq k\}$. Hence,

$$P\{T_k \leq n\} = P\{N_n \geq k\} = \sum_{j=k}^{n} P\{N_n = j\},$$

which implies (3.4) when $P\{N_n = j\}$ is replaced by its value, using Theorem (2.7).

By Lemma (3.2b) the events $\{T_k = n\}$ and $\{N_{n-1} = k - 1, X_n = 1\}$ are the same, and by Definition (2.2.24) of independence, X_n and N_{n-1} are independent. Hence,

$$P\{T_k = n\} = P\{N_{n-1} = k - 1\} \, P\{X_n = 1\},$$

which yields the formula (3.5) when the two probabilities on the right are evaluated using (2.7) and the definition of the X_n.   ☐

The preceding theorem, in principle, enables us to compute probabilities and expectations concerning only one of the T_k. In order to consider more complex events involving several of the T_k, we need to know more about the structure of the process $\{T_k; k = 1, 2, \ldots\}$. In fact this will further yield theorems which considerably simplify the computations of quantities such as $E(T_k)$, $\mathrm{Var}(T_k)$, etc.

Fix $\omega \in \Omega$ and suppose $T_1(\omega) = 3$, $T_2(\omega) = 4$, $T_3(\omega) = 5$, $T_4(\omega) = 12$. Then, in order that $T_5(\omega) = 17$, we must have $X_{13}(\omega) = 0$, $X_{14}(\omega) = 0$, $X_{15}(\omega) = 0$, $X_{16}(\omega) = 0$, $X_{17}(\omega) = 1$. More generally, in order that $T_5(\omega) = 12 + m$, we must have $X_{12+1}(\omega) = 0, \ldots, X_{12+m-1}(\omega) = 0, X_{12+m}(\omega) = 1$. We note that, given T_4, the information concerning T_1, T_2, and T_3 had no value in predicting T_5. In fact the same is true in predicting $T_5 - T_4$. These observations lead to the following two theorems. (Throughout the following we have $T_0 = 0$, and $T_1, T_2, \ldots$ as the times of the first, second, $\ldots$ successes in a Bernoulli process $\{X_n; n \geq 1\}$ with probability p of success in any one trial.)

(3.6)   THEOREM.   For any $k \in \mathbb{N}$ and $n \geq k$,

$$P\{T_{k+1} = n \mid T_0, \ldots, T_k\} = P\{T_{k+1} = n \mid T_k\}$$

independent of $T_0, \ldots, T_{k-1}$.

Proof. Fix k and n and consider the conditional probability $f(T_0, \ldots, T_k)$ on the left, that is, for integers $0 = a_0 < a_1 < \cdots < a_k = a$ we have

$$f(a_0, \ldots, a_k) = P\{T_{k+1} = n \mid T_0 = a_0, \ldots, T_k = a_k\}.$$

If $n \leq a_k = a$, then this conditional probability is zero. Otherwise, if $n > a_k = a$, in order that $T_k = a$, $T_{k+1} = n$, we must have $X_{a+1} = 0, \ldots, X_{n-1} = 0$, $X_n = 1$. Hence,

$$f(a_0, \ldots, a_k) = P\{X_{a+1} = 0, \ldots, X_{n-1} = 0, X_n = 1\} = pq^{n-1-a}.$$

We have shown that

$$(3.7) \qquad P\{T_{k+1} = n \,|\, T_0, \ldots, T_k\} = \begin{cases} 0 & \text{on } \{T_k \geq n\}, \\ pq^{n-1-T_k} & \text{on } \{T_k < n\}, \end{cases}$$

and the right hand side is independent of $T_0, \ldots, T_{k-1}$ in both cases. □

The preceding theorem shows that given the time T_k of the kth success, the time of the $(k + 1)$th success is conditionally independent of $T_0, \ldots, T_{k-1}$. In fact, the proof has shown something stronger: from (3.7) we have

$$P\{T_{k+1} = T_k + m \,|\, T_0, \ldots, T_k\} = pq^{T_k+m-1-T_k} = pq^{m-1},$$

which implies the following

(3.8) PROPOSITION. For any $k \in \mathbb{N}$,

$$P\{T_{k+1} - T_k = m \,|\, T_0, \ldots, T_k\} = P\{T_{k+1} - T_k = m\} = pq^{m-1},$$

for all $m \in \{1, 2, \ldots\}$.

In words, then, the time between two successes is independent of the times of previous successes. The distribution appearing here is called the *geometric distribution*. For purposes of increased articulation we now restate this proposition.

(3.9) PROPOSITION. $T_1, T_2 - T_1, T_3 - T_2, \ldots$ are independent and identically distributed random variables with the geometric distribution

$$P\{T_{k+1} - T_k = m\} = pq^{m-1}, \quad m = 1, 2, \ldots,$$

as the common distribution.

Thus, each T_k is the sum of k independent and identically distributed random variables. In this regard, there is no difference between the processes $\{N_n\}$ and $\{T_n\}$. The difference lies in the distribution of the increments $N_{n+1} - N_n$ and $T_{n+1} - T_n$. During the discussion following Theorem (2.12) and Corollary (2.13) we mentioned that Corollary (2.13) holds for any process $\{Z_n\}$ where Z_n is the sum of n independent and identically distributed random variables. Hence it also holds for the process $\{T_n\}$:

(3.10) COROLLARY. Let $0 = n_0 < n_1 < \cdots < n_j$ be integers. Then the random variables $T_{n_1} - T_{n_0}, T_{n_2} - T_{n_1}, \ldots, T_{n_j} - T_{n_{j-1}}$ are independent. Furthermore, the distribution of $T_{m+n} - T_m$ is independent of m.

Of course, the distribution of $T_{m+n} - T_m$ is the same as the distribution of T_n, and this was computed in Theorem (3.3). As we had mentioned after Theorem (3.3), we are now in a much better position to compute expectations and complex-looking probabilities.

It follows from the distribution of $T_{k+1} - T_k$ given in Proposition (3.9) that

$$(3.11) \qquad E[T_{k+1} - T_k] = \sum_{m=1}^{\infty} mpq^{m-1} = \frac{1}{p}$$

(see Example (2.1.20) for this computation and the footnote there to help in the next computation). Similarly,

$$(3.12) \quad E[(T_{k+1} - T_k)^2] = \sum_{m=1}^{\infty} m^2 pq^{m-1}$$

$$= \sum_{m=2}^{\infty} m(m-1)pq^{m-1} + \sum_{m=1}^{\infty} mpq^{m-1}$$

$$= pq\frac{2}{(1-q)^3} + \frac{1}{p} = \frac{2q}{p^2} + \frac{1}{p} = \frac{1+q}{p^2},$$

and using the formula (2.1.28),

$$(3.13) \qquad \operatorname{Var}(T_{k+1} - T_k) = \frac{q}{p^2}.$$

We can use these results to obtain similar ones about T_k without using the distribution (given in (3.3)) of T_k directly. Writing $T_k = T_1 + (T_2 - T_1) + \cdots + (T_k - T_{k-1})$, since the expectation of a sum is the sum of the expectations (cf. Corollary (2.1.25)), (3.11) yields

$$(3.14) \qquad E[T_k] = \frac{k}{p}.$$

Coupling the result of (2.3.9) with the independence of $T_1, T_2 - T_1, \ldots, T_k - T_{k-1}$ (cf. Proposition (3.9) for this), we get

$$(3.15) \qquad \operatorname{Var}(T_k) = \frac{kq}{p^2}$$

from (3.13). The reader might want to compute $\operatorname{Var}(T_k)$ directly from the distribution of T_k given in Theorem (3.3) in order to appreciate the values of Propositions (3.8) and (3.9). The following could not be computed without such propositions.

(3.16) EXAMPLE. Consider $P\{T_1 = 3, T_5 = 9, T_7 = 17\}$. The event in question is the same as the event $\{T_1 = 3, T_5 - T_1 = 6, T_7 - T_5 = 8\}$. Since, by Corollary (3.10), $T_1, T_5 - T_1, T_7 - T_5$ are independent, we can write

$$P\{T_1 = 3, T_5 - T_1 = 6, T_7 - T_5 = 8\}$$
$$= P\{T_1 = 3\}P\{T_5 - T_1 = 6\}P\{T_7 - T_5 = 8\}$$
$$= P\{T_1 = 3\}P\{T_4 = 6\}P\{T_2 = 8\},$$

where the last equality is again obtained from (3.10). Now using Theorem (3.3) to evaluate these, we obtain

$$P\{T_1 = 3, T_5 = 9, T_7 = 17\} = P\{T_1 = 3\}P\{T_4 = 6\}P\{T_2 = 8\}$$
$$= \binom{3-1}{1-1}p^1q^2\binom{6-1}{4-1}p^4q^2\binom{8-1}{2-1}p^2q^6$$
$$= (pq^2)(10p^4q^2)(7p^2q^6) = 70p^7q^{10}. \qquad \square$$

(3.17) EXAMPLE. A certain component in a large system has a lifetime whose distribution can be approximated by $\pi(m) = pq^{m-1}$, $m \geq 1$. When the component fails, it is replaced by an identical one. Let $T_1, T_2, \ldots$ denote the times of failure; then $U_k = T_k - T_{k-1}$ is the lifetime of the kth item replaced. Since the components are identical,

$$P\{U_k = m\} = pq^{m-1}, \qquad m \geq 1,$$

and the items have independent lifetimes. Thus $\{T_k\}$ can be looked upon as the times of "successes" in a Bernoulli process. Past records indicate that the first three failures occurred at times 3, 12, and 14. We would like to estimate the time of the fifth failure.

By Theorem (3.6),

$$E[T_5 \mid T_1, T_2, T_3] = E[T_5 \mid T_3].$$

Writing $T_5 = T_3 + (T_5 - T_3)$, using Corollary (2.2.12) to write the conditional expectation as the sum of the conditional expectations, noting that $E[T_3 \mid T_3] = T_3$ by Proposition (2.2.15) and that $E[T_5 - T_3 \mid T_3] = E[T_5 - T_3]$ by the independence of $T_5 - T_3$ from T_3 (this is by Corollary (3.10)), we have

$$E[T_5 \mid T_3] = E[T_3 \mid T_3] + E[T_5 - T_3 \mid T_3] = T_3 + \frac{2}{p}.$$

Hence, the quantity of interest is

$$E[T_5 \mid T_1 = 3, T_2 = 12, T_3 = 14] = 14 + \frac{2}{p}.$$

If $p = 0.60$, this is 17.33. $\square$

(3.18) EXAMPLE. In the preceding example suppose that each item costs c dollars when new. Assuming that α dollars today is worth 1 dollar during the next period, we now compute the expected total discounted value of all future expenditures on this component.

It is assumed that the discount rate remains the same throughout. Therefore, c dollars spent at time n is worth $c\alpha^n$ dollars at present. Note that α is less than 1; if the "inflation" rate is 0.06, for example, then $\alpha = 1/(1 + 0.06)$.

Fix $\omega \in \Omega$; the kth replacement occurs at time $T_k(\omega)$. That kth item costs c dollars, and therefore its discounted value is

$$C_k(\omega) = c\alpha^{T_k(\omega)}.$$

Thus, the total discounted cost is

$$C(\omega) = \sum_{k=1}^{\infty} C_k(\omega) = \sum_{k=1}^{\infty} c\alpha^{T_k(\omega)}, \quad \omega \in \Omega.$$

The expected value of a sum being the sum of the expectations (cf. Theorem (2.1.34)), we have

$$E[C] = \sum_{k=1}^{\infty} cE[\alpha^{T_k}].$$

Writing $T_k = T_1 + (T_2 - T_1) + \cdots + (T_k - T_{k-1})$, noting that $T_1, T_2 - T_1, \ldots, T_k - T_{k-1}$ are independent by Proposition (3.9), and applying Proposition (2.1.26) on the expected value of such products, we obtain

$$E[\alpha^{T_k}] = E[\alpha^{T_1} \cdot \alpha^{T_2 - T_1} \cdots \alpha^{T_k - T_{k-1}}]$$
$$= E[\alpha^{T_1}]E[\alpha^{T_2 - T_1}] \cdots E[\alpha^{T_k - T_{k-1}}].$$

Since $T_1, T_2 - T_1, \ldots$ have the same geometric distribution,

$$E[\alpha^{T_1}] = E[\alpha^{T_2 - T_1}] = \cdots = \sum_{n=1}^{\infty} \alpha^n pq^{n-1} = \frac{\alpha p}{1 - \alpha q}.$$

Hence,

$$E[C] = \sum_{k=1}^{\infty} c\left(\frac{\alpha p}{1 - \alpha q}\right)^k = c \cdot \frac{\alpha p}{1 - \alpha q} \frac{1}{1 - \dfrac{\alpha p}{1 - \alpha q}} = c\frac{\alpha p}{1 - \alpha}. \qquad \Box$$

We close this section with a generalization of the computations appearing in Example (3.18). Suppose that the value of the amount spent at time n is equal to $f(n)$ in terms of present worth. In the preceding example we had $f(n) = c\alpha^n$. Now we only assume that $f(n) \geq 0$ for all n. In fact, the result (3.19) below holds for any f for which $\sum f(n)$ is absolutely convergent.

We would like to compute

$$E\left[\sum_{k=1}^{\infty} f(T_k)\right], \quad f \geq 0.$$

Without further information on f, it looks as if we have to use Theorem (3.3). But we can do better if we examine the sum $S = \sum_{k=1}^{\infty} f(T_k)$. If, for $\omega \in \Omega$, $T_1(\omega) = 2$, $T_2(\omega) = 5$, $T_3(\omega) = 10$, $T_4(\omega) = 11$, etc., then $S(\omega) = f(2) + f(5) + f(10) + f(11) + \cdots$. Note that 2, 5, 10, 11, ... are precisely those indices n for which $X_n(\omega) = 1$, and that $S(\omega) = f(1) X_1(\omega) + f(2) X_2(\omega) + f(3) X_3(\omega) + \cdots$. This is true in general, and we can write

$$\sum_{k=1}^{\infty} f(T_k) = \sum_{n=1}^{\infty} f(n)X_n.$$

The rest is easy:

(3.19) $$E\left[\sum_{k=1}^{\infty} f(T_k)\right] = E\left[\sum_{n=1}^{\infty} f(n)X_n\right]$$

$$= \sum_{n=1}^{\infty} f(n)E[X_n] = p \sum_{n=1}^{\infty} f(n).$$

4. Sums of Independent Random Variables

In comparing the stochastic structures of the process $\{N_n; n \in \mathbb{N}\}$ of Section 2 and the process $\{T_n; n \in \mathbb{N}\}$ of Section 3, as revealed for example by Corollaries (2.13) and (3.10) respectively, we see that they are very much alike. As we mentioned before, this is due to the facts that $N_1 - N_0, N_2 - N_1, \ldots$ are independent and identically distributed and that $T_1 - T_0, T_2 - T_1, \ldots$ are also independent and identically distributed. It is this property that we now will put in a general setting.

Let $Y_1, Y_2, \ldots$ be a sequence of independent and identically distributed random variables, and define for each $n \in \mathbb{N}$

(4.1) $$Z_n = \begin{cases} 0 & \text{if } n = 0, \\ Y_1 + \cdots + Y_n & \text{if } n \geq 1. \end{cases}$$

We are interested in the stochastic process $\{Z_n; n \in \mathbb{N}\}$ and, especially, its limiting behavior as n approaches infinity. These limit theorems have applications in data collection and statistical estimation, and we shall point out the way they are used.

The following property is immediate from the definition of the Z_n and Definition (2.2.24) of independence.

(4.2) PROPOSITION.   The stochastic process $\{Z_n; n \in \mathbb{N}\}$ has stationary and independent increments; that is,

(a) $Z_{n_1} - Z_{n_0}, Z_{n_2} - Z_{n_1}, \ldots, Z_{n_k} - Z_{n_{k-1}}$ are independent for any integers $0 \leq n_0 < n_1 < \cdots < n_k$; and

(b)   the distribution of $Z_{n+m} - Z_n$ is independent of n.

Conversely, if a process $\{Z_n\}$ has stationary independent increments, then taking $n_0 = 0, n_1 = 1, \ldots, n_k = k$ shows that $Z_1 - Z_0, Z_2 - Z_1, \ldots$ are independent and identically distributed. Hence, all *discrete parameter* processes with stationary independent increments are obtained in this manner.

Proposition (4.2) can be used to prove, just as in Theorem (2.14), the following result: in words, as far as predicting the value of a random variable W depending on the future $Z_n, Z_{n+1}, \ldots$ of the process $\{Z_n\}$ is concerned, all past information concerning $Z_0, \ldots, Z_{n-1}$ becomes worthless once the present value Z_n is given.

(4.3) PROPOSITION.   Suppose $W = g(Z_n, Z_{n+1}, \ldots)$ for some function $g \geq 0$ and some integer $n \in \mathbb{N}$. Then

$$E[W \,|\, Z_0, \ldots, Z_n] = E[W \,|\, Z_n].$$

The following computations are easily made by using the definition of Z_n given by (4.1). Suppose

(4.4) $E[Y_n] = a, \qquad \mathrm{Var}(Y_n) = b^2.$

Then (4.1) and Corollary (2.1.25) give

(4.5) $E[Z_n] = na,$

and (4.1), the independence of $Y_1, Y_2, \ldots$, and (2.3.9) imply

(4.6) $\mathrm{Var}(Z_n) = nb^2.$

(4.7) EXAMPLE.   Successive customers arriving at a gas station for fuel spend independent amounts. Let $Y_1, Y_2, \ldots$ be the amounts (measured in dollars) demanded by the first, second, ... customers. Then Z_n is the amount demanded by the first n customers. If an average demand is 4.87 dollars, then the amount demanded by the first 9 customers has the expected value $9 \times 4.87 = 43.83$ dollars.   □

In the preceding example it is easy to convince ourselves (by considering the independence of the actions of different customers) that $Y_1, Y_2, \ldots$ are independent and identically distributed. But how do we find out the expected value a or, more generally, the common distribution function φ? We start considering the problem of estimating the expected value $E(Y_n) = a$ first.

Noting that $E[Z_n] = na$, it seems reasonable to take the average amount demanded by the first n customers, namely Z_n/n, as an estimate for a. Since by (4.6) and (2.3.8)

$$\operatorname{Var}\left(\frac{Z_n}{n}\right) = \frac{b^2}{n}, \qquad E\left(\frac{Z_n}{n}\right) = a,$$

we see that this estimate gets better, as far as the variance is concerned, as n gets larger. In fact, using Chebyshev's inequality (2.1.27), we have

$$(4.8) \qquad\qquad P\left\{\left|\frac{Z_n}{n} - a\right| > \varepsilon\right\} \leq \frac{b^2}{n\varepsilon^2}$$

for any $\varepsilon > 0$. Hence, by taking the number n of our observations large enough, we can make the estimate Z_n/n as close to its intended value as we desire. Taking limits as $n \longrightarrow \infty$ in (4.8), we obtain

(4.9) PROPOSITION. For any $\varepsilon > 0$,

$$\lim_{n\to\infty} P\left\{\left|\frac{Z_n}{n} - a\right| > \varepsilon\right\} = 0.$$

This is called the *weak law of large numbers*. Returning to our estimation problem, as we observe the process, we are getting a possible realization ω in the sample space. If we observe a total of n customers, then the values we obtain are $Y_1(\omega), Y_2(\omega), \ldots, Y_n(\omega)$ for some $\omega \in \Omega$. So our estimate of a is $Z_n(\omega)/n = (1/n)(Y_1(\omega) + \cdots + Y_n(\omega))$. Though the preceding proposition indicates the reasonableness of using this estimate, it is too weak for the real purpose we had. We need a stronger result which assures us that the observed average $Z_n(\omega)/n$ is close to a as $n \to \infty$. This indeed is true and is called the *strong law of large numbers*.

(4.10) THEOREM. For almost all $\omega \in \Omega$,

$$\lim_{n\to\infty} \frac{Z_n(\omega)}{n} = a.$$

(4.11) EXAMPLE. Suppose, in Example (1.3), that we do not know the probability p that a vehicle turns left. Since $X_1, X_2, \ldots$ are independent and identically distributed with $E[X_n] = p$, the theorem above applies to the number N_n of vehicles turning left to give

$$\lim_{n\to\infty} \frac{N_n(\omega)}{n} = p.$$

for almost all ω. Thus, if we observe the number of vehicles turning left and divide that number by the total number of vehicles arriving at the junction, we obtain an estimate of p. If the observation lasts long enough, our estimate will be very close to the actual value p. Of course, this is as we would expect, and it is nice to have the assurance of Theorem (4.10).   □

The simple-looking estimation of the expected value a in Theorem (4.10) also settles our second problem, namely, the estimation of the "unknown" distribution function φ. Having observed the values $Y_1(\omega)$, $Y_2(\omega)$, ..., $Y_n(\omega)$, it seems reasonable to estimate $\varphi(t)$ by the proportion of the $Y_k(\omega)$ observed to be less than or equal to t. This proportion is

(4.12)
$$F_n(t, \omega) = \frac{1}{n} \sum_{k=1}^{n} I_{(-\infty, t]}(Y_k(\omega)).$$

Note that $I_B(x) = 1$ or 0 according as $x \in B$ or $x \notin B$, so the sum on the right is the number of $Y_k(\omega)$ falling to the left of t; see Figure 3.4.1. Next we show that $F_n(t, \omega)$ is very close to $\varphi(t)$ if n is large enough.

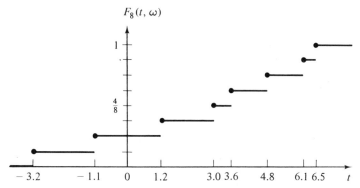

Figure 3.4.1   The empirical distribution for eight observations giving $Y_1(\omega) = 4.8$, $Y_2(\omega) = -1.1$, $Y_3(\omega) = 1.2$, $Y_4(\omega) = 6.5$, $Y_5(\omega) = 3.6$, $Y_6(\omega) = 6.1$, $Y_7(\omega) = -3.2$, and $Y_8(\omega) = 3.0$.

(4.13) PROPOSITION.   For almost all ω,

$$\lim_{n \to \infty} F_n(t, \omega) = \varphi(t)$$

for every $t \in \mathbb{R}$.

Proof. Fix t; let $X_k = I_{(-\infty, t]}(Y_k)$. Then $X_1, X_2, \ldots$ are independent and identically distributed random variables taking only the values 1 and 0, that is, $\{X_k\}$ is a Bernoulli process with success at trial k meaning $Y_k \leq t$. So the probability of success is

$$p = P\{X_k = 1\} = P\{Y_k \le t\} = \varphi(t).$$

Now, note that $F_n(t, \omega) = N_n(\omega)/n$ is the average number of successes in the first n trials. Applying Theorem (4.10) to the process $\{N_n\}$, we get (as in Example (4.11))

$$\lim_{n \to \infty} F_n(t, \omega) = \lim_{n \to \infty} \frac{N_n(\omega)}{n} = p = \varphi(t). \qquad \square$$

For fixed n and ω, $t \to F_n(t, \omega)$ is a step function increasing at each number $Y_k(\omega)$ by a jump of magnitude $1/n$. It is called the *empirical distribution* based on n observations. Proposition (4.13) shows that for large n it approximates the distribution function φ.

We have seen in Proposition (4.9) that for large n the average Z_n/n is close to the expected value a of the increments Y_k. Somewhat better information is contained in the following famous result. It is called the *central limit theorem*.

(4.14) THEOREM. For any $t \in \mathbb{R}$,

$$\lim_{n \to \infty} P\left\{\frac{Z_n - na}{b\sqrt{n}} \le t\right\} = \int_{-\infty}^{t} \frac{1}{\sqrt{2\pi}} \exp\left[-\frac{1}{2}x^2\right] dx.$$

We omit the proof. In words, for large n, the distribution of the sum Z_n is approximately the normal distribution with mean na and variance nb^2. As with the *weak law* and *strong law* of large numbers, this theorem also holds for any distribution for the Y_n provided that the variance b^2 is finite.

The standard normal distribution φ is defined by

$$(4.15) \qquad \varphi(x) = \int_{-\infty}^{x} \frac{1}{\sqrt{2\pi}} e^{-y^2/2} dy, \qquad -\infty < x < \infty.$$

The corresponding density function f (defined by $f(y) = (1/\sqrt{2\pi})\, e^{-y^2/2}$) is drawn in Figure 3.4.2. Since $f(y) = f(-y)$ for all y, we have

$$(4.16) \qquad \varphi(-x) = 1 - \varphi(x), \qquad -\infty < x < \infty.$$

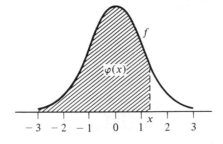

Figure 3.4.2 The standard normal density function f. In the table of Figure 3.4.3, the value $\varphi(x)$ is the area under f to the left of x.

In view of this relation, it is sufficient to list the values of $\varphi(x)$ for x positive only. This is done in the table in Figure 3.4.3.

x	$\varphi(x)$	x	$\varphi(x)$	x	$\varphi(x)$	x	$\varphi(x)$
0.0	0.5000	1.0	0.8413	2.0	0.9772	3.0	0.9987
0.1	0.5398	1.1	0.8643	2.1	0.9821	3.1	0.9990
0.2	0.5793	1.2	0.8849	2.2	0.9861	3.2	0.9993
0.3	0.6179	1.3	0.9032	2.3	0.9893	3.3	0.9995
0.4	0.6554	1.4	0.9192	2.4	0.9918	3.4	0.9997
0.5	0.6915	1.5	0.9332	2.5	0.9938	3.5	0.9998
0.6	0.7257	1.6	0.9452	2.6	0.9953	3.6	0.9998
0.7	0.7580	1.7	0.9554	2.7	0.9965	3.7	0.9999
0.8	0.7881	1.8	0.9641	2.8	0.9974	3.8	0.9999
0.9	0.8159	1.9	0.9713	2.9	0.9981	3.9	1.0000

Figure 3.4.3   Values of the standard normal distribution φ for positive x. For negative x use this table along with (4.16).

(4.17) EXAMPLE. Consider a Bernoulli process with success probability $p = 0.5$. Then applying (4.14) to the process $\{N_n; n \in \mathbb{N}\}$ of success counts, we get

$$\lim_{n \to \infty} P\left\{\frac{N_n - 0.5n}{0.5\sqrt{n}} \le t\right\} = \int_{-\infty}^{t} \frac{1}{\sqrt{2\pi}} e^{-x^2/2} dx.$$

Next, applying (4.14) to the process $\{T_n; n \in \mathbb{N}\}$ of the times of successes, noting that now $a = E(T_{n+1} - T_n) = 1/p = 2$ and $b^2 = \mathrm{Var}(T_{n+1} - T_n) = q/p^2 = 2$ by (3.11) and (3.13),

$$\lim_{n \to \infty} P\left\{\frac{T_n - 2n}{\sqrt{2n}} \le t\right\} = \int_{-\infty}^{t} \frac{1}{\sqrt{2\pi}} e^{-x^2/2} dx.$$

For $n = 50$, for example,

$$P\{T_{50} \le 112\} = P\left\{\frac{T_{50} - 100}{10} \le 1.2\right\}$$

is approximately equal to (using the table in Figure 3.4.3)

$$\int_{-\infty}^{1.2} \frac{1}{\sqrt{2\pi}} e^{-x^2/2} \, dx = 0.8849.$$

Similarly,

$$P\{T_{50} \le 77\} = P\left\{\frac{T_{50} - 100}{10} \le -2.3\right\}$$

is approximately equal to

$$\int_{-\infty}^{-2.3} \frac{1}{\sqrt{2\pi}} e^{-x^2/2} dx = \int_{2.3}^{\infty} \frac{1}{\sqrt{2\pi}} e^{-x^2/2} dx$$

$$= 1 - \int_{-\infty}^{2.3} \frac{1}{\sqrt{2\pi}} e^{-x^2/2} dx$$

$$= 1 - 0.9893 = 0.0107. \qquad \square$$

5. Exercises

(5.1)   Consider a possible realization $\omega = (S, F, F, F, S, S, F, S, \ldots)$ of a sequence of independent trials with two possible outcomes, S and F. For this particular ω, what are the values of the random variables

   (a) $X_1, X_2, \ldots, X_8$;
   (b) $N_0, N_1, \ldots, N_8$;
   (c) T_0, T_1, T_2, T_3, T_4?

(5.2)   In Example (1.2) let the defective rate be $p = 0.05$. What is the probability that the first, second, and third items inspected are all defective? What is the probability that exactly one of the first, second, or third items is defective?

(5.3)   For the process described in Example (1.3), compute and interpret the following quantities.

   (a) $P\{N_1 = 0, N_2 = 0, N_3 = 1, N_4 = 1\}$
   (b) $P\{N_1 = 1, N_3 = 2, N_4 = 2\}$
   (c) $P\{N_8 = 6, N_{15} = 12\}$.

(5.4)   For the problem introduced in Example (1.4), what is the expected number of bearings, among the first 400 produced, which do not meet the specifications?

(5.5)   For $p = 0.8$, compute
   (a) $E[N_3], E[N_7], E[N_3 + 4N_7]$
   (b) $\text{Var}(N_3), \text{Var}(N_7 - N_3)$
   (c) $E[6N_4 + N_7 | N_2]$.

(5.6)   Consider Bernoulli trials with probability $p = 0.8$ for success. Suppose that the first five trials resulted in, respectively, S, F, F, S, S. What is the expected value of $N_3 + 2N_7$ given this past history?

(5.7)   Generalize the result of Example (2.16) by showing that for any $n, m \in \mathbb{N}$,

$$E[N_{n+m} | N_n] = N_n + mp.$$

(5.8)   Repeat the steps of the proof of Theorem (2.14) to show the truth of the particular result that

$$E[3N_5^4 + N_8^3 | N_0, N_1, N_2] = E[3N_5^4 + N_8^3 | N_2].$$

(5.9)   *Hypergeometric distribution.* Let k, m, n be integers with $k \leq m + n$. Show that for any integer j with $0 \leq j \leq m$, $0 \leq k - j \leq n$,

$$P\{N_m = j \mid N_{m+n} = k\} = \frac{\binom{m}{j}\binom{n}{k-j}}{\binom{m+n}{k}}.$$

This defines a probability distribution in j; it is called the hypergeometric distribution. Note that N_m can be replaced by the sum $X_{n_1} + \cdots + X_{n_m}$ of any m variables selected from $X_1, \ldots, X_{m+n}$.

(5.10) In Example (1.2), suppose the items are packaged in boxes of 100 each. A sampling inspection plan calls for rejecting a box if a sample of 5 drawn from that box contained one or more defectives. What is the probability that a box containing exactly 4 defectives is rejected?

(5.11) In Example (3.1) suppose the rate of crossings is 4 vehicles per minute. Compute and interpret the following quantities.

    (a) $p = P\{X_n = 1\}$,
    (b) $P\{T_4 - T_3 = 12\}$
    (c) $E[T_4 - T_3], E[T_{13} - T_3]$
    (d) $\mathrm{Var}(T_2 + 5T_3)$.

(5.12) Repeat the proof of Theorem (3.6) to show that

$$P\{T_8 = 17 \mid T_0, \ldots, T_7\} = pq^{16-T_7}$$

if the event $\{T_7 \leq 16\}$ occurs.

(5.13) Show, by following the steps of Example (3.17), that

$$E[T_{n+m} \mid T_n] = T_n + \frac{m}{p}.$$

(5.14) Compute the following (and compare with (2.2.14)):

    (a) $P\{T_1 = k, T_2 = m, T_3 = n\}$
    (b) $P\{T_3 = n \mid T_1 = k, T_2 = m\}$
    (c) $E[T_3 \mid T_1 = k, T_2 = m]$
    (d) $E[g(T_3) \mid T_1, T_2]$ for $g(b) = \alpha^b, \alpha \in [0, 1]$.

(5.15) If a random variable T has the geometric distribution, then

$$P\{T > n + m \mid T > n\} = q^m = P\{T > m\}$$

for all n and m. Show that the converse is also true: if a discrete random variable T is such that

$$P\{T > n + m \mid T > n\} = P\{T > m\}$$

for all $m, n \in \mathbb{N}$, then T has a geometric distribution.

(5.16) The probability that a given driver stops to pick up a hitchhiker is $p = 0.04$; different drivers, of course, make their decisions to stop or not independently of each other. Given that our hitchhiker counted 30 cars passing her without stopping, what is the probability that she will be picked up by the 37th car or before?

(5.17) *Continuation.* Suppose the arrivals of cars themselves are as described in Example (3.1) with 4 vehicles per minute. Then "success" for the hitchhiker occurs at time n provided that both an arrival occurs at n *and* that car stops to pick her up. Let T be the time (in seconds) she finally gets a ride.

    (a)   Find the distribution of T; compute $E(T)$, $\text{Var}(T)$.

    (b)   Given that after 5 minutes (during which 12 cars pass by) she is still there, compute the expected value of T.

(5.18) A young man and an irate husband are to fight a duel. Each has a large supply of ammunition, and at stated intervals each fires once at the other. On any given shot, the probability that the first man kills the second is p_1 and the probability that the second kills the first is p_2 (it is possible that they both get killed). We assume that the succeeding rounds of the duel are independent and the probabilities remain constant.

    (a)   Find the probability that the duel lasts exactly 13 rounds.

    (b)   Find the probability that the young man comes out alive.

    (c)   Find the probability that the duel lasts exactly 13 rounds with the young man alive at the end.

(5.19) Consider a computer equipped for time-sharing (simultaneous use by more than one person). Suppose each customer uses time in one-minute blocks, and the probability that a given customer is using the computer during any minute is 0.30.

    (a)   Suppose there are 30 customers and the computer can handle only 10 at a time. What is the probability that it is overloaded during any minute?

    (b)   Suppose the computer can handle 10 at a time and it is desired to have the probability of overload during any minute to be less than 0.10. How many customers should there be?

(5.20) In Example (4.7), the number of customers arriving at the station during a certain hour is a random variable N with distribution

$$P\{N = n\} = \frac{e^{-30}30^n}{n!}, \quad n = 0, 1, \ldots.$$

Compute the expected value and variance of the total sales Z_N during that hour if the expected value and variance of a single sale are 3.67 and 2.24, respectively.

(5.21) Let Y_1, Y_2, $\ldots$ be independent and identically distributed random variables taking values in $\mathbb{R}_+ = [0, \infty)$. Then if we put $Z_0 = 0$, $Z_1 = Y_1$, $Z_2 = Y_1 + Y_2, \ldots$, we see that $Z_0 \leq Z_1 \leq Z_2 \leq \cdots$. We think of Z_n as the time of the nth arrival into a store (the stochastic process $\{Z_n; n \in \mathbb{N}\}$ is called a *renewal process*). Let N_t be the number of arrivals during $(0, t]$.

    (a)   Show that

$$P\{N_t \geq n\} = P\{Z_n \leq t\}$$

for all $n \in \mathbb{N}$ and $t \in \mathbb{R}_+$.

    (b)   Show that for almost all ω,

$$\lim_{t \to \infty} N_t(\omega) = +\infty.$$

    (c)   Note that Z_{N_t} is the time of the last arrival before time t and that Z_{N_t+1} is the time of the next arrival after time t.

    (d)   Use the strong law of large numbers on Z_n/n and the result of (b) above to show that, almost surely,

$$\lim_{t \to \infty} Z_{N_t} / N_t = a,$$

where a is the expected value of Y_n.

    (e)   It follows from (c) that $Z_{N_t} \leq t < Z_{N_t+1}$. From this we have

$$\frac{Z_{N_t}}{N_t} \leq \frac{t}{N_t} < \frac{Z_{N_t+1}}{N_t + 1} \cdot \frac{N_t + 1}{N_t}.$$

Using (d) and (b), now show that

$$\lim_{n \to \infty} \frac{N_t(\omega)}{t} = \frac{1}{a}$$

for almost all ω.

$$* \qquad * \qquad *$$

In addition to Bernoulli processes, this chapter tried to cover various limit theorems which by now have become part of the lore of probability theory. Our object was to show the motivation behind the early work in probability theory rather than to give a systematic coverage of it. The weak and strong laws of large numbers and the central limit theorems are the glorious chapters of the classical theory. For two elegant introductions to these subjects we refer the reader to the books by CHUNG *[2] and* LAMPERTI *[1]. For further limit theorems on sums of random variables the complete reference is the book by* GNEDENKO *and* KOLMOGOROV *[1].*

CHAPTER 4

Poisson Processes

The Poisson process arises in many applications, especially as a model for the arrival process at a store, for the arrivals of calls at a telephone exchange, for the arrivals of radioactive particles at a Geiger counter, etc. Just as we employed the terminology of "successes" in discussing the Bernoulli process, we will now employ the term "arrival" with the understanding that its real meaning will vary, depending on the application involved.

Let Ω be a sample space and P a probability. Suppose that for each realization $\omega \in \Omega$ and each $t \geq 0$, we have a non-negative integer $N_t(\omega)$. In applications, $N_t(\omega)$ will stand for the number of arrivals in the interval $[0, t]$ for the realization ω. Thus, for fixed $\omega \in \Omega$, the mapping $t \longrightarrow N_t(\omega)$ is a step function; the times of its jumps correspond to instants of arrivals, and each jump is of size one (namely, if there are group arrivals, we agree to call the arrival of a group one arrival). The resulting family $\{N_t; t \geq 0\}$ of random variables is an arrival process. This is a stochastic process with continuous time parameter and discrete state space $E = \mathbb{N} = \{0, 1, 2, \ldots\}$.

Suppose our observations of the process in the past have convinced us that the distribution of the number of arrivals $N_{t+s} - N_t$ in the interval $(t, t + s]$ depends only on s and not on t. Suppose further that we have been able to find no correlation between the number of arrivals $N_{t+s} - N_t$ in the interval $(t, t + s]$ and the arrivals during $[0, t]$. This second property can be given a justification by noting that an arrival process is the result of the actions of very large numbers of people; and if these actions are independent, as is usually the case, the number of arrivals in $(t, t + s]$ should be independent of the arrivals in $[0, t]$.

In Sections 1, 2, and 3 different aspects of such an arrival process will be examined. In particular, it will be shown that the simple qualitative observations above completely specify the probabilistic structure of the process. Sections 4 and 5 will consider the superposition (merging together) and decomposition of arrival processes. Finally, in Sections 6 and 7 we will discuss the modifications that are necessary when the hypotheses outlined above do not all hold. For example, in most "real life" applications, the distribution of $N_{t+s} - N_t$ depends both on s *and t.* We then will show how to reduce the study of such a process to the original case, and conversely (for use in simulation studies), how to obtain such a process from the simpler ones.

This is an important chapter and is worth studying carefully.

1. Arrival Counting Process

Let Ω be a sample space and P a probability measure on it. Throughout the following, by an *arrival process* we will mean a stochastic process $N = \{N_t; t \geq 0\}$ defined on Ω such that for any $\omega \in \Omega$, the mapping $t \to N_t(\omega)$ is non-decreasing, increases by jumps only, is right continuous, and has $N_0(\omega) = 0$. See, for example, Figure 4.1.1.

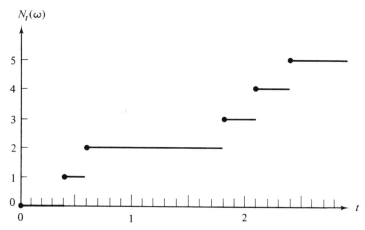

Figure 4.1.1 The function $N_t(\omega)$ for the realization $\omega \in \Omega$ for which the arrival times are $T_1(\omega) = 0.4$, $T_2(\omega) = 0.6$, $T_3(\omega) = 1.8$, $T_4(\omega) = 2.1$, $T_5(\omega) = 2.4, \ldots$.

(1.1) DEFINITION.   An arrival process $N = \{N_t; t \geq 0\}$ is called a Poisson process provided that the following axioms hold:

(a)   for almost all ω, each jump of $t \to N_t(\omega)$ is of unit magnitude;
(b)   for any $t, s \geq 0$, $N_{t+s} - N_t$ is independent of $\{N_u; u \leq t\}$;
(c)   for any $t, s \geq 0$, the distribution of $N_{t+s} - N_t$ is independent of t.

$\square$

In the definition, axiom (b) expresses the independence of the number of arrivals in $(t, t + s]$ from the past history of the process until time t. In particular, it implies that $N_{t+s} - N_t$ is independent of $N_{t_1}, N_{t_2}, \ldots, N_{t_n}$ provided that $t_1, t_2, \ldots, t_n \leq t$ (cf. Definition (2.2.24) of independence). As the knowledge of $N_{t_1}, \ldots, N_{t_n}$ is equivalent to that of $N_{t_1}, N_{t_2} - N_{t_1}, \ldots,$ $N_{t_n} - N_{t_{n-1}}$, Theorem (2.2.23) implies that $N_{t+s} - N_t$ is independent of $N_{t_1},$ $N_{t_2} - N_{t_1}, \ldots, N_{t_n} - N_{t_{n-1}}$ when $t_1, \ldots, t_n \leq t$. Taking $t_1 < t_2 < \cdots < t_n$ in these arguments, and iterating, we reach the conclusion that $N_{t_1}, N_{t_2} - N_{t_1},$ $\ldots, N_{t_n} - N_{t_{n-1}}$ are independent; that is, the numbers of arrivals in disjoint time intervals are independent. Comparing this with Corollary (3.2.13), we see th' a Poisson process is the continuous time-parameter version of the counting process for Bernoulli trials. Or somewhat more generally, what we have is an example of a continuous time-parameter process with stationary and independent increments; here axiom (1.1c) gives the stationarity, and axiom (1.1b) gives the independent increments property.

Our definition of the Poisson process is completely qualitative. What is surprising is that these qualitative axioms completely specify the distribution of N_t and, thereby, the probability law of the whole process. The following results are directed at obtaining this distribution. Throughout this section $N = \{N_t; t \geq 0\}$ is a Poisson process.

(1.2) LEMMA. For all $t \geq 0$,

$$P\{N_t = 0\} = e^{-\lambda t}$$

for some constant $\lambda \geq 0$.

Proof. The number of arrivals in $[0, t + s]$ is zero if and only if there are no arrivals in either $[0, t]$ or $(t, t + s]$; that is, the event $\{N_{t+s} = 0\}$ is equal to the event $\{N_t = 0, N_{t+s} - N_t = 0\}$. Therefore, by the independence of $N_{t+s} - N_t$ from N_t,

$$P\{N_{t+s} = 0\} = P\{N_t = 0\}P\{N_{t+s} - N_t = 0\}.$$

On the other hand, by the stationarity axiom (1.1c),

$$P\{N_{t+s} - N_t = 0\} = P\{N_s = 0\}.$$

Hence, the function $f(t) = P\{N_t = 0\}$ satisfies the equations

$$(1.3) \qquad f(t + s) = f(t)f(s), \qquad 0 \leq f(t) \leq 1,$$

for all $t, s \geq 0$.

Then it is known that either $f(t) = 0$ for all $t \geq 0$ or else $f(t)$ has the form

$$(1.4) \qquad\qquad f(t) = e^{-\lambda t}$$

for some constant $\lambda \geq 0$. This completes the proof once we show that $f(t)$ does not vanish identically. If $f(t)$ were 0 for all $t > 0$, then for any t and almost all ω, $N_{t+s}(\omega) - N_t(\omega) \geq 1$ no matter how small s is. This in turn implies, by writing $N_t = N_{t_1} + (N_{t_2} - N_{t_1}) + \cdots + (N_{t_n} - N_{t_{n-1}})$ with $t_1 < t_2 < \cdots < t_n = t$, that $N_t(\omega) \geq n$ for almost all ω and any n. Hence, if $f(t) = 0$ for all $t \geq 0$, then $N_t(\omega) = +\infty$ for all t for almost all ω, thus contradicting the finiteness of $N_t(\omega)$.   □

The case $\lambda = 0$ corresponds to the degenerate case where $N_t = 0$ identically for all t. This case will be excluded from further consideration. Within the next proof we will need the fact that $E[N_t] < \infty$; this can be proved using Proposition (2.2), to be given later, which will use only the definition and (1.2).

(1.5) LEMMA. We have

$$\lim_{t \downarrow 0} \frac{1}{t} P\{N_t \geq 2\} = 0.$$

Proof. Let $h(t) = P\{N_t \geq 2\}$. Since $N_t(\omega) \geq 2$ implies $N_{t+s}(\omega) \geq 2$, by Proposition (1.1.8), $h(t) \leq h(t + s)$, that is, h is non-decreasing. Let $n_t = \sup\{n \in \mathbb{N}: n \leq 1/t\}$; then $t \leq 1/n_t$ and $1/t < n_t + 1$. Hence $h(t) \leq h(1/n_t)$, and

$$0 \leq \frac{1}{t} h(t) \leq (n_t + 1) h\Big(\frac{1}{n_t}\Big) = \frac{n_t + 1}{n_t} n_t h\Big(\frac{1}{n_t}\Big).$$

As $t \downarrow 0$, $n_t \uparrow \infty$ and $(n_t + 1)/n_t \downarrow 1$. Therefore, to show that $h(t)/t \to 0$ as $t \to 0$, it is sufficient to show that $nh(1/n) \to 0$ as $n \to \infty$.

Divide the unit interval $[0, 1]$ into n subintervals of length $1/n$ each, and let $S_n(\omega)$ be the number of these subintervals during which there were two or more arrivals for the realization ω. We can think of S_n as the number of successes in n Bernoulli trials, where by "success at the kth trial" is meant having two or more arrivals during the kth subinterval. Then the probability of success at any one trial is $p = h(1/n)$, and hence

(1.6) $$E[S_n] = np = nh\Big(\frac{1}{n}\Big).$$

By axiom (1.1a), for almost all $\omega \in \Omega$, the minimum time between two arrivals in $[0, 1]$ is some number $\delta(\omega) > 0$. If n is large enough to have $1/n < \delta(\omega)$, then no subinterval can contain two arrivals and we must have $S_n(\omega) = 0$. Hence, for almost all ω

(1.7) $$\lim_{n \to \infty} S_n(\omega) = 0.$$

Finally, S_n cannot be greater than the number of arrivals in $[0, 1]$; namely, the S_n are bounded by N_1, whose expected value $E[N_1]$ is finite. Therefore the bounded convergence theorem (2.1.35) applies, and from (1.7) we get

$$\lim_{n \to \infty} E[S_n] = 0.$$

In view of (1.6), this completes the proof.   □

(1.8) LEMMA. We have

$$\lim_{t \downarrow 0} \frac{1}{t} P\{N_t = 1\} = \lambda$$

where λ is the constant appearing in Lemma (1.2).

Proof. We have $P\{N_t = 1\} = 1 - P\{N_t = 0\} - P\{N_t \geq 2\}$. Thus, by Lemmas (1.2) and (1.5),

$$\lim_{t \downarrow 0} \frac{1}{t} P\{N_t = 1\} = \lim_{t \downarrow 0} \frac{1 - e^{-\lambda t}}{t} - \lim_{t \downarrow 0} \frac{1}{t} P\{N_t \geq 2\} = \lambda. \qquad □$$

The next theorem gives the distribution of N_t and, in view of axioms (1.1b) and (1.1c), specifies the joint distribution of $N_{t_1}, N_{t_2}, \ldots, N_{t_n}$ for any $n \geq 1$ and $t_1, t_2, \ldots, t_n \in \mathbb{R}_+$.

(1.9) THEOREM. If $\{N_t; t \geq 0\}$ is a Poisson process, then for any $t \geq 0$,

$$P\{N_t = k\} = \frac{e^{-\lambda t}(\lambda t)^k}{k!}, \quad k = 0, 1, \ldots,$$

for some constant $\lambda \geq 0$.

Proof. Let $G(t) = E[\alpha^{N_t}]$. Writing $N_{t+s} = N_t + (N_{t+s} - N_t)$, using the independence of $N_{t+s} - N_t$ from N_t by axiom (1.1b) to apply Proposition (2.1.26), and using the stationarity axiom (1.1c), we get

$$G(t + s) = E[\alpha^{N_{t+s}}] = E[\alpha^{N_t} \cdot \alpha^{N_{t+s} - N_t}]$$
$$= E[\alpha^{N_t}]E[\alpha^{N_{t+s} - N_t}] = G(t)G(s).$$

Since $G(t) = \sum \alpha^n P\{N_t = n\} \geq P\{N_t = 0\} = e^{-\lambda t}$, G does not vanish for any t, and $G(t + s) = G(t)G(s)$ can be satisfied only if

(1.10) $G(t) = e^{tg(\alpha)}, \quad t \geq 0.$

Note that $g(\alpha)$ is the derivative of G at $t = 0$, that is, since $G(0) = 1$ and

$$G(t) = \sum_0^\infty \alpha^n P\{N_t = n\},$$

(1.11) $$g(\alpha) = \lim_{t \downarrow 0} \frac{1}{t}(G(t) - 1)$$

$$= \lim_{t \downarrow 0} \frac{1}{t}[P\{N_t = 0\} - 1] + \lim_{t \downarrow 0} \frac{1}{t}\alpha P\{N_t = 1\}$$

$$+ \lim_{t \downarrow 0} \frac{1}{t} \sum_{n=2}^\infty \alpha^n P\{N_t = n\}.$$

By Lemma (1.2), the first limit is $-\lambda$; by Lemma (1.8), the second limit is $\alpha\lambda$; and by Lemma (1.5), for $\alpha \in [0, 1]$,

$$0 \le \lim_{t \downarrow 0} \frac{1}{t} \sum_2^\infty \alpha^n P\{N_t = n\}$$

$$\le \lim_{t \downarrow 0} \frac{1}{t} \sum_2^\infty P\{N_t = n\} = \lim_{t \downarrow 0} \frac{1}{t} P\{N_t \ge 2\} = 0.$$

Putting these results in (1.11), we see that $g(\alpha) = -\lambda + \lambda\alpha$, and (1.10) becomes

(1.12) $$G(t) = e^{-\lambda t + \lambda t \alpha}.$$

We thus have shown that for all $\alpha \in [0, 1]$,

$$\sum_{n=0}^\infty \alpha^n P\{N_t = n\} = \sum_{n=0}^\infty \frac{e^{-\lambda t}(\lambda t)^n}{n!} \alpha^n.$$

The equality of these two power series in α implies that the corresponding coefficients are equal to each other, that is,

$$P\{N_t = n\} = \frac{e^{-\lambda t}(\lambda t)^n}{n!}, \quad n = 0, 1, \ldots,$$

as was to be shown. $\qquad\qquad\qquad\qquad\qquad\qquad\qquad\qquad\qquad\qquad$ □

So far, λ is only a constant without any particular meaning attached to it. We now note that for any $t \ge 0$,

(1.13) $$E[N_t] = \sum_{n=0}^\infty n P\{N_t = n\}$$

$$= \sum_{n=0}^\infty n \frac{e^{-\lambda t}(\lambda t)^n}{n!} = \lambda t.$$

Thus λ is the expected number of arrivals in an interval of unit length, or in other words, λ is the *arrival rate*. Incidentally, a computation similar to the one in Example (2.1.29) shows that

(1.14) $$\mathrm{Var}(N_t) = \lambda t$$

also. The distribution appearing in the preceding theorem is called the *Poisson distribution with parameter λt*.

The next corollary combines Theorem (1.9) with the axioms on stationarity and the independence of increments.

(1.15) COROLLARY. Let $N = \{N_t; t \geq 0\}$ be a Poisson process with rate λ. Then for any $s, t \geq 0$,

$$P\{N_{t+s} - N_t = k \mid N_u; u \leq t\} = P\{N_{t+s} - N_t = k\}$$
$$= \frac{e^{-\lambda s}(\lambda s)^k}{k!}, \quad k = 0, 1, \dots.$$

Proof. By axiom (1.1b), $N_{t+s} - N_t$ is independent of $\{N_u; u \leq t\}$; this implies the first equality. By axiom (1.1c), $P\{N_{t+s} - N_t = k\} = P\{N_s = k\}$. By Theorem (1.9), this last number is $e^{-\lambda s}(\lambda s)^k/k!$. $\square$

(1.16) EXAMPLE. Let N be a Poisson process with rate $\lambda = 8$. We would like to compute

$$P\{N_{2.5} = 17, N_{3.7} = 22, N_{4.3} = 36\}.$$

The event in question is equal to the event $\{N_{2.5} = 17, N_{3.7} - N_{2.5} = 5, N_{4.3} - N_{3.7} = 14\}$. By the independence of increments, $N_{2.5}, N_{3.7} - N_{2.5}$, and $N_{4.3} - N_{3.7}$ are independent; and by Corollary (1.15), they have the Poisson distributions with respective parameters $8 \times 2.5 = 20$, $8(3.7 - 2.5) = 9.6$, $8(4.3 - 3.7) = 4.8$. Hence, the probability desired is

$$\frac{e^{-20}\, 20^{17}}{17!} \cdot \frac{e^{-9.6}\, 9.6^5}{5!} \cdot \frac{e^{-4.8}\, 4.8^{14}}{14!}. \qquad \square$$

From the practical point of view, checking to see if the axioms (1.1b) and (1.1c) hold for a particular process can be quite difficult. The following theorem reduces the checks involved considerably.

(1.17) THEOREM. $N = \{N_t; t \geq 0\}$ is a Poisson process with rate λ if and only if
  (a) for almost all ω, each jump of $N_t(\omega)$ is of unit magnitude, and
  (b) for any $t, s \geq 0$,

$$E[N_{t+s} - N_t \mid N_u; u \leq t] = \lambda s.$$

Proof is omitted. The theorem states the following. Suppose that, as far as predicting the expected number of arrivals during $(t, t + s]$ is concerned, the past history of arrivals before t has no value and that expectation is a constant λ times the length of the interval in question. Then that process is Poisson. Of course, if N is a Poisson process, then (1.17a) and (1.17b) are quite readily seen to hold. What is surprising is that the seemingly mild condition (1.17b) implies the axioms (1.1b) and (1.1c).

The preceding theorem gives the simplest qualitative characterization of the Poisson process. It is easy to use, especially in situations where there is strong *a priori* evidence to suggest the independence of $N_{t+s} - N_t$ from the past history of arrivals before t. The next theorem is useful in the opposite situation, where the independence axiom is hard to check but there are enough data to obtain the distribution of the number of arrivals falling in various time sets. Before stating this theorem we extend our notation somewhat.

Let N be a Poisson process with rate λ. For any subset B of $\mathbb{R}_+ = [0, \infty)$ we write $N_B(\omega)$ for the number of arrivals falling in B for the realization ω. For example, if $B = (t, t + s]$, then $N_B = N_{t+s} - N_t$; if $B = [0, t]$, then $N_B = N_t - N_0 = N_t$; if B is the union of two disjoint intervals $(t, u]$ and $(s, v]$, then $N_B = N_{(t,u]} + N_{(s,v]} = N_u - N_t + N_v - N_s$.

If B is an interval of length b, then by Theorem (1.9), the distribution of N_B is Poisson with parameter λb. Next let A and B be two disjoint intervals with respective lengths a and b, say, $A = (t, t + a]$, $B = (s, s + b]$. Then N_A and N_B are independent random variables having the Poisson distribution with respective parameters λa and λb. From the stationarity axiom, N_B has the same distribution as N_C where $C = (t + a, t + a + b]$, and N_A and N_C are still independent. Therefore, $N_A + N_B$ has the same distribution as $N_A + N_C$. But the latter is the number of arrivals in $(t, t + a + b]$ and hence has the Poisson distribution with parameter $\lambda(a + b)$. Thus we have shown that $N_A + N_B$ has the Poisson distribution with parameter $\lambda(a + b)$. The same argument goes through for any number of intervals and shows the necessity part of the following

(1.18)   THEOREM.   *N is a Poisson process with rate λ if and only if*

$$P\{N_B = k\} = \frac{e^{-\lambda b}(\lambda b)^k}{k!}, \quad k = 0, 1, \ldots,$$

for any subset B of $\mathbb{R}_+$ which is the union of a finite number of disjoint intervals whose lengths sum up to b.

Proof of necessity was already outlined, and we omit the proof of sufficiency. As we mentioned before, the strength of the theorem lies in its avoidance of conditional expectations and distributions in characterizing the Poisson process. This is useful in situations where there are no intuitive reasons to expect axiom (1.1b) to hold. In words, this theorem states that an arrival process is Poisson with rate λ if the distribution of the number of arrivals occurring during any time set B is Poisson with parameter λ times the "length" of B.

Suppose next that the number of arrivals in an interval $B = (t, t + b]$ is known to be k but the exact times of these k arrivals are not known. The next proposition shows that given $N_B = k$, each one of those k arrival times has the uniform distribution over the interval B independent of the times of the other $k - 1$ arrivals. In this sense, knowing the number of arrivals in an interval B yields no information on the times of those arrivals (other than the trivial information that they fall in B).

(1.19) PROPOSITION. Let $A_1, \ldots, A_n$ be disjoint intervals with union B, let $a_1, \ldots, a_n$ be their respective lengths, and set $b = a_1 + \cdots + a_n$. Then for $k_1 + \cdots + k_n = k$, $k_1, \ldots, k_n \in \mathbb{N}$,

$$P\{N_{A_1} = k_1, \ldots, N_{A_n} = k_n | N_B = k\} = \frac{k!}{k_1! \cdots k_n!} \left(\frac{a_1}{b}\right)^{k_1} \cdots \left(\frac{a_n}{b}\right)^{k_n}.$$

Proof. The event $\{N_{A_1} = k_1, \ldots, N_{A_n} = k_n\}$ implies $\{N_B = k\}$; therefore, the conditional probability in question is equal to

$$\alpha = \frac{P\{N_{A_1} = k_1, \ldots, N_{A_n} = k_n\}}{P\{N_B = k\}}.$$

Since $A_1, \ldots, A_n$ are disjoint, $N_{A_1}, \ldots, N_{A_n}$ are independent, and

$$\alpha = \frac{e^{-\lambda a_1}(\lambda a_1)^{k_1}}{k_1!} \cdots \frac{e^{-\lambda a_n}(\lambda a_n)^{k_n}}{k_n!} \cdot \frac{k!}{e^{-\lambda b}(\lambda b)^k},$$

which becomes the desired result after rearrangement.   □

We close this section with a few remarks on the estimation of the parameter λ and the long-run behavior of N. We already computed $E[N_t]$ to be λt. For any integer n, we can write N_n as the sum of n independent and identically distributed random variables (namely, $N_n = N_1 + (N_2 - N_1) + \cdots + (N_n - N_{n-1})$), each one of which has the expected value λ. Thus, by the strong law of large numbers (3.4.10), for almost all ω,

$$\lim_{n \to \infty} \frac{N_n(\omega)}{n} = \lambda.$$

From this it is easy to see that

(1.20)
$$\lim_{t \to \infty} \frac{N_t(\omega)}{t} = \lambda$$

for almost all ω. This is the strong law of large numbers for the Poisson process. By this result, we are justified in estimating λ by the average number of arrivals $N_t(\omega)/t$ we observed during the course of our observations.

A similar reasoning based on writing N_t as the sum of a large number of independent and identically distributed random variables also yields a central limit theorem:

(1.21)
$$\lim_{t \to \infty} P\left\{ \frac{N_t - \lambda t}{\sqrt{\lambda t}} \le x \right\} = \int_{-\infty}^{x} \frac{1}{\sqrt{2\pi}} e^{-y^2/2} \, dy$$

for any $x \in \mathbb{R}$. Namely, if λt is large enough, N_t has the normal distribution with mean λt and variance λt. For practical purposes, the approximation is quite good if $\lambda t \ge 10$.

2. Times of Arrivals

Let N be a Poisson process with rate λ. For almost all $\omega \in \Omega$, the function $t \longrightarrow N_t(\omega)$ is non-decreasing and right continuous and increases by jumps of size one. Such a function is completely determined by the times of its jumps; let these be $T_1(\omega), T_2(\omega), \ldots$. Then $T_1, T_2, \ldots$ are the successive instants of arrivals.

We have seen in Lemma (1.2) that if t is an arbitrary point in time, the probability that there are no arrivals in $(t, t + s]$ is $e^{-\lambda s}$, independent of the history of arrivals before t. The same result happens to hold even when the arbitrary time point t is replaced by a time of arrival, say T_n; namely,

(2.1)
$$P\{N_{T_n+s} - N_{T_n} = 0 \,|\, N_u; \, u \le T_n\} = e^{-\lambda s},$$

where $\{N_u; \, u \le T_n\}$ is the history of the process until the time T_n of the nth arrival. Noting that the event $\{N_{T_n+s} - N_{T_n} = 0\}$ is equal to the event $\{T_{n+1} - T_n > s\}$, and that the information contained in the history $\{N_u; \, u \le T_n\}$ is the same as that contained in $\{T_1, \ldots, T_n\}$, what we have can be restated as follows (we write $T_0 = 0$ for convenience):

(2.2) PROPOSITION. For any $n \ge 0$,
$$P\{T_{n+1} - T_n \le t \,|\, T_0, \ldots, T_n\} = 1 - e^{-\lambda t}, \quad t \ge 0.$$

In other words, the interarrival times $T_1, T_2 - T_1, T_3 - T_2, \ldots$ are independent and identically distributed random variables with the common distribution being

(2.3) $1 - e^{-\lambda t}, \qquad t \geq 0.$ □

The distribution (2.3) above is called the *exponential distribution* with parameter λ. It is differentiable, and the probability density function corresponding to it is

$$\lambda e^{-\lambda t}, \qquad t \geq 0.$$

Note that this density function is monotone decreasing. As a result, an interarrival time is more likely to have a length in $[0, s]$ than a length in $[t, t + s]$ for any t. Thus, a Poisson process has more short intervals than long ones. Therefore, a plot of the times of arrivals on a line looks, to the naive eye, as if the arrivals occur in clusters.

Another property, usually referred to as the *memorylessness* of the exponential distribution, is the prime reason for the pleasant independence properties of the Poisson process: if a random variable X has an exponential distribution, then

(2.4) $P\{X > t + s \,|\, X > t\} = P\{X > s\}, \qquad t, s \geq 0;$

and conversely, if X has this property, then it must have an exponential distribution. That is, knowing that an interarrival time has already lasted t units does not alter the probability of its lasting another s units.

In certain applied situations it is more efficient to work with the times of arrivals. Then it is of value to find conditions on interarrival times which guarantee that the counting process will be Poisson. The answer turns out to be that the conditions of (2.2) above are sufficient. We omit the proof of this result:

(2.5) THEOREM.   Let $T_1, T_2, \ldots$ be the successive times of arrivals of an arrival process $N = \{N_t; t \geq 0\}$. Then N is a Poisson process with rate λ if and only if the interarrival times $T_1, T_2 - T_1, \ldots$ are independent and identically distributed exponential random variables with parameter λ.

(2.6) EXAMPLE.   An item has a random lifetime whose distribution is exponential with parameter λ. When it fails, it is immediately replaced by an identical item; and when that fails, it is replaced by another identical item; etc. What this means is that the lifetimes $X_1, X_2, \ldots$ of the successive items in use are independent and identically distributed random variables with distribution

$$P\{X_n \leq t\} = 1 - e^{-\lambda t}, \qquad t \geq 0.$$

If T_1, T_2, ... are the times of successive failures, then $T_1 = X_1$, $T_2 = X_1 + X_2$, ... ; and we see that the conditions of Theorem (2.5) are satisfied by the sequence T_1, T_2, ... Hence, if N_t is the number of failures in $[0, t]$, we may conclude by Theorem (2.5) that the process of failures $\{N_t; t \geq 0\}$ is Poisson with rate λ. □

It follows from the fact that $T_{n+1} - T_n$ has an exponential distribution with parameter λ that the expected value of an interarrival time in a Poisson process is (we are using Theorem (2.1.9))

$$(2.7) \qquad E[T_{n+1} - T_n] = \int_0^\infty P\{T_{n+1} - T_n > t\}\, dt = \int_0^\infty e^{-\lambda t}\, dt = \frac{1}{\lambda};$$

and the variance is (see Example (2.1.30) for an illustration)

$$(2.8) \qquad \text{Var}(T_{n+1} - T_n) = \frac{1}{\lambda^2}.$$

By writing $T_n = T_1 + (T_2 - T_1) + \cdots + (T_n - T_{n-1})$ and using the fact that the expectation of a sum is the sum of the expectations, we obtain from (2.7) that

$$(2.9) \qquad E[T_n] = \frac{n}{\lambda}.$$

Using the independence of T_1, $T_2 - T_1$, ..., $T_n - T_{n-1}$ proved in (2.2) to apply (2.3.9) (namely, that the variance of a sum of independent random variables is the sum of the variances), we obtain from (2.8) that

$$(2.10) \qquad \text{Var}(T_n) = \frac{n}{\lambda^2}.$$

Similarly, various higher moments of T_n may be obtained by utilizing Proposition (2.2) once such moments are obtained for one interarrival time.

(2.11) EXAMPLE. In Example (2.6), suppose $\lambda = 0.0002$ (time is measured in hours). Then the expected lifetime of an item is $E[X_n] = 1/\lambda = 5,000$ hours, and the variance is $\text{Var}(X_n) = 1/\lambda^2 = 25 \times 10^6$ hours². Hence, if a piece of equipment contains 3 such items connected so that the second starts functioning as soon as the first fails, and the third starts functioning as soon as the second fails, then the lifetime of this equipment is $T_3 = X_1 + X_2 + X_3$. It has mean $E[T_3] = 3/\lambda = 15,000$ hours and the variance $\text{Var}(T_3) = 3/\lambda^2 = 75 \times 10^6$ hours². □

(2.12) EXAMPLE. In Example (2.6), again suppose that the items are replaced as soon as they fail. Suppose the cost of a replacement is β dollars,
and suppose the discount rate of money is $\alpha > 0$, so that one dollar spent at
time t has the present value $e^{-\alpha t}$ (typically, α is the interest rate). Then for
the realization $\omega \in \Omega$, time of the nth failure is $T_n(\omega)$, and the present value
of the cost of replacement is $\beta \exp(-\alpha T_n(\omega))$. Summing this over all n, we
obtain the present value of all future replacement costs; this is

$$C(\omega) = \sum_{n=1}^{\infty} \beta e^{-\alpha T_n(\omega)}, \quad \omega \in \Omega.$$

We are interested in the expected value of C. Since the expected value of a
sum is the sum of the expectations,

$$E[C] = \beta \sum_{n=1}^{\infty} E[e^{-\alpha T_n}].$$

For fixed n we can write $T_n = T_1 + (T_2 - T_1) + \cdots + (T_n - T_{n-1})$ where
$T_1, T_2 - T_1, \ldots, T_n - T_{n-1}$ are independent and identically distributed.
Hence, using Proposition (2.1.26),

$$\begin{aligned}
E[e^{-\alpha T_n}] &= E[e^{-\alpha T_1} e^{-\alpha(T_2-T_1)} \cdots e^{-\alpha(T_n-T_{n-1})}] \\
&= E[e^{-\alpha T_1}] E[e^{-\alpha(T_2-T_1)}] \cdots E[e^{-\alpha(T_n-T_{n-1})}] \\
&= E[e^{-\alpha T_1}]^n.
\end{aligned}$$

Since the distribution of T_1 is exponential with parameter λ,

$$E[e^{-\alpha T_1}] = \int_0^{\infty} e^{-\alpha t} \lambda\, e^{-\lambda t}\, dt = \frac{\lambda}{\alpha + \lambda}.$$

Hence, remembering that $1 + x + x^2 + \cdots = 1/(1 - x)$ for $x \in [0, 1)$,

$$E[C] = \beta \sum_{n=1}^{\infty} \left(\frac{\lambda}{\alpha + \lambda}\right)^n = \beta \frac{\lambda}{\alpha + \lambda}\left(1 - \frac{\lambda}{\alpha + \lambda}\right)^{-1} = \frac{\beta\lambda}{\alpha}.$$

In particular, if the mean lifetime is 5,000 hours, replacement cost 800 dollars, and the interest rate 24 percent per year, then $\beta = 800$, $\lambda = 1/5,000$,
$\alpha = 0.24/(365 \times 24) = 0.01/365$, and hence $E[C] = 800 \times 36,500/5,000 =$
5840 dollars.   ☐

Though in many computations we can do without the distribution of T_n,
there are situations when it is helpful to know it. One way of obtaining the
distribution of T_n would be to invert the Laplace transform of T_n, which is
$E[e^{-\alpha T_n}] = [\lambda/(\alpha + \lambda)]^n$. However, it is easier (and smarter) to observe first

that for any realization ω, the time $T_5(\omega)$ of the fifth arrival is before t if and only if the number of arrivals $N_t(\omega)$ in $[0, t]$ is equal to five or more. This reasoning gives, for any n and t,

(2.13) $$\{T_n \leq t\} = \{N_t \geq n\}.$$

From this equality between two events we obtain, by (1.9),

(2.14) PROPOSITION. For any $n \in \{1, 2, \ldots\}$,

$$P\{T_n \leq t\} = 1 - \sum_{k=0}^{n-1} \frac{e^{-\lambda t}(\lambda t)^k}{k!}, \quad t \geq 0. \qquad \square$$

The distribution of T_n is called the *Erlang-n distribution*. It is a member of the gamma family of distributions. It is differentiable, and the derivative is

(2.15) $$\frac{\lambda(\lambda t)^{n-1}e^{-\lambda t}}{(n-1)!}, \quad t \geq 0.$$

It is important to distinguish this probability density function $t \longrightarrow \lambda(\lambda t)^{n-1}$ $e^{-\lambda t}/(n-1)!$ of the random variable T_n taking values in $\mathbb{R}_+$ from the distribution $n \longrightarrow e^{-\lambda t}(\lambda t)^n/n!$ of the discrete random variable N_t.

(2.16) EXAMPLE. Taking a certain fixed point on a highway, let U_1, $U_2, \ldots$ be the successive interarrival times of vehicles at this point. Suppose $U_1, U_2, \ldots$ are independent and identically distributed random variables with the distribution

$$P\{U_k \leq t\} = 1 - e^{-\lambda t} - \lambda t e^{-\lambda t}, \quad t \geq 0.$$

We are interested in the distribution of the number M_t of vehicles crossing this fixed point during $[0, t]$.

First we observe that the distribution of U_k given is the Erlang-2 distribution. Thus, we may think of each U_k as being the sum of two interarrival times in a Poisson process with rate λ. That is, the times vehicles cross the given point may be thought of as $U_1 = T_2$, $U_1 + U_2 = T_4$, $U_1 + U_2 + U_3 = T_6, \ldots$ where $T_1, T_2, \ldots$ are the times of arrivals in a Poisson process $\{N_t\}$ with rate λ. Then the number $M_t(\omega)$ of vehicles crossing is equal to 6 if and only if the Poisson process N has, for that realization ω, 12 or 13 arrivals in $[0, t]$. So for any $t \geq 0$,

$$P\{M_t = k\} = P\{N_t = 2k\} + P\{N_t = 2k + 1\}$$
$$= \frac{e^{-\lambda t}(\lambda t)^{2k}}{2k!} + \frac{e^{-\lambda t}(\lambda t)^{2k+1}}{(2k+1)!}, \quad k = 0, 1, \ldots. \qquad \square$$

Next is a computational result which extends that of Example (2.12) to situations where the interest rate, instead of remaining constant over time as in (2.12), may vary as a function of time. Again we assume that $T_1, T_2, \ldots$ are the times of arrivals of a Poisson process with rate λ.

(2.17) PROPOSITION. For any non-negative function f on $\mathbb{R}_+$,

$$E\left[\sum_{n=1}^{\infty} f(T_n)\right] = \lambda \int_0^{\infty} f(t)\, dt.$$

Proof. Using Proposition (2.1.23), noting that the probability density function of T_n is given by (2.15),

$$E[f(T_n)] = \int_0^{\infty} f(t) \frac{\lambda e^{-\lambda t}(\lambda t)^{n-1}}{(n-1)!}\, dt.$$

Hence,

$$E\left[\sum_{n=1}^{\infty} f(T_n)\right] = \sum_{n=1}^{\infty} \int_0^{\infty} f(t) \frac{\lambda e^{-\lambda t}(\lambda t)^{n-1}}{(n-1)!}\, dt$$

$$= \int_0^{\infty} \lambda f(t) \sum_{n=1}^{\infty} \frac{e^{-\lambda t}(\lambda t)^{n-1}}{(n-1)!}\, dt$$

$$= \int_0^{\infty} \lambda f(t)\, dt. \qquad \square$$

We can now re-do the computations of Example (2.12): $E[C] = \lambda \int f$, where $f(t) = \beta e^{-\alpha t}$. So $\int f = \beta/\alpha$, and $E[C] = \lambda\beta/\alpha$.

We close this section by giving a more general version of the property (2.1) with which we started. In certain situations, one is interested in the number of arrivals occurring in an interval $(T, T+s]$ where T is a random variable instead of a fixed number. It turns out that for a certain class of random times T, the independence of $N_{T+s} - N_T$ from the past history $\{N_u; u \leq T\}$ until T is still preserved, and furthermore, the distribution of $N_{T+s} - N_T$ is again Poisson with parameter λs. Such "good" random times T are characterized by the property that for any number $t \geq 0$, one can determine whether the event $\{T \leq t\}$ has occurred or not by knowing the history $\{N_u; u \leq t\}$ of the arrival process until the time t. Such random times are called *stopping times*. For example, the time T_6 of the sixth arrival is a stopping time, since we can tell whether $T_6 \leq 3.7$ or not by knowing what the arrival process looked like until the time $t = 3.7$ (and similarly for any other t). The first time T when an interarrival time exceeds a certain critical value c is a stopping time (this is, for example, the random time when a pedestrian completes crossing a highway if he needs a gap of c time units in traffic in order to cross safely).

(2.18) THEOREM. For any number $s \geq 0$ and any stopping time T,

$$P\{N_{T+s} - N_T = k \,|\, N_u; \; u \leq T\} = \frac{e^{-\lambda s}(\lambda s)^k}{k!}, \quad k = 0, 1, \dots .$$

Proof is omitted. Note that Proposition (2.2) is a special case, where $T = T_n$ and $k = 0$.

(2.19) EXAMPLE. The times vehicles cross a fixed point on a highway form a Poisson process with rate λ vehicles per second. A pedestrian arrives at this point at time 0 wanting to cross the highway. He needs 4 seconds to cross it. Therefore, for the realization ω, if $T_1(\omega) \leq 4$, $T_2(\omega) - T_1(\omega) \leq 4, \dots$, $T_6(\omega) - T_5(\omega) \leq 4$, and $T_7(\omega) - T_6(\omega) > 4$, then the time $T(\omega)$ he completes crossing is $T_6(\omega) + 4$; if the first interval greater than 4 were $T_{13}(\omega) - T_{12}(\omega)$, then $T(\omega)$ would have been $T(\omega) = T_{12}(\omega) + 4$. The time T is a stopping time (because the picture of N in the interval $[0, t]$ is sufficient for us to tell whether $T \leq t$ or not).

After he completes crossing, he will have escaped his doom by less than 2 seconds with probability

$$P\{N_{T+2} - N_T \geq 1\} = 1 - e^{-2\lambda}. \qquad \square$$

(2.20) EXAMPLE. Arrivals of buses at a certain stop form a Poisson process with rate $\lambda = 0.2$ buses per minute. (Though the arrival times are scheduled, when the average time between arrivals is short as is the case here, traffic delays, etc., introduce enough variation to make the arrival process become a Poisson process.) An inspector from the bus company arrives at the stop with the fifth bus after 9:00 A.M. We are interested in the distribution of the number of buses he will count during the one hour he plans to stay there. If 9:00 A.M. corresponds to time $t_0 = 120$ (i.e., if the time origin is taken to be 7:00 A.M.), then the time the inspector arrives is

$$T(\omega) = T_{N_{120}(\omega)+5}(\omega), \quad \omega \in \Omega.$$

This random time T is a stopping time; therefore, the distribution of the number of buses he will count in the next 60 minutes is

$$P\{N_{T+60} - N_T = k\} = \frac{e^{-12} \, 12^k}{k!}, \quad k = 0, 1, \dots . \qquad \square$$

3. Forward Recurrence Times

Let $\{N_t; t \geq 0\}$ be a Poisson process with rate λ, let $T_1, T_2, \dots$ be the succesive arrival times, and set $T_0 = 0$. Fix $\omega \in \Omega$ and $t \geq 0$. For the realization ω, the number of arrivals in $[0, t]$ is $N_t(\omega)$, the time of the last arrival before t is $T_{N_t(\omega)}(\omega)$, and the time of the next arrival is $T_{N_t(\omega)+1}(\omega)$. Then the time since the last arrival is $t - T_{N_t(\omega)}(\omega)$, and the length of the time from t until the next arrival is

$$V_t(\omega) = T_{N_t(\omega)+1}(\omega) - t.$$

In a different context, in Example (2.6), V_t is the remaining lifetime of the item in use at time t.

(3.1) THEOREM. We have

$$P\{V_t \le u \,|\, N_s, \ s \le t\} = 1 - e^{-\lambda u}, \quad u \ge 0,$$

independent of t.

   Proof. $\{V_t \le u\} = \{T_{N_t+1} - t \le u\} = \{T_{N_t+1} - t > u\}^c = \{T_{N_t+1} > t + u\}^c$ $= \{N_{t+u} - N_t = 0\}^c$. Thus,

$$
\begin{aligned}
P\{V_t \le u \,|\, N_s, \ s \le t\} &= P(\{N_{t+u} - N_t = 0\}^c \,|\, N_s, \ s \le t) \\
&= 1 - P\{N_{t+u} - N_t = 0 \,|\, N_s, \ s \le t\} \\
&= 1 - P\{N_{t+u} - N_t = 0\} \\
&= 1 - P\{N_u = 0\} = 1 - e^{-\lambda u}. \qquad \square
\end{aligned}
$$

In words, then, the time from t until the next arrival is independent of the past history of the process until t. Furthermore, it has the same distribution as any interarrival time. In particular, no matter how long it has been since the last arrival, the probability of waiting u more units for the next arrival to occur remains the same.

   As a result, $E[V_t] = 1/\lambda$. Note that $E[T_{n+1} - T_n]$ is also equal to $1/\lambda$ and that

$$E[T_{N_t+1} - T_{N_t}] = E[T_{N_t+1} - t + t - T_{N_t}] = E[T_{N_t+1} - t] + E[t - T_{N_t}]$$

$$= \frac{1}{\lambda} + E[t - T_{N_t}].$$

Thus, the interval $[T_{N_t}, T_{N_t+1})$ which covers the instant t is longer, on the average, than an ordinary interval $[T_n, T_{n+1})$ between two arrivals.

(3.2) EXAMPLE. Suppose the buses arrive at a bus stop according to a Poisson process at the rate of one every five minutes. You arrive at the bus stop in the evening at time $t = 617.23$ (measured in minutes starting from 7:00 in the morning). Theorem (3.1) implies that the waiting time until the next bus arrives has the same distribution as if you just missed the last one. The expected waiting time until the next bus is 5 minutes. In fact we can show that the expected time since the arrival of the last one is also 5 minutes (this latter depends, in general, on t, but becomes 5 for large t). Comparing the expected length of the interval within which you have happened to arrive (namely, 10 minutes) with the bus company's claim that their buses arrive on the average every 5 minutes, you may be annoyed. You are right in feeling annoyed; so is the company in their claim. If it is any comfort, look at it this way: it takes a larger than average interval to contain an instant distinguished enough to be your arrival time.   □

4. Superposition of Poisson Processes

Let $L = \{L_t; t \geq 0\}$ and $M = \{M_t; t \geq 0\}$ be two Poisson processes, independent of each other, with respective rates λ and μ. For each $\omega \in \Omega$ and $t \geq 0$, let

(4.1) $$N_t(\omega) = L_t(\omega) + M_t(\omega).$$

Then the process $N = \{N_t; t \geq 0\}$ is called the *superposition* of the processes L and M (see also Figure 4.4.1).

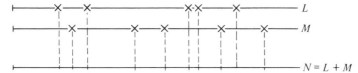

Figure 4.4.1 Times of arrivals of the superposition process N are obtained by putting together the arrival times of the processes L and M.

(**4.2**) THEOREM. N is a Poisson process with rate $\nu = \lambda + \mu$.

(4.3) EXAMPLE.   Let L_t be the number of vehicles arriving at an intersection in $(0, t]$ from one direction, and let M_t be the number of vehicles arriving in $(0, t]$ from the other direction. Then N_t is the number of vehicles arriving at the intersection from both directions during $(0, t]$. If the arrivals from the two directions are independent Poisson processes with respective rates λ and μ, then the total arrival process is also a Poisson process and has rate $\nu = \lambda + \mu$.   □

Proof of (4.2). By Theorem (1.18), it is sufficient to show that, for any time set B which is the union of a finite number of disjoint intervals of total length b, the number N_B of arrivals during B has the Poisson distribution with parameter νb. By definition, $N_B = L_B + M_B$, and L_B and M_B are independent

Poisson distributed random variables with respective parameters λb and μb. Hence,

$$
\begin{aligned}
P\{N_B = n\} &= \sum_{k=0}^{n} P\{L_B = k, M_B = n - k\} \\
&= \sum_{k=0}^{n} \frac{e^{-\lambda b}(\lambda b)^k}{k!} \frac{e^{-\mu b}(\mu b)^{n-k}}{(n-k)!} \\
&= \frac{e^{-(\lambda+\mu)b}(\lambda+\mu)^n b^n}{n!} \sum_{k=0}^{n} \frac{n!}{k!(n-k)!} \left(\frac{\lambda}{\lambda+\mu}\right)^k \left(\frac{\mu}{\lambda+\mu}\right)^{n-k} \\
&= \frac{e^{-\nu b}(\nu b)^n}{n!}
\end{aligned}
$$

for any $n \in \mathbb{N}$ (as a reminder, the last sum is 1 since it is the sum of all the terms of the distribution of the number of successes in n trials where the probability of success in one trial is $p = \lambda/(\lambda + \mu)$). □

Poisson processes are unique in this regard. For example, if L and M are independent renewal processes (that is, the times between arrivals of the L process are independent and identically distributed but with a distribution which is not necessarily exponential, and similarly for M), and if their super-position $N = L + M$ turns out to be a renewal process, then all three processes must be Poisson.

When one considers how restrictive the axioms for a Poisson process are, it looks rather unreasonable that it should come up in real life as often as it does. An explanation for this phenomenon can be based on the observation that, quite often, an arrival process being examined is the superposition of a large number of uniformly sparse arrival processes. For example, the arrivals of phone calls at a telephone exchange is the superposition of the times of calls of the individual subscribers. As there are a large number of subscribers, compared with the total telephone traffic, the contributions of the individual subscribers are each infinitesimal. It can then be shown that the superposition process is approximately a Poisson process.

5. Decomposition of a Poisson Process

Let $N = \{N_t; t \geq 0\}$ be a Poisson process with rate λ. Let $\{X_n; n = 1, 2, \ldots\}$ be a Bernoulli process, independent of N, with p for probability of success; and let S_n be the number of successes in the first n trials. Suppose that the nth trial is performed at the time T_n of the nth arrival. For a realization $\omega \in \Omega$, the number of trials performed during $[0, t]$ is $N_t(\omega)$, and therefore the number of successes obtained during $[0, t]$ is

$$
(5.1) \qquad\qquad M_t(\omega) = S_{N_t(\omega)}(\omega),
$$

and the number of failures is

(5.2) $L_t(\omega) = N_t(\omega) - M_t(\omega).$

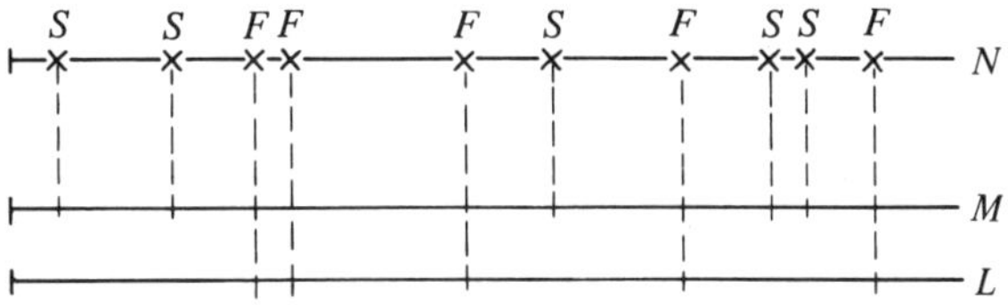

Figure 4.5.1   Those arrival times of N at which successes occur form
the process M and those at which failures occur form the process L.

(5.3)   THEOREM.   The processes $M = \{M_t; t \geq 0\}$ and $L = \{L_t; t \geq 0\}$ are
Poisson with respective rates λp and λq ($q = 1 - p$). Furthermore, L and M
are independent of each other.

(5.4)   EXAMPLE.   Vehicles arriving at a junction turn left or right with
probabilities 0.60 and 0.40 independent of each other. Arrivals at the junction
form a Poisson process with $\lambda = 30$ vehicles per minute. Then the vehicles
turning left form a Poisson process with parameter $30 \times 0.60 = 18$, and the
vehicles turning right form a Poisson process with parameter $30 \times 0.40 = 12$.
The two traffic streams thus created are independent of each other.   □

(5.5)   EXAMPLE.   Vehicles stopping at a roadside restaurant form a Poisson
process N with rate $\lambda = 20$ per hour. A vehicle has 1, 2, 3, 4, or 5 persons in
it with respective probabilities 0.30, 0.30, 0.20, 0.10, and 0.10. We would like
to find the expected number of persons arriving at the restaurant within one
hour.
   Let $N^1, N^2, \ldots,$ and N^5 be the number of vehicles arriving (within one
hour) with 1, 2, $\ldots,$ and 5 passengers. By Theorem (5.3), $N^1, \ldots, N^5$ are
Poisson distributed with respective parameters 6, 6, 4, 2, 2. Hence, $E[N^1]$,
$\ldots, E[N^5]$ are 6, 6, 4, 2, 2, and the expected value of the number of persons
arriving is

$E[N^1 + 2N^2 + 3N^3 + 4N^4 + 5N^5]$
$$= 6 + 2 \times 6 + 3 \times 4 + 4 \times 2 + 5 \times 2 = 48. \qquad □$$

   Proof of Theorem (5.3). In view of Theorem (1.17), it is sufficient to
show that

(5.6) $P\{M_{t+s} - M_t = m, L_{t+s} - L_t = k \mid M_u, L_u; u \leq t\}$
$$= \frac{e^{-\lambda ps}(\lambda ps)^m}{m!} \frac{e^{-\lambda qs}(\lambda qs)^k}{k!}; \quad k, m = 0, 1, \ldots$$

for any $s, t \geq 0$.

The event $\{M_{t+s} - M_t = m,\ L_{t+s} - L_t = k\}$ is equal to the event $\{N_{t+s} - N_t = m + k,\ M_{t+s} - M_t = m\}$, which is in turn equal to the event

$$A = \{N_{t+s} - N_t = m + k,\ S_{N_{t+s}} - S_{N_t} = m\}.$$

On the other hand, knowledge of the random variables $\{M_u, L_u;\ u \leq t\}$ is equivalent to that of

$$\mathcal{K} = \{N_u;\ u \leq t;\ X_1, \ldots, X_{N_t}\}.$$

Since N is a Poisson process, and since the processes N and S are independent, the random variable $N_{t+s} - N_t$ is independent of $\mathcal{K}$; similarly, since the number of successes in the trials numbered $N_t + 1, \ldots, N_{t+s}$ is independent of the past history $X_1, \ldots, X_{N_t}$, and since S and N are independent, $S_{N_{t+s}} - S_{N_t}$ is independent of $\mathcal{K}$. Hence, event A is independent of $\mathcal{K}$, and we have that the left-hand side of (5.6) is equal to

$$P(A) = P\{N_{t+s} - N_t = m + k,\ S_{N_{t+s}} - S_{N_t} = m\}$$

$$= \sum_{n=0}^{\infty} P\{N_t = n,\ N_{t+s} - N_t = m + k,\ S_{N_{t+s}} - S_{N_t} = m\}$$

$$= \sum_{n=0}^{\infty} P\{N_t = n,\ N_{t+s} = m + k + n,\ S_{m+k+n} - S_n = m\}$$

$$= \sum_{n=0}^{\infty} P\{N_t = n,\ N_{t+s} = n + m + k\}P\{S_{m+k+n} - S_n = m\}$$

by the independence of S and N. By (3.2.8), the distribution of $S_{n+m+k} - S_n$ is the same as that of S_{m+k}; thus,

$$P(A) = \sum_{n=0}^{\infty} P\{N_t = n,\ N_{t+s} - N_t = m + k\}P\{S_{m+k} = m\}$$

$$= P\{N_{t+s} - N_t = m + k\}P\{S_{m+k} = m\}$$

$$= \frac{e^{-\lambda s}(\lambda s)^{m+k}}{(m+k)!}\frac{(m+k)!}{m!k!}p^m q^k,$$

which is equal to the right-hand side of (5.6). $\qquad\square$

6.　Compound Poisson Processes

In this section we will remove the restriction (1.1a) in the definition of a Poisson process, and thus allow jumps of any size.

Throughout we will take $Z = \{Z_t;\ t \geq 0\}$ to be a stochastic process such that for any $\omega \in \Omega$, the mapping $t \longrightarrow Z_t(\omega)$ is right continuous and real-valued, has $Z_0(\omega) = 0$, and increases or decreases by jumps only.

(6.1) DEFINITION. *Z* is said to be a *compound Poisson process* provided that
(a) for almost all $\omega \in \Omega$, the function $t \longrightarrow Z_t(\omega)$ has only finitely many jumps in any finite interval;
(b) for any t and $s \geq 0$, $Z_{t+s} - Z_t$ is independent of the past history $\{Z_u; u \leq t\}$ until t;
(c) for any $t, s \geq 0$, the distribution of $Z_{t+s} - Z_t$ depends only on s (independent of t). ☐

Let Z be a compound Poisson process. By axiom (6.1a) it has only finitely many jumps in any finite interval; therefore, the jump times can be ordered. For each $\omega \in \Omega$, let $T_1(\omega), T_2(\omega), \ldots$ be the times of the first jump, the second jump, $\ldots$, and let $X_1(\omega), X_2(\omega), \ldots$ be the magnitudes of the corresponding jumps.

It follows from (6.1b) that the number of jumps in an interval $(t, t + s]$ is independent of the past history in $[0, t]$, and by (6.1c), the distribution of the number of jumps in $(t, t + s]$ is independent of t. Hence, $T_1, T_2, \ldots$ must be the arrival times in a Poisson process $N = \{N_t; t \geq 0\}$; that is, if N_t is the number of jumps of Z in $(0, t]$, then N is a Poisson process. The process Z differs from N by the fact that the magnitudes of the jumps of Z are random variables instead of being all equal to one. Moreover, it follows again from (6.1b) and (6.1c) that the jump sizes $X_1, X_2, \ldots$ are independent and identically distributed random variables which are, furthermore, independent of $T_1, T_2, \ldots$.

Conversely, if $T_1, T_2, \ldots$ are the arrival times in a Poisson process and if $X_1, X_2, \ldots$ are independent and identically distributed random variables which are also independent of the T_n, then the process Z obtained by summing all the X_i for which $T_i \leq t$ to make up Z_t is a compound Poisson process. Thus, we have the following qualitative characterization. See Figure 4.6.1 for a possible realization.

(6.2) PROPOSITION. *Z* is a compound Poisson process if and only if its jump times form a Poisson process and the magnitudes of its successive jumps are independent and identically distributed random variables independent of the jump times.

(6.3) EXAMPLE. Arrivals of customers into a store form a Poisson process N. The amount of money spent by the nth customer is a random variable X_n which is independent of all the arrival times (including his own) and all the amounts spent by others. The total amount spent by the first n customers is $Y_n = X_1 + \cdots + X_n$ if $n \geq 1$, and we set $Y_0 = 0$. Since the number of customers arriving in $(0, t]$ is N_t, the sales to customers arriving in $(0, t]$ total $Z_t = Y_{N_t}$. Our hypotheses on N and the X_n are such that the process $Z = \{Z_t; t \geq 0\}$ is a compound Poisson process. ☐

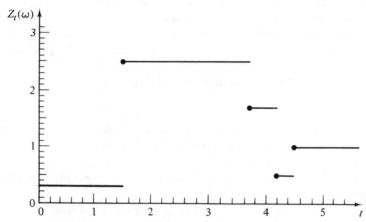

Figure 4.6.1 A compound Poisson process increases or decreases, by random amounts, at the times of arrivals of a Poisson process. For the realization ω pictured, arrival times are $T_1(\omega) = 1.5$, $T_2(\omega) = 3.7$, $T_3(\omega) = 4.2$, $T_4(\omega) = 4.5, \ldots$, and the jump sizes associated are $X_1(\omega) = 2.2$, $X_2(\omega) = -0.8$, $X_3(\omega) = -1.2$, $X_4(\omega) = 0.5, \ldots$.

(6.4) EXAMPLE. The times between the successive failures of a computer are independent and identically distributed exponential variables. With each failure there is associated a cost of repair. The costs associated with different failures are independent and identically distributed, and furthermore, the cost of a repair is independent of the times of failures. If Z_t is the cumulative cost in $(0, t]$, then $Z = \{Z_t; t \geq 0\}$ is a compound Poisson process. $\square$

Suppose Z is a compound Poisson process whose jumps occur at rate λ and by discrete amounts, that is, the random variables $X_1, X_2, \ldots$ take values in a countable set $E = \{a, b, \ldots\} \subset \mathbb{R}$. For each $\omega \in \Omega$, let $N_t^a(\omega), N_t^b(\omega), \ldots$ be the number of jumps of Z in $(0, t]$ whose magnitudes were exactly equal to a, equal to $b, \ldots$ (see Example (5.5) for the approach we are following). Then Theorem (5.3) applies to show that the processes $N^a, N^b, \ldots$ are independent Poisson processes with rates $\lambda(a) = \lambda P\{X_i = a\}$, $\lambda(b) = \lambda P\{X_i = b\}, \ldots$. This yields the following representation of Z: for any $t \geq 0$ and $\omega \in \Omega$,

(6.5) $$Z_t(\omega) = aN_t^a(\omega) + bN_t^b(\omega) + \cdots,$$

where $N^a, N^b, \ldots$ are independent Poisson processes with rates $\lambda(a), \lambda(b), \ldots$.

The representation (6.5) can be used to obtain any quantity concerning Z from the results already obtained for Poisson processes. For example, suppose $a, b, \ldots \geq 0$, and consider the Laplace transform of Z_t. By the independence of $N^a, N^b, \ldots$ in (6.5), Proposition (2.1.26) applies, and

$$E[e^{-\alpha Z_t}] = E[e^{-\alpha a N_t^a}]E[e^{-\alpha b N_t^b}] \cdots.$$

For a Poisson process M with rate μ, for any $\beta \geq 0$,

$$E[e^{-\beta M_t}] = \sum_{k=0}^{\infty} e^{-\beta k} \frac{e^{-\mu t}(\mu t)^k}{k!} = \exp[-\mu t(1 - e^{-\beta})].$$

Using this result repeatedly with $\beta = \alpha a, \alpha b, \ldots$ and $\mu = \lambda(a), \lambda(b), \ldots$, we get

$$(6.6) \qquad E[e^{-\alpha Z_t}] = \exp[-t \sum_{i \in E} (1 - e^{-\alpha i})\lambda(i)].$$

Conversely, if a process Z has the Laplace transform (6.6) for Z_t for all $t \geq 0$, then Z is a weighted sum of independent Poisson processes with rates $\lambda(i), i \in E$.

A somewhat more general approach makes use of the structure of Z outlined in Proposition (6.2) by conditioning on the number of jumps, as was done previously in Example (2.2.27).

If the number of jumps N_t of Z in $(0, t]$ is n, then Z_t is the sum of n independent and identically distributed random variables. Hence, if $E[X_i] = \mu$ and the rate of jumps is λ, then

$$(6.7) \qquad E[Z_t \mid N_t] = \mu N_t;$$

which gives, by Proposition (2.2.16),

$$(6.8) \qquad E[Z_t] = \mu \lambda t.$$

This approach can also be used to compute the Laplace transform of Z_t in the case where all the jumps are upward. We then get

(6.9) PROPOSITION. Suppose the jump times of Z form a Poisson process with rate λ and that the magnitudes of jumps are independent and identically distributed non-negative random variables, independent of the jump times, with distribution φ. Then for any $\alpha \geq 0$ and $t \geq 0$,

$$(6.10) \qquad E[e^{-\alpha Z_t}] = \exp[-\lambda t(1 - f(\alpha))],$$

where

$$f(\alpha) = E[e^{-\alpha X_t}] = \int_0^{\infty} e^{-\alpha u} \, d\varphi(u).$$

Proof. By the independence of the X_i from N, since $Z_t = X_1 + \cdots + X_n$ when $N_t = n, n \geq 1$,

$$E[e^{-\alpha Z_t} \mid N_t = n] = E[e^{-\alpha(X_1 + \cdots + X_n)}].$$

Since the X_i are independent of each other, by Proposition (2.1.26),

$$E[e^{-\alpha(X_1 + \cdots + X_n)}] = E[e^{-\alpha X_1}] \cdots E[e^{-\alpha X_n}] = f(\alpha)^n.$$

We have thus shown that

$$E[e^{-\alpha Z_t} | N_t] = [f(\alpha)]^{N_t};$$

and by Proposition (2.2.16),

$$E[e^{-\alpha Z_t}] = E[f(\alpha)^{N_t}] = e^{-\lambda t(1-f(\alpha))},$$

since for any $\beta \geq 0$,

$$E[\beta^{N_t}] = \sum_{k=0}^{\infty} \beta^k \frac{e^{-\lambda t}(\lambda t)^k}{k!} = e^{-\lambda t(1-\beta)}. \qquad \square$$

Note that we can write

$$(1 - f(\alpha))\lambda = \int_0^{\infty} (1 - e^{-\alpha u}) \, d\lambda(u)$$

by letting $\lambda(u) = \lambda \varphi(u)$. Then (6.10) becomes

(6.11) $$E[e^{-\alpha Z_t}] = \exp\left[-t \int_0^{\infty} (1 - e^{-\alpha u}) \, d\lambda(u)\right],$$

which is the continuous analog of (6.6). Analogous to the case where the jump sizes are discrete random rariables, we have the following result. Consider those jumps of Z whose magnitudes fall in the interval $(a, b]$; then the instants at which those jumps occur form a Poisson process with rate

$$\lambda(b) - \lambda(a) = \lambda \, P\{a < X_i \leq b\},$$

independent of the jumps with other sizes.

It follows from axiom (6.1c) that the distribution of any increment $Z_{t+s} - Z_t$ is the same as that of Z_s. By axiom (6.1b), the increments $Z_{t_1} - Z_{t_0}, \ldots,$ $Z_{t_n} - Z_{t_{n-1}}$ are independent of each other for any $n \geq 1$ and $0 \leq t_0 < t_1 < \cdots < t_n$. Thus, once the distribution of Z_t is known for all t, the joint distribution of any number m of variables $Z_{s_1}, \ldots, Z_{s_m}$ can be computed easily. Hence, the probability law of the process Z is completely specified once we are given the rate of jumps λ and the distribution φ of the jump sizes.

7. Non-Stationary Poisson Processes

In Definition (1.1), axioms (a) and (b) are quite applicable to many real-life arrival processes, but (c) is not. Usually, the rate of arrivals will depend on the time of day, etc. In this section we will study those arrival processes which satisfy only (1.1a) and (1.1b).

Consider an arrival process N; that is, for any $\omega \in \Omega$, the mapping $t \rightarrow N_t(\omega)$ is non-decreasing, increases by jumps only, is right continuous, and has $N_0(\omega) = 0$.

(7.1) DEFINITION. N is said to be a possibly non-stationary Poisson process provided that

(a)   for almost all ω, $t \rightarrow N_t(\omega)$ has jumps of size one only;

(b)   for any $t, s \geq 0$, the random variable $N_{t+s} - N_t$ is independent of the past history $\{N_u; u \leq t\}$ until t. $\square$

We will refer to N as a *stationary* Poisson process if it further satisfies (1.1c). Let N be a non-stationary Poisson process and let

$$(7.2) \qquad a(t) = E[N_t], \qquad t \geq 0.$$

Since $N_{t+s} \geq N_t$, $E[N_{t+s}] \geq E[N_t]$; that is, a is a non-decreasing function. If a sequence of numbers t_n decreases to t, then N_{t_n} decreases to N_t by the right continuity of N. By the monotone convergence theorem (2.1.34), this implies that, then, $a(t_n)$ decreases to $a(t)$. Hence a is a non-decreasing right continuous function. We will, for the time being, assume that a is, in fact, continuous. We define the *time inverse* of a by

$$(7.3) \qquad \tau(t) = \inf\{s : a(s) > t\}, \quad t \geq 0.$$

Then a look at Figure 4.7.1 will show that τ is also non-decreasing. Suppose that we have constructed a clock which reads t when the actual time is $\tau(t)$. The next theorem states that if one were to reckon time according to this clock, then the arrival process he is observing would be a stationary Poisson process.

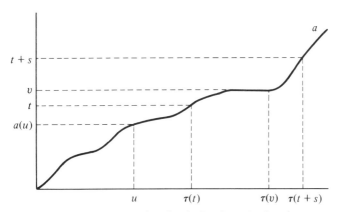

Figure 4.7.1   Constructing the clock τ from the function a.

(7.4) THEOREM. Let N be a non-stationary Poisson process, suppose a is continuous, and define

$$M_t(\omega) = N_{\tau(t)}(\omega)$$

for all $t \geq 0$ and $\omega \in \Omega$. Then $M = \{M_t; t \geq 0\}$ is a stationary Poisson process with rate 1.

Proof. Fix $t, s \geq 0$ and put $t' = \tau(t)$, $t' + s' = \tau(t + s)$. Then

$$(7.5) \qquad E[M_{t+s} - M_t \,|\, M_u; u \leq t] = E[N_{t'+s'} - N_{t'} \,|\, N_u; u \leq t'].$$

By axiom (7.1b), $N_{t'+s'} - N_{t'}$ is independent of the past history $\{N_u; u \leq t'\}$. Hence, the right-hand side of (7.5) is equal to

$$(7.6) \qquad E[N_{t'+s'} - N_{t'}] = a(t' + s') - a(t').$$

Since a is continuous, $a(t') = a(\tau(t)) = t$, and similarly $a(t' + s') = t + s$. Putting these in (7.6), we obtain

$$(7.7) \qquad E[M_{t+s} - M_t \,|\, M_u \leq t] = t + s - t = s.$$

By Theorem (1.17), (7.7) implies that M is a Poisson process with rate equal to one. $\qquad\square$

The following corollary restates (7.4) in terms of the times of arrivals. The proof is evident from Figure 4.7.2.

(7.8) COROLLARY. Let a be a continuous non-decreasing function. Then $T_1, T_2, \ldots$ are the arrival times in a non-stationary Poisson process N with

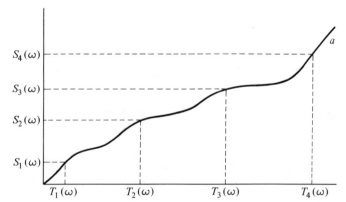

Figure 4.7.2 $T_1, T_2, \ldots$ form a non-stationary Poisson process with expectation function a if and only if $S_1, S_2, \ldots$ form a stationary Poisson process with rate 1.

$E[N_t] = a(t)$ if and only if $a(T_1)$, $a(T_2)$, ... are the arrival times in a stationary Poisson process with rate 1.

Let N be a non-stationary Poisson process with expectation function a. Almost surely, an arrival T_n falls in the interval $(t, t + s]$ if and only if $a(T_n)$ falls in the interval $(a(t), a(t + s)]$. Hence the number of arrivals in $(t, t + s]$ is equal to the number of $a(T_n)$ falling in the interval $(a(t), a(t + s)]$. By Corollary (7.8), the latter random number has the Poisson distribution with parameter $1 \times (a(t + s) - a(t))$. We have thus proved the following quantitative result.

(7.9) PROPOSITION. If N is a non-stationary Poisson process with a continuous expectation function a, then

$$P\{N_{t+s} - N_t = k\} = \frac{e^{-b(t,s)}b(t, s)^k}{k!}, \quad k = 0, 1, \ldots,$$

for any $t, s \geq 0$; here

$$b(t, s) = a(t + s) - a(t).$$

The preceding proposition is an example of how to use Theorem (7.4) and Corollary (7.8) to reduce a computation concerning a non-stationary Poisson process to a computation about a stationary one. Next is another result, again obtained by the same method, but now on the times of arrivals.

(7.10) PROPOSITION. Let $T_1, T_2, \ldots$ be the successive arrival times of a non-stationary Poisson process with a continuous expectation function a. Then for any n,

$$P\{T_{n+1} - T_n > t \mid T_1, \ldots, T_n\} = e^{-[a(T_n+t)-a(T_n)]}.$$

Before proving this, we note that the interarrival times are no longer independent as in a stationary Poisson process. However, they still have conditionally exponential distributions.

Proof. The event $A = \{a(T_{n+1}) > a(T_n + t)\}$ implies the event $\{T_{n+1} - T_n > t\}$, which in turn implies the event $B = \{a(T_{n+1}) \geq a(T_n + t)\}$. Thus,

$$P\{A \mid T_1, \ldots, T_n\} \leq P\{T_{n+1} - T_n > t \mid T_1, \ldots, T_n\} \leq P\{B \mid T_1, \ldots, T_n\}.$$

The $T_1, \ldots, T_n$ determine $a(T_1), \ldots, a(T_n)$, and $a(T_n + t)$. Therefore, by Corollary (7.8), $a(T_{n+1}) - a(T_n)$ is independent of $T_1, \ldots, T_n$, and has the exponential distribution with parameter one. Hence,

$$P\{A \mid T_1, \ldots, T_n\} = P\{a(T_{n+1}) - a(T_n) > a(T_n + t) - a(T_n) \mid T_1, \ldots, T_n\}$$
$$= \exp[-(a(T_n + t) - a(T_n))].$$

By the same reasoning,

$$P\{B \mid T_1, \ldots, T_n\} = \exp[-(a(T_n + t) - a(T_n))]$$

also, since $P\{X \geq t\} = P\{X > t\}$ for any random variable X having an exponential distribution. This completes the proof in view of the double inequality above.   □

(7.11) EXAMPLE. The times when successive demands occur for an expensive "fad" item (such as a certain model computer) form a non-stationary Poisson process N with the arrival rate at time t being

$$\lambda(t) = \frac{1875 \, e^{-3t}(3t)^2}{2}, \quad t \geq 0;$$

that is,

$$E[N_t] = a(t) = \int_0^t \lambda(s) \, ds = 625(1 - e^{-3t} - 3te^{-3t}).$$

The total demand N_∞ for this item throughout $[0, \infty)$ has the Poisson distribution with parameter 625. If, based on this analysis, the company has manufactured exactly 700 such items, the probability of a shortage would be

$$P\{N_\infty > 700\} = \sum_{k=701}^{\infty} \frac{e^{-625}(625)^k}{k!}.$$

Using the normal distribution to approximate this, we have

$$P\{N_\infty > 700\} = P\left\{ \frac{N_\infty - 625}{\sqrt{625}} > 3 \right\} \simeq \int_3^\infty \frac{1}{\sqrt{2\pi}} e^{-x^2/2} \, dx = 0.0013.$$

The time of sale of the last item is T_{700}; since $T_{700} \leq t$ if and only if $N_t \geq 700$, its distribution is

$$P\{T_{700} \leq t\} = P\{N_t > 700\} = 1 - \sum_{k=0}^{699} \frac{e^{-a(t)}a(t)^k}{k!}. \qquad \square$$

(7.12) EXAMPLE. In considering stochastic systems with non-stationary Poisson arrivals, simulation is used quite often (more often than needed, in fact) as a quick method of "solution." It seems desirable, therefore, to mention how Corollary (7.8) can be used to generate a possible realization of arrival times from a non-stationary Poisson process with a given continuous expectation function a.

First, draw the function a as in Figure 4.7.2. To get one possible realization, we do the following:

(a)   set $s_0 = 0, t_0 = 0$;
(b)   for $n = 1, 2, \ldots$
      (i)   pick a "random number" ω_n from the tables,
     (ii)   compute $x_n = -\log \omega_n$,
    (iii)   compute $s_n = s_{n-1} + x_n$,
    (iv)   enter s_n on the vertical axis in the figure drawn above, and read the corresponding number t_n on the horizontal axis (as in Figure 4.7.2).

The sequence of numbers $t_1, t_2, t_3, \ldots$ obtained gives a possible realization of the arrival process desired.

In the language of probability theory, what we have done is the following. We have taken a sample space Ω which is the set of all possible sequences $\omega = (\omega_1, \omega_2, \ldots)$ of numbers ω_i in [0, 1]. A probability P is constructed on Ω in such a way that for any n and numbers $a_1, \ldots, a_n$ in [0, 1],

$$P(\{\omega: \omega_1 \leq a_1, \ldots, \omega_n \leq a_n\}) = a_1 \cdots a_n.$$

A sequence $\{X_n; n = 1, 2, \ldots\}$ of random variables is defined by

$$X_n(\omega) = -\log \omega_n, \quad \omega \in \Omega.$$

Then for any $b_1, \ldots, b_n \geq 0$,

$$\begin{aligned}
P\{X_1 > b_1, \ldots, X_n > b_n\} &= P(\{\omega: -\log \omega_1 > b_1, \ldots, -\log \omega_n > b_n\}) \\
&= P(\{\omega: \omega_1 < e^{-b_1}, \ldots, \omega_n < e^{-b_n}\}) \\
&= e^{-b_1} \cdots e^{-b_n};
\end{aligned}$$

that is, $X_1, X_2, \ldots$ are independent and identically distributed with the common distribution being the exponential one with parameter 1. Then by Theorem (2.5), $S_1 = X_1, S_2 = X_1 + X_2, S_3 = X_1 + X_2 + X_3, \ldots$ are the arrival times of a stationary Poisson process with rate 1. By entering the S_n on the vertical axis to read off the T_n on the horizontal one (on a graph of the function a) we obtain a sequence $T_1, T_2, \ldots$ which are the arrival times of the non-stationary Poisson process N with the expectation function a, by Corollary (7.8). In the simulation procedure outlined above, the sequence $\omega = (\omega_1, \omega_2, \ldots)$ of "random numbers" ω_n picked is a possible outcome in the sample space Ω we have constructed. The numbers $x_1, x_2, \ldots$ computed were the actual values $X_1(\omega), X_2(\omega), \ldots$ of the random variables $X_1, X_2, \ldots$ corresponding to this outcome ω. Similarly, the numbers $t_1, t_2, \ldots$ that were obtained are the actual values $T_1(\omega), T_2(\omega), \ldots$ corresponding to this particular ω.   □

Finally, before ending this section, we briefly consider the case where the expectation function a is not continuous. We have shown that a is necessarily non-decreasing and right continuous. Since it is non-decreasing, the left-hand limit $a(t-) = \lim_{s \uparrow t} a(s)$ exists for all t. Suppose now that for a particular point t, $a(t-) \neq a(t)$, and let $\alpha = a(t) - a(t-)$. Then the number of arrivals $N_t - N_{t-}$ occurring at exactly t has the expected value $\alpha > 0$. By axiom (7.1a), this random variable $N_t - N_{t-}$ is either 0 or 1; therefore, its expectation is also the probability that it is equal to 1, that is,

(7.13) $P\{N_t - N_{t-} = 1\} = a(t) - a(t-) = \alpha.$

Thus, we may think of such a point t as the time at which an arrival is scheduled but might fail to materialize with probability $1 - \alpha$.

If a has jumps of magnitudes $\alpha_1, \alpha_2, \ldots$ at the fixed times $t_1, t_2, \ldots$, then the above analysis shows that there will be an arrival at exactly t_i with probability α_i, $i = 1, 2, \ldots$.

A general non-stationary Poisson process is, then, composed of two streams of arrivals:

(7.14) $N_t(\omega) = N_t^f(\omega) + N_t^c(\omega), \quad \omega \in \Omega, t \geq 0.$

The possible jump times of N^f are all *fixed;* these times are the points of discontinuity of a; and the probability of an arrival at such a fixed time t is $a(t) - a(t-)$. If we write $a^f(t)$ for the sum of all the jumps of a in $[0, t]$, that is, if

(7.15) $a^f(t) = \sum_{s \leq t} [a(s) - a(s-)],$

then

(7.16) $a^c(t) = a(t) - a^f(t), \quad t \geq 0,$

is a continuous non-decreasing function. The second component N^c in the decomposition (7.14) is a non-stationary Poisson process with the expectation function a^c. To this component, then, our previous propositions are applicable.

(7.17) EXAMPLE. Arrivals of customers at a single-barber shop form a non-stationary Poisson process. On a particular day, there are appointments made for 12: 00, 12: 20, 1: 00, 3: 40, 4: 20, 4: 40. Past experience indicates that an appointment is kept with probability $\alpha = 2/3$. Customers without appointments arrive at the rate of 1 per hour during the first 3 hours, at the rate of 0.4 during the next 2 hours, and at the rate of 0.2 per hour during the last hour. The expectation function a is the sum of a^f and a^c shown in Figure 4.7.3.

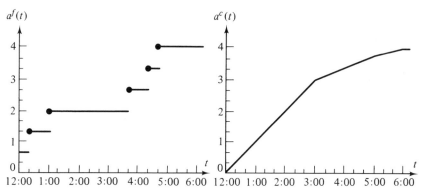

Figure 4.7.3   Expectation functions for scheduled and unscheduled arrivals at a barber shop.

The number of customers arriving with appointments during the first 4 hours is N_4^f; it has the distribution (noting that there are 3 scheduled during that interval)

$$P\{N_4^f = k\} = \binom{3}{k} 0.6^k \, 0.4^{3-k}, \qquad k = 0, 1, 2, 3.$$

The distribution of the number arriving without appointments during this same time is (noting that $a^c(4) = 3.4$)

$$P\{N_4^c = k\} = \frac{e^{-3.4} \, 3.4^k}{k!}, \qquad k = 0, 1, 2, \ldots.$$

The probability that the total number N_4 to arrive is 8 is, since N^f and N^c are independent,

$$P\{N_4 = 8\} = \sum_{k=0}^{3} P\{N_4^f = k\} P\{N_4^c = 8 - k\}$$

$$= (0.4)^3 \frac{e^{-3.4}(3.4)^8}{8!} + 3(0.6)(0.4)^2 \frac{e^{-3.4}(3.4)^7}{7!}$$

$$+ 3(0.6)^2(0.4) \frac{e^{-3.4}(3.4)^6}{6!} + (0.6)^3 \frac{e^{-3.4}(3.4)^5}{5!}. \qquad \square$$

8.   Exercises

(8.1)   Let $N = \{N_t; t \geq 0\}$ be a Poisson process with rate $\lambda = 15$. Compute
(a) $P\{N_6 = 9\}$,
(b) $P\{N_6 = 9, N_{20} = 13, N_{56} = 27\}$,
(c) $P\{N_{20} = 13 \,|\, N_6 = 9\}$,
(d) $P\{N_6 = 9 \,|\, N_{20} = 13\}$.

(8.2) Let N be a Poisson process with rate $\lambda = 2$. Compute
   (a) $E[N_t]$, $\text{Var}(N_t)$,
   (b) $E[N_{t+s} \mid N_t]$.

(8.3) Arrivals of passengers at a bus stop form a Poisson process N with rate $\lambda = \frac{1}{3}$ per minute. Assume that a bus has left at time $t = 0$ leaving no customers behind. Let T denote the time of arrival for the next bus; then the number of passengers present when it arrives will be N_T. We suppose that T is independent of N and has the distribution φ.
   (a)   Compute $E[N_T \mid T]$ and $E[N_T^2 \mid T]$.
   (b)   Compute $E[N_T]$ and $\text{Var}(N_T)$ for

$$d\varphi(t) = \begin{cases} \dfrac{1}{2}\,dt & \text{if } 9 \le t \le 11, \\ 0 & \text{otherwise.} \end{cases}$$

(8.4)   A store promises to give a small gift to every thirteenth customer to arrive. If the arrivals of customers form a Poisson process with rate λ,
   (a)   find the probability density function of the times between the lucky arrivals;
   (b)   find $P\{M_t = k\}$ for the number of gifts M_t given in the interval $[0, t]$.

(8.5)   Arrivals of customers into a store form a Poisson process N with rate $\lambda = 20$ per hour. Find the expected number of sales made during an eight-hour business day if the probability that a customer buys something is 0.30.

(8.6)   Consider the road network pictured in Figure 4.8.1. The inputs are Poisson processes with the rates indicated, and the probabilities of a vehicle choosing the indicated directions are written on the arrows. Describe the traffic flow on each branch of the network.

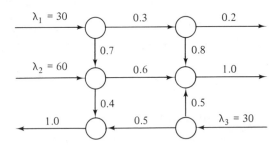

Figure 4.8.1   Network of Exercise (8.6).

(8.7)   A department store has 3 doors. Arrivals at each door form Poisson processes with rates $\lambda_1 = 110$, $\lambda_2 = 90$, $\lambda_3 = 160$ customers per hour. 30% of all customers are male. The probability that a male customer buys something is 0.80, and the probability of a female customer buying something is 0.10. An average purchase is worth $4.50.
   (a)   What is the average worth of total sales made in a 10 hour day?

(b)   What is the probability that the third female customer to purchase anything arrives during the first 15 minutes? What is the expected time of her arrival?

(8.8)   Considering the traffic on the road pictured in Figure 4.8.2, the following is known. The number of vehicles passing the point A in an hour follows the Poisson

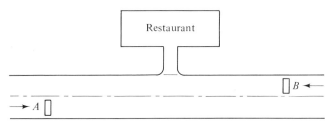

Figure 4.8.2   Picture for Exercise (8.8).

distribution with mean 60; 20% of these vehicles are trucks. The number of vehicles passing B in an hour is also Poisson distributed with mean 80; 30% of these are trucks. In general, 10% of all vehicles stop at the restaurant. The number of persons in a truck is one; the number of passengers in a car is equal to 1, 2, 3, 4, or 5 with respective probabilities 0.30, 0.30, 0.20, 0.10, and 0.10.

(a)   Find the expected value $E[Z]$ of the number of persons Z arriving at the restaurant within that one hour.

(b)   Compute $E[\alpha^Z]$ for $\alpha \in [0, 1]$.

(8.9)   A device is subject to shocks which occur according to a Poisson process N with rate λ. The device can fail only due to a shock, and the probability that a given shock causes failure is p independent of the number and times of previous shocks. Let K be the total number of shocks the device takes before failure, and let $T = T_K$ be the time of failure.

(a)   Compute $E[T]$, $\text{Var}(T)$.

(b)   Compute $E[T|K]$.

(c)   Compute $E[T|K > 9]$.

(8.10)   *Mixtures of Poisson Processes.* Let X be a random variable taking values in $E = \{a, b, \ldots\}$ with distribution $\pi = (\pi(a), \pi(b), \ldots)$. For each $i \in E$, let $N(i)$ be a Poisson process with rate $\lambda(i)$. Suppose the processes $N(a), N(b), \ldots$ are independent of each other and of X. Define, for each $t \in [0, \infty)$,

$$N_t = N_t(X).$$

The process $N = \{N_t; t \geq 0\}$ is then called the mixture of $N(a), N(b), \ldots$ according to the distribution π. (This is a misnomer, since N is in fact one of the $N(a), N(b), \ldots$ picked at random; the computation below explains the name better).

(a)   Show that for any $t \geq 0$,

$$P\{N_t = k\} = \sum_{i \in E} \pi(i) \frac{e^{-\lambda(i)t}(\lambda(i)t)^k}{k!}, \quad k = 0, 1, \ldots.$$

(b)   Show that in general, N does not have independent increments.

(8.11) Suppose $T_1, T_2, \ldots$ are the arrival times of a Poisson process with rate λ, and let f be a non-negative function. Show that

$$E\left[\exp\left[-\sum_{n=1}^{\infty} f(T_n)\right]\right] = \exp\left[-\lambda \int_0^{\infty} [1 - e^{-f(t)}]\, dt\right].$$

(*Hint:* First show this to be true for simple functions f, then use the fact that any function $f \geq 0$ is the increasing limit of a sequence of simple functions.)

(8.12) *Continuation.* Compute the expected value and variance of $Z = \sum f(T_n)$ by using (8.11). (*Hint:* Use (8.11) to obtain $E[e^{-\alpha Z}]$ first for $\alpha \geq 0$.)

(8.13) *Continuation.* Suppose $T_1, T_2, \ldots$ are the arrival times of a non-stationary Poisson process N with a continuous expectation function a. Show that for any function $f \geq 0$,

$$(8.14) \qquad E\left[\exp\left[-\sum_{n=1}^{\infty} f(T_n)\right]\right] = \exp\left[-\int_0^{\infty} [1 - e^{-f(t)}]\, da(t)\right].$$

Use this to compute

$$E[\textstyle\sum f(T_n)], \qquad \mathrm{Var}(\textstyle\sum f(T_n)).$$

(8.15) It can be shown that if (8.14) holds for all functions $f \geq 0$, then $T_1, T_2, \ldots$ form a non-stationary Poisson process with the continuous function a as the expectation function. Use this result to show that the superposition of two independent non-stationary Poisson processes without fixed discontinuities is again a non-stationary Poisson process.

$$*\qquad *\qquad *$$

The learned Bishop Hall, I mean the famous Dr. Joseph Hall, who was Bishop of Exeter in King James the First's reign, tells us in one of his Decads, at the end of his devine art of meditation, imprinted at London, in the year 1610, by John Beal, dwelling in Aldersgate-street, "That it is an abominable thing for a man to commend himself;"—and I really think it is so.

And yet, on the other hand, when a thing is executed in a masterly kind of fashion, which thing is not likely to be found out;—I think it is full as abominable, that a man should lose the honour of it, and go out of the world with the conceit of it rotting in his head.

This is precisely my situation.

For in this long chapter which by accident became too long, as in all my chapters, there is a skillful blending of the abstract with the practical, the merit of which has all along, I fear, been overlooked by my reader,—not for want of penetration in him,—but because 'tis an excellence seldom looked for, or expected indeed, in a work on stochastic processes.

—This is vile work.—For which reason, from the beginning, you see, I have constructed the main work and the adventitious parts of it with such intersections, *and have so complicated and involved the practical and the abstract that the two, which were thought to be at variance with each other, are reconciled as one wheel within another, and the whole machine, in general, has been kept a-going.*[1]

I have given a qualitative definition which enables the practitioner to decide, using his intuition, whether or not the Poisson process provides a plausible model for the phenomenon he is interested in. The method used for the proof of Theorem (1.9) is close to that of KHINCHINE [1]. *The proof of Lemma (1.5) is the neatest known to us; it is based on a recent proof by* CHUNG [3]. *Most textbooks include this lemma as an axiom within the definition, with claims of it being intuitively clear. Somehow, why it should be intuitive has always escaped us.*

Theorem (1.17) is due to WATANABE [1]; *it is a nice example of how abstract mathematics, honestly executed, can yield the most intuitive and practical results. Theorem (1.18) is due to* RÉNYI [1]; *its proof is a rare example of using quantitative data to prove stochastic independence. In most instances, such things as independence are matters of assumption to be checked by statistical means.*

[1]Fashioned after *Tristram Shandy* by Laurence Sterne, Chap. XXII.

CHAPTER 5

Markov Chains

This and the next three chapters are devoted to stochastic processes whose futures are conditionally independent of their pasts provided that their present values be known. The present chapter is to introduce the fundamental notions and terminology of such processes.

The theory of Markovian processes comprises the largest and the most important chapter in the theory of stochastic processes; this importance is further enhanced by the many applications it has found in both the physical, biological, and social sciences and in engineering and commerce.

The subject matter and the notations to be used are introduced in Section 1. It is usual to think of a Markov chain as the sequence of states entered by a system evolving in time, or the sequence of positions occupied by a moving particle. In Section 2, we consider the process as an observer at a fixed position would see it: that is, the successive times at which that position is visited by the moving particle. It turns out that much can be learned about the motion as a whole from such fragmentary data. Section 3 introduces a classification of states which provides the tools for accomplishing this.

1. Introduction

Let Ω be a sample space and P a probability measure on it. Consider a stochastic process $X = \{X_n; n \in \mathbb{N}\}$ with a countable state space E; that is, for each $n \in \mathbb{N} = \{0, 1, \ldots\}$ and $\omega \in \Omega$, $X_n(\omega)$ is an element of the countable set E. It is customary to say "the process is in state j at time n" to mean $X_n = j$. Thus, X_n is referred to as the state of the process X at time n, and the set E is called the state space of the process X.

(1.1) DEFINITION. The stochastic process $X = \{X_n; n \in \mathbb{N}\}$ is called a *Markov chain* provided that

$$P\{X_{n+1} = j \mid X_0, \ldots, X_n\} = P\{X_{n+1} = j \mid X_n\}$$

for all $j \in E$ and $n \in \mathbb{N}$. □

A Markov chain, then, is a sequence of random variables such that for any n, X_{n+1} is conditionally independent of $X_0, \ldots, X_{n-1}$ given X_n. That is, the "next" state X_{n+1} of the process is independent of the "past" states $X_0, \ldots, X_{n-1}$ provided that the "present" state X_n be known.

Throughout the discussion we will restrict ourselves to Markov chains for which the conditional probability

$$(1.2) \qquad P\{X_{n+1} = j \mid X_n = i\} = P(i, j), \qquad i, j \in E,$$

is independent of n. This restriction is convenient from the computational point of view and, as Exercise (4.13) shows, does not cause any loss of generality from the theory. The probabilities $P(i, j)$ are then called the *transition probabilities* for the Markov chain X, and a Markov chain X satisfying (1.2) is said to be *time-homogeneous* if emphasis is needed. It is customary to arrange the $P(i, j)$ into a square array and to call the resulting matrix P the *transition matrix* of the Markov chain X. If $E = \{0, 1, \ldots\}$, for example, the transition matrix is

$$P = \begin{bmatrix} P(0, 0) & P(0, 1) & P(0, 2) & \cdots \\ P(1, 0) & P(1, 1) & P(1, 2) & \cdots \\ P(2, 0) & P(2, 1) & P(2, 2) & \cdots \\ \cdot & \cdot & \cdot & \cdots \\ \cdot & \cdot & \cdot & \cdots \\ \cdot & \cdot & \cdot & \cdots \end{bmatrix}.$$

Following are the principles of notation we will follow regarding matrices and vectors.

(1.3) NOTATIONS. If $M(i, j)$ are real numbers defined for all i, j in some countable set E, then by M we denote the matrix whose (i, j)-entry is $M(i, j)$. Conversely, if M is a matrix, by $M(i, j)$ we mean the (i, j)-entry of M. We will denote the column vectors by lower case letters such as $f, g, h, \ldots$, and the row vectors by lower case Greek characters such as $\pi, \nu, \ldots$. If f is a column vector, $f(i)$ is its i-entry; if π is a row vector, $\pi(j)$ is its j-entry. Column vectors are to be thought of as *functions* defined on E, and when it is clear that we are talking of a function f we will take the liberty of displaying it as $f = (a, b, \ldots)$ for typographical ease, even though we are still thinking of it as a column vector.

The *identity matrix* will always be denoted by I; so $I(i, j) = 1$ or 0 according as $i = j$ or $i \neq j$. Any vector or matrix whose every entry is zero will be denoted by 0. The column vector of ones will be denoted by 1; so $1(i) = 1$ for all i. For any j, we will write 1_j for the column vector whose every entry is zero except the j-entry, which is one; so $1_j(i) = I_{\{j\}}(i) = 1$ or 0 according as $i = j$ or $i \neq j$.

Equalities and inequalites between vectors (or between matrices) are always term by term. So $M \geq 0$ means $M(i, j) \geq 0$ for all i, j; $f \leq g$ means $f(i) \leq g(i)$ for all i.   ☐

Every entry of a transition matrix P is non-negative, and the terms in any one row sum to unity. Such matrices are given a special name.

(1.4)   DEFINITION.   Let P be a square matrix of entries $P(i, j)$ defined for all $i, j \in E$. Then P is called a *Markov matrix over E* provided that
   (a) for any $i, j \in E$, $P(i, j) \geq 0$, and
   (b) for each $i \in E$, $\sum_{j \in E} P(i, j) = 1$.   ☐

Thus, the transition matrix of a Markov chain is a Markov matrix. Conversely, for any given Markov matrix over a countable set E, it is possible to construct a sample space Ω, a probability P on all subsets of Ω, and random variables $X_0, X_1, \ldots$ on Ω taking values in E such that $X = \{X_n\}$ is a Markov chain whose transition matrix is the given matrix. In such constructions, it is usual to take Ω to be the set of all sequences $\omega = (\omega_0, \omega_1, \omega_2, \ldots)$ with $\omega_i \in E$ for every i and to define $X_n(\omega) = \omega_n$. Then every realization ω is the sequence of states being observed corresponding to that realization.

Let X be a Markov chain with transition matrix P and state space E, and let $i, j, k \in E$ be fixed. From the elementary definition (1.3.1) of conditional probabilities, we can write

$$P\{X_6 = j, X_7 = k \mid X_5 = i\} = P\{X_7 = k \mid X_5 = i, X_6 = j\}P\{X_6 = j \mid X_5 = i\}.$$

On the other hand, by Definition (1.1),

$$P\{X_7 = k \mid X_5 = i, X_6 = j\} = P\{X_7 = k \mid X_6 = j\};$$

and by the time-homogeneity condition (1.2),

$$P\{X_7 = k \mid X_6 = j\} = P(j, k),$$
$$P\{X_6 = j \mid X_5 = i\} = P(i, j).$$

Hence,

$$P\{X_6 = j, X_7 = k \mid X_5 = i\} = P(i, j)P(j, k).$$

This method of proof gives, by induction, the following more general result.

(1.5) THEOREM. For any $n, m \in \mathbb{N}$ with $m \geq 1$ and $i_0, \ldots, i_m \in E$,

$$P\{X_{n+1} = i_1, \ldots, X_{n+m} = i_m \mid X_n = i_0\} = P(i_0, i_1)P(i_1, i_2) \cdots P(i_{m-1}, i_m).$$

Taking $n = 0$ and using Definition (1.3.1) of conditional probabilities yields the following

(1.6) COROLLARY. Let π be a probability distribution on E, and suppose

$$P\{X_0 = i\} = \pi(i)$$

for all $i \in E$. Then for any $m \in \mathbb{N}$ and $i_0, \ldots, i_m \in E$,

$$P\{X_0 = i_0, X_1 = i_1, \ldots, X_m = i_m\} = \pi(i_0)P(i_0, i_1) \cdots P(i_{m-1}, i_m).$$

This corollary shows that the joint distribution of $X_0, \ldots, X_m$ is completely specified for every m once the initial distribution π and the transition matrix P are known. By using Theorem (1.1.10) as many times as necessary, we can then get the joint distribution of $X_{n_1}, \ldots, X_{n_k}$ for any integer $k \geq 1$ and $n_1, \ldots, n_k \in \mathbb{N}$. The following are some results in this direction.

For any $h, i, j, k \in E$ we can write, from Theorem (1.5),

$$P\{X_{n+1} = i, X_{n+2} = j, X_{n+3} = k \mid X_n = h\} = P(h, i)P(i, j)P(j, k);$$

from this, it follows through Theorem (1.1.10) that

$$P\{X_{n+3} = k \mid X_n = h\} = \sum_{i \in E} P(h, i) \sum_{j \in E} P(i, j)P(j, k).$$

We now note that summing the terms $P(i, j)P(j, k)$ over all $j \in E$ yields the (i, k)-entry of the square of the matrix P, which is $P^2(i, k)$ in our system of notation. Putting this in the equation above, we get

$$P\{X_{n+3} = k \mid X_n = h\} = \sum_{i \in E} P(h, i)P^2(i, k).$$

But the sum on the right is the (h, k)-entry of the product of the matrices P and P^2, that is, the (h, k)-entry of the third power of the matrix P, which is written as $P^3(h, k)$ in our system. Hence, we have shown that

$$P\{X_{n+3} = k \mid X_n = h\} = P^3(h, k)$$

for all $h, k \in E$ and $n \in \mathbb{N}$. By iteration, the same proof yields the following general result.

(1.7) Proposition. For any $m \in \mathbb{N}$,

$$P\{X_{n+m} = j \mid X_n = i\} = P^m(i, j)$$

for all $i, j \in E$ and $n \in \mathbb{N}$ (in particular, for $m = 0$, $P^0 = I$).

In words, the probability that the chain moves from state i to state j in m steps is the (i, j)-entry of the mth power of the transition matrix P. Obviously, for any $m, n \in \mathbb{N}$

$$P^{m+n} = P^m P^n,$$

which in open form becomes

(1.8) $$P^{m+n}(i, j) = \sum_{k \in E} P^m(i, k) P^n(k, j); \quad i, j \in E.$$

Equation (1.8) is called the *Chapman–Kolmogorov equation*; it states that starting at state i, in order for the process X to be in state j after $m + n$ steps, it must be in some intermediate state k after the mth step and then move from that state k into state j during the remaining n steps.

In some applications, the powers P^m of the transition matrix P are desired for a large number of m. Then it may be worth considering the various matrix-theoretic methods available for such computations. Some such results may be found in the appendix at the end of this book.

(1.9) Example. Let $X = \{X_n; n \in \mathbb{N}\}$ be a Markov chain with state space $E = \{a, b, c\}$ and transition matrix

$$P = \begin{bmatrix} \frac{1}{2} & \frac{1}{4} & \frac{1}{4} \\ \frac{2}{3} & 0 & \frac{1}{3} \\ \frac{3}{5} & \frac{2}{5} & 0 \end{bmatrix}.$$

Then

$$P\{X_1 = b, X_2 = c, X_3 = a, X_4 = c, X_5 = a, X_6 = c, X_7 = b \mid X_0 = c\}$$
$$= P(c, b)P(b, c)P(c, a)P(a, c)P(c, a)P(a, c)P(c, b)$$
$$= \frac{2}{5} \cdot \frac{1}{3} \cdot \frac{3}{5} \cdot \frac{1}{4} \cdot \frac{3}{5} \cdot \frac{1}{4} \cdot \frac{2}{5} = \frac{3}{2500}.$$

The two-step transition probabilities are given by

$$P^2 = \begin{bmatrix} \dfrac{17}{30} & \dfrac{9}{40} & \dfrac{5}{24} \\[2mm] \dfrac{8}{15} & \dfrac{3}{10} & \dfrac{1}{6} \\[2mm] \dfrac{17}{30} & \dfrac{3}{20} & \dfrac{17}{60} \end{bmatrix}$$

so that, for example, $P\{X_{n+2} = c \mid X_n = b\} = P^2(b, c) = \frac{1}{6}$. □

(1.10) EXAMPLE. Let X be a Markov chain with state space $E = \{1, 2\}$, initial distribution $\pi = (\frac{1}{3}, \frac{2}{3})$, and transition matrix

$$P = \begin{bmatrix} 0.5 & 0.5 \\ 0.3 & 0.7 \end{bmatrix}.$$

Then

$$P^2 = \begin{bmatrix} 0.40 & 0.60 \\ 0.36 & 0.64 \end{bmatrix}, \quad P^3 = \begin{bmatrix} 0.380 & 0.620 \\ 0.372 & 0.628 \end{bmatrix}, \quad P^4 = \begin{bmatrix} 0.3760 & 0.6240 \\ 0.3744 & 0.6256 \end{bmatrix},$$

and in general, using the techniques explained in the appendix,

$$P^m = \begin{bmatrix} \frac{3}{8} + \frac{5}{8}(0.2)^m & \frac{5}{8} - \frac{5}{8}(0.2)^m \\[1mm] \frac{3}{8} - \frac{3}{8}(0.2)^m & \frac{5}{8} + \frac{3}{8}(0.2)^m \end{bmatrix}.$$

We have

$$P\{X_1 = 2, X_4 = 1, X_6 = 1, X_{18} = 1 \mid X_0 = 1\}$$
$$= P(1, 2)P^3(2, 1)P^2(1, 1)P^{12}(1, 1) = (0.5)(0.372)(0.40)(\tfrac{3}{8} + \tfrac{5}{8}(0.2)^{12});$$
$$P\{X_2 = 1, X_7 = 2, X_9 = 2 \mid X_0 = 1\} = P^2(1, 1)P^5(1, 2)P^2(2, 2)$$
$$= (0.40)(\tfrac{5}{8} - \tfrac{5}{8}(0.2)^5)(0.64);$$
$$P\{X_2 = 1, X_7 = 2\} = \sum_i P\{X_0 = i\}P\{X_2 = 1, X_7 = 2 \mid X_0 = i\}$$
$$= \pi(1)P^2(1, 1)P^5(1, 2) + \pi(2)P^2(2, 1)P^5(1, 2)$$
$$= \tfrac{1}{3}(0.40)(\tfrac{5}{8} - \tfrac{5}{8}(0.2)^5) + \tfrac{2}{3}(0.36)(\tfrac{5}{8} - \tfrac{5}{8}(0.2)^5).$$ □

(1.11) EXAMPLE. *Number of successes in Bernoulli process.* Let N_n denote the number of successes in n Bernoulli trials, where the probability of a success in any one trial is p. Then Theorem (3.2.11) implies that $\{N_n; n \geq 0\}$

is a Markov chain. Here the state space is $\{0, 1, 2, \ldots\}$, the initial distribution is $\pi(0) = 1$, $\pi(j) = 0$, $j \geq 1$; and the transition probabilities are

$$P(i, j) = P\{N_{n+1} = j \mid N_n = i\} = \begin{cases} p & \text{if } j = i + 1, \\ q & \text{if } j = i, \\ 0 & \text{otherwise.} \end{cases}$$

The transition matrix is

$$P = \begin{bmatrix} p & q & & & & 0 \\ & p & q & & & \\ & & p & q & & \\ & & & p & q & \\ & & & & \cdot & \cdot \\ 0 & & & & & \cdot \end{bmatrix}.$$

We then have, by Theorem (3.2.11) again,

$$P^n(i, j) = \binom{n}{j - i} p^{j-i} q^{n-j+i}, \quad j = i, \ldots, n + i. \qquad \square$$

(1.12) EXAMPLE. Let T_n be the time of the nth success in a Bernoulli process. Then Theorem (3.6) of Chapter 3 shows that $\{T_n; n \geq 0\}$ is a Markov chain. The state space is $\{0, 1, 2, \ldots\}$; $T_0 = 0$, so that the initial distribution is $\pi(0) = 1$, $\pi(1) = \pi(2) = \cdots = 0$; and the transition probabilities are

$$P(i, j) = P\{T_{n+1} = j \mid T_n = i\} = P\{T_{n+1} - T_n = j - i\}$$
$$= \begin{cases} pq^{j-i-1} & \text{if } j \geq i + 1, \\ 0 & \text{otherwise} \end{cases}$$

by Proposition (3.3.8). The transition matrix then is

$$P = \begin{bmatrix} 0 & p & pq & pq^2 & pq^3 & \cdot & \cdot & \cdot \\ & 0 & p & pq & pq^2 & \cdot & \cdot & \cdot \\ & & 0 & p & pq & \cdot & \cdot & \cdot \\ & & & 0 & p & \cdot & \cdot & \cdot \\ & & & & 0 & \cdot & \cdot & \cdot \\ & & & & & \cdot & & \\ 0 & & & & & & \cdot \end{bmatrix}.$$

The m-step transition probabilities can be obtained by multiplying P by itself m times. But we have already computed these in Theorem (3.3.3):

$$P^m(i, j) = P\{T_{n+m} = j \mid T_n = i\}$$
$$= P\{T_{n+m} - T_n = j - i\}$$
$$= P\{T_m = j - i\} = \begin{cases} \binom{j - i - 1}{m - 1} p^m q^{j-i-m} & \text{if } j \geq i + m, \\ 0 & \text{otherwise.} \end{cases}$$

(Note that the second equality is justified by the independence of $T_{n+m} - T_n$ and T_n; see Corollary (3.3.10).) $\qquad\square$

(1.13) EXAMPLE. *Independent trials process.* Let $X_0, X_1, \ldots$ be independent discrete random variables with the common distribution

$$P\{X_n = k\} = \begin{cases} \pi(k) & k = 0, 1, 2, \ldots; \\ 0 & k < 0. \end{cases}$$

Then since X_{n+1} is independent of $X_0, \ldots, X_n$,

$$P\{X_{n+1} = j \mid X_0, \ldots, X_n\} = P\{X_{n+1} = j\} = \pi(j),$$

and

$$P\{X_{n+1} = j \mid X_n\} = P\{X_{n+1} = j\} = \pi(j).$$

Hence, $\{X_n; n \geq 0\}$ is a Markov chain. Its initial distribution is π, and its transition matrix is

$$P = \begin{bmatrix} \pi(0) & \pi(1) & \pi(2) & \cdots \\ \pi(0) & \pi(1) & \pi(2) & \cdots \\ \pi(0) & \pi(1) & \pi(2) & \cdots \\ \cdot & \cdot & \cdot & \cdots \\ \cdot & \cdot & \cdot & \cdots \\ \cdot & \cdot & \cdot & \cdots \end{bmatrix}.$$

Note that all the rows of P are identical and that $P^m = P$ for all $m \geq 1$. Conversely, if $\{X_n; n \geq 0\}$ is a Markov chain whose transition matrix has all the rows identical, then $X_0, X_1, \ldots$ are independent and identically distributed. $\qquad\square$

(1.14) EXAMPLE. *Sums of independent random variables.* Let $Y_1, Y_2, \ldots$ be independent and identically distributed discrete random variables with probability distribution $\{p_k; k = 0, 1, 2, \ldots\}$. Put

$$X_n = \begin{cases} 0 & n = 0, \\ Y_1 + Y_2 + \cdots + Y_n & n \geq 1. \end{cases}$$

Then since $X_{n+1} = X_n + Y_{n+1}$,

$$P\{X_{n+1} = j \,|\, X_0, \ldots, X_n\} = P\{Y_{n+1} = j - X_n \,|\, X_0, \ldots, X_n\} = p_{j-X_n}$$

by the independence of Y_{n+1} and $X_0, \ldots, X_n$. Thus, $\{X_n; n \geq 0\}$ is a Markov chain whose transition probabilities are

$$P(i, j) = P\{X_{n+1} = j \,|\, X_n = i\} = p_{j-i};$$

or in matrix form,

$$P = \begin{bmatrix} p_0 & p_1 & p_2 & p_3 & \cdot & \cdot & \cdot \\ & p_0 & p_1 & p_2 & \cdot & \cdot & \cdot \\ & & p_0 & p_1 & \cdot & \cdot & \cdot \\ & & & p_0 & \cdot & \cdot & \cdot \\ & & & & \cdot & & \\ 0 & & & & & \cdot & \\ & & & & & & \cdot \end{bmatrix}.$$

Comparing this with the transition matrix in Example (1.12), we note that the latter is a special case with $p_k = pq^{k-1}$, $k = 1, 2, \ldots$. □

(1.15) EXAMPLE. Let $Y_1, Y_2, \ldots$ be independent identically distributed random variables each taking values in $\{0, 1, 2, 3, 4\}$ with the common distribution $(p_0, p_1, p_2, p_3, p_4)$.

Let $X_0 = 0$ and define

$$X_{n+1} = X_n + Y_{n+1} \quad (\text{modulo } 5).$$

(For example, if for some $\omega \in \Omega$, $Y_1(\omega) = 2$, $Y_2(\omega) = 4$, $Y_3(\omega) = 4$, $Y_4(\omega) = 1$, $Y_5(\omega) = 3$, $Y_6(\omega) = 4, \ldots$, then we have $X_0(\omega) = 0$, $X_1(\omega) = 2$, $X_2(\omega) = 1$, $X_3(\omega) = 0$, $X_4(\omega) = 1$, $X_5(\omega) = 4$, $X_6(\omega) = 3, \ldots$.)

Then $\{X_n; n \geq 0\}$ is a Markov chain with state space $\{0, 1, 2, 3, 4\}$ and transition matrix

$$P = \begin{bmatrix} p_0 & p_1 & p_2 & p_3 & p_4 \\ p_4 & p_0 & p_1 & p_2 & p_3 \\ p_3 & p_4 & p_0 & p_1 & p_2 \\ p_2 & p_3 & p_4 & p_0 & p_1 \\ p_1 & p_2 & p_3 & p_4 & p_0 \end{bmatrix}.$$

Note that in this example, not only each row, but also each column sums to one. Such matrices are called *doubly Markov*. □

(1.16) EXAMPLE. *Inventory Theory.* A commodity is being stocked to satisfy a continuing demand. The inventory policy is described by specifying two non-negative critical values s and S, with $s < S$. The inventory level is checked periodically at fixed times $t_0, t_1, t_2, \ldots$. If the stock on hand at t_n is less than or equal to s, then by immediate procurement the stock level is brought up to the level S. On the other hand, if the stock level is greater than s, then no replenishment is undertaken.

Let Z_n be the total demand for the commodity during the interval $[t_{n-1}, t_n)$, and let X_n be the stock on hand just before the time t_n. Then for any $\omega \in \Omega$,

$$X_{n+1}(\omega) = \begin{cases} X_n(\omega) - Z_{n+1}(\omega) & \text{if } s < X_n(\omega) \leq S, Z_{n+1}(\omega) \leq X_n(\omega), \\ S - Z_{n+1}(\omega) & \text{if } X_n(\omega) \leq s, Z_{n+1}(\omega) \leq S, \\ 0 & \text{otherwise.} \end{cases}$$

Now, then,

$$P\{X_{n+1} = j \mid X_0, \ldots, X_n\} = P\{X_{n+1} = j \mid X_n\}$$

if and only if

$$P\{Z_{n+1} = k \mid X_0, \ldots, X_n\} = P\{Z_{n+1} = k \mid X_n\}.$$

Thus, if the demand Z_{n+1} for the commodity in the interval $[t_n, t_{n+1})$ is independent of $X_0, \ldots, X_{n-1}$ given the value of X_n, then $\{X_n, n \geq 0\}$ is a Markov chain with state space $E = \{0, 1, \ldots, S\}$.   □

(1.17) EXAMPLE. *Remaining Lifetime.* Consider some piece of equipment which is now in use. When it fails, it is replaced immediately by an identical one. When that one fails it is again replaced by an identical one, and so on. Let p_k denote the probability that a new item lasts for k units of time, $k = 1, 2, \ldots$. Let X_n be the remaining lifetime of the item in use at time n. Then for fixed ω,

$$(1.18) \qquad X_{n+1}(\omega) = \begin{cases} X_n(\omega) - 1 & \text{if } X_n(\omega) \geq 1, \\ Z_{n+1}(\omega) - 1 & \text{if } X_n(\omega) = 0, \end{cases}$$

where $Z_{n+1}(\omega)$ is the lifetime of the item installed at n. Since the lifetimes of successive items installed are independent, Z_{n+1} is independent of $X_0, X_1, \ldots, X_n$. Thus, (1.18) implies that $\{X_n, n \geq 0\}$ is a Markov chain with state space $\{0, 1, 2, \ldots\}$.

Let $P(i, j)$ be the transition probabilities. For $i \geq 1$,

$$P(i, j) = P\{X_{n+1} = j \mid X_n = i\} = P\{X_n - 1 = j \mid X_n = i\}$$

$$= P\{X_n = j + 1 \mid X_n = i\} = \begin{cases} 1 & \text{if } j = i - 1, \\ 0 & \text{if } j \neq i - 1; \end{cases}$$

and for $i = 0$,

$$P(0, j) = P\{X_{n+1} = j \mid X_n = 0\}$$
$$= P\{Z_{n+1} - 1 = j \mid X_n = 0\} = P\{Z_{n+1} = j + 1\} = p_{j+1}$$

for any $j = 0, 1, \ldots$. Thus, the transition matrix is

$$P = \begin{bmatrix} p_1 & p_2 & p_3 & \cdot & \cdot & \cdot \\ 1 & 0 & 0 & \cdot & \cdot & \cdot \\ & 1 & 0 & \cdot & \cdot & \cdot \\ & & 1 & & \cdot & \\ & & & & & \\ 0 & & & & \cdot & \cdot \end{bmatrix}.$$

□

The definition of a Markov chain stated that the next state of the process is independent of the past provided that the present state be known. Next we will show that a stronger statement is also true: the past and the future are conditionally independent given the present.

(1.19) THEOREM. Fix $n \in \mathbb{N}$, and let Y be a bounded function of the random variables $X_n, X_{n+1}, \ldots$. Then

$$E[Y \mid X_0, \ldots, X_n] = E[Y \mid X_n].$$

Proof. It is sufficient to prove this for a random variable Y which is a function of finitely many of the $X_n, X_{n+1}, \ldots$, say $Y = f(X_n, X_{n+1}, \ldots, X_{n+m})$ for some $m \in \mathbb{N}$ and some bounded function f from E^{m+1} into $\mathbb{R}$. If $m = 0$, then the claim holds trivially by Proposition (2.2.15). Now we make the induction hypothesis that the claim holds for any bounded random variable of the form $f(X_n, \ldots, X_{n+m})$, and proceed to show that the claim then holds for any bounded $Y = g(X_n, \ldots, X_{n+m}, X_{n+m+1})$.

By Theorem (2.2.19),

$$(1.20) \quad E[Y \mid X_0, \ldots, X_n] = E[E[Y \mid X_0, \ldots, X_{n+m}] \mid X_0, \ldots, X_n].$$

On the other hand, using the Markov property at time $n + m$,

$$E[Y \mid X_0, \ldots, X_{n+m}] = \sum_{j \in E} g(X_n, \ldots, X_{n+m}, j) P\{X_{n+m+1} = j \mid X_0, \ldots, X_{n+m}\}$$
$$= \sum_j g(X_n, \ldots, X_{n+m}, j) P(X_{n+m}, j)$$
$$= f(X_n, \ldots, X_{n+m})$$

for some f. The last expression being a function of only $X_n, \ldots, X_{n+m}$, the induction hypothesis applies and we have

(1.21)
$$E[E[Y|X_0, \ldots, X_{n+m}]|X_0, \ldots, X_n]$$
$$= E[E[Y|X_0, \ldots, X_{n+m}]|X_n] = E[Y|X_n],$$

where we used Theorem (2.2.19) in writing the last equality. The proof now follows from (1.20) and (1.21).   □

A closer look at the proof of the preceding theorem yields also a nice computational formula.

(1.22) PROPOSITION. For any $n \in \mathbb{N}$ and any bounded function f on $E \times E \times \cdots$, we have

$$E[f(X_n, X_{n+1}, \ldots)|X_n = i] = E[f(X_0, X_1, \ldots)|X_0 = i].$$

Proof. It is sufficient to show this for a function f whose values depend only on a finite number of its arguments, say $f(a_0, a_1, \ldots) = g(a_0, \ldots, a_m)$. Then, using Theorem (1.5),

$$E[g(X_n, \ldots, X_{n+m})|X_n = i] = \sum_{i_1} \cdots \sum_{i_m} g(i, i_1, \ldots, i_m)P(i, i_1) \ldots P(i_{m-1}, i_m),$$

which is also the expression for $E[g(X_0, \ldots, X_m)|X_0 = i]$.   □

Putting the results of the two preceding propositions together, we obtain the following useful

(1.23) COROLLARY. Let f be a bounded function on $E \times E \times \cdots$, and let

$$g(i) = E[f(X_0, X_1, \ldots)|X_0 = i].$$

Then for any $n \in \mathbb{N}$,

$$E[f(X_n, X_{n+1}, \ldots)|X_0, \ldots, X_n] = g(X_n).$$

* * *

We have shown in Theorem (1.19) that the past and the future of a Markov chain are independent provided that the "present" state be known; here, by "present" was meant a fixed time n. In various applications a similar concept of independence is needed with "present" replaced by a random time T. It turns out that for certain random times T, the past $\{X_m; m \leq T\}$ and the future $\{X_m; m \geq T\}$ are conditionally independent given the "present" state X_T. When this holds for a random time T, the *strong Markov property* is said to hold at time T. In particular, the strong Markov property holds for random times T having the property that for every $n \in \mathbb{N}$, the occurrence or non-occurrence of the event $\{T \leq n\}$ can be determined by looking at the values of $X_0, \ldots, X_n$; in other words, for every $n \in \mathbb{N}$, there is a function f_n such that $f_n(X_0(\omega), \ldots, X_n(\omega)) = 1$ or 0 according as $T(\omega) \leq n$ or $T(\omega) > n$. Such random times are called *stopping times*. (See also (4.2.18) for the same concept.)

For example: the first time X visits a certain state is a stopping time; the first time X visits a certain set of states is a stopping time; the kth time X visits a fixed set of states is a stopping time; the last time a state is visited is *not* a stopping time. The sum of two stopping times is a stopping time. The smaller (or the larger) of two stopping times is again a stopping time. Every constant non-negative integer is a stopping time.

A stopping time T may equal $+\infty$ for certain ω; in that case $X_{T(\omega)}(\omega)$ is not well defined, since we have not yet defined $X_\infty(\omega)$. We now introduce the following conventions to handle such situations. First we append an extra state, which we denote by Δ, to the state space. We then define $X_\infty(\omega) = \Delta$ for all ω. Finally, we make the convention that any function f defined on E is automatically extended to $E \cup \{\Delta\}$ by setting $f(\Delta) = 0$. This way, if $T(\omega) = \infty$, then $f(X_{T(\omega)}(\omega)) = f(\Delta) = 0$. These conventions are useful in simplifying the statements of the following theorems.

(1.24) THEOREM.   For any stopping time T,

$$E[f(X_T, X_{T+1}, \ldots) \mid X_n; n \leq T] = E[f(X_T, X_{T+1}, \ldots) \mid X_T]$$

for all bounded functions f on $E \times E \times \cdots$.

Proof is omitted. Computation of the quantity on the right of the expression in (1.24) is quite simple because of time-homogeneity. Just as in Proposition (1.22) and Corollary (1.23), we have the following

(1.25) PROPOSITION.   Let f be a bounded function on $E \times E \times \cdots$, and let

$$g(i) = E[f(X_0, X_1, \ldots) \mid X_0 = i].$$

Then for any stopping time T,

$$E[f(X_T, X_{T+1}, \ldots) \mid X_n; n \leq T] = g(X_T).$$

A particular case of interest is when $f(a_0, a_1, \ldots) = 1$ or 0 according as $a_m = j$ or $a_m \neq j$, where $j \in E$ is fixed and $m \in \mathbb{N}$ is fixed. Then

$$E[f(X_0, X_1, \ldots) \mid X_0 = i] = P\{X_m = j \mid X_0 = i\} = P^m(i, j),$$

and

$$E[f(X_T, X_{T+1}, \ldots) \mid X_n; n \leq T] = P\{X_{T+m} = j \mid X_n; n \leq T\}.$$

Now Proposition (1.25) yields the following alternative statement of the strong Markov property at T.

(1.26) COROLLARY.   For any stopping time T,

$$P\{X_{T+m} = j \mid X_n; n \le T\} = P^m(X_T, j)$$

for all $m \in \mathbb{N}$ and $j \in E$.

In other words, the randomness of the "present" time does not alter our computations. In particular,

(1.27) $$P\{X_{T+m} = j \mid X_T = i\} = P^m(i, j)$$

as before for the case of constant times T.

2.   Visits to a Fixed State

Throughout this section $X = \{X_n; n \in \mathbb{N}\}$ is a Markov chain with state space E and transition matrix P. Instead of writing $P\{A \mid X_0 = i\}$ we simply write $P_i\{A\}$; and similarly, instead of $E[Y \mid X_0 = i]$ we write $E_i[Y]$. For fixed $i \in E$, P_i is a probability measure on Ω and E_i is the corresponding expectation operator.

Let $j \in E$ be fixed, and for each $\omega \in \Omega$, let $N_j(\omega)$ be the number of times j appears in the sequence $X_0(\omega), X_1(\omega), X_2(\omega), \dots$. Then $N_j(\omega)$ is the total number of visits to state j by the process X for the realization ω. If $N_j(\omega)$ is finite, then X eventually leaves state j never to return; that is, there is an integer n such that $X_n(\omega) = j$ and $X_m(\omega) \ne j$ for $m > n$ (this n will, of course, depend on ω). On the other hand, if $N_j(\omega) = \infty$ for a realization ω, then X visits j again and again (and there is no integer n with $X_m(\omega) \ne j$ for all $m > n$). For purposes of predicting the future, it is a matter of some importance to determine whether an ω of the former type (that yields only finitely many visits to j) or an ω of the latter type (that yields infinitely many visits to j) is to come up. Our objective here is to obtain criteria for settling this matter. In this section and the next, a number of reasonably simple criteria will be obtained in terms of the given data (which are the transition probabilities $P(i, k)$).

For each $\omega \in \Omega$, the realization $X_0(\omega), X_1(\omega), \dots$ of the process X is a sequence of elements of E; let $T_1(\omega), T_2(\omega), \dots$ be the successive indices $n \ge 1$ for which $X_n(\omega) = j$ as long as there are such n. It may be that there is no such n: then we put $T_1(\omega) = T_2(\omega) - T_1(\omega) = \cdots = +\infty$; it may be that j appears only a finite number m of times: then we let $T_1(\omega), \dots, T_m(\omega)$ be the successive n for which $X_n(\omega) = j$ and let $T_{m+1}(\omega) - T_m(\omega) =$

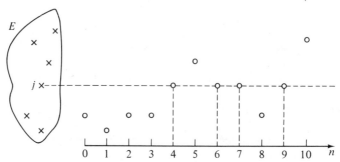

Figure 5.2.1 For the realization ω pictured here, the times of visits to j are 4, 6, 7, 9, ...

$T_{m+2} - T_{m+1}(\omega) = \cdots = \infty$. Then $T_1, T_2, \ldots$ are the times of successive visits to state j; see Figure 5.2.1 for a realization.

For any $n \in \mathbb{N}$, $T_m(\omega) \leq n$ if and only if j appears in $\{X_1(\omega), \ldots, X_n(\omega)\}$ at least m times. Thus, every T_m is a stopping time, and by Theorem (1.24), the strong Markov property holds at T_m; namely, the future after T_m is independent of the past before T_m given X_{T_m}. On the other hand, by the way the T_m are defined, $X_{T_m} = j$ on $\{T_m < \infty\}$. Hence, every time the process X visits j, the past loses all its influence upon the future. In particular we have the following.

(2.1) PROPOSITION.   For any $i \in E$ and finite integers $k, m \geq 1$

$$P_i\{T_{m+1} - T_m = k \mid T_1, \ldots, T_m\} = \begin{cases} 0 & \text{on } \{T_m = \infty\}, \\ P_j\{T_1 = k\} & \text{on } \{T_m < \infty\}. \end{cases}$$

Proof. For fixed j and m, $T = T_m$ is a stopping time. The event $\{T_{m+1} - T_m = k\}$ is the same as the event $\{X_{T+1} \neq j, \ldots, X_{T+k-1} \neq j, X_{T+k} = j\}$. Hence, by Proposition (1.25),

(2.2) $$P_i\{T_{m+1} - T_m = k \mid X_n, n \leq T\} = g(X_T),$$

where $g(\Delta) = 0$ and $g(j) = P_j\{X_1 \neq j, \ldots, X_{k-1} \neq j, X_k = j\} = P_j\{T_1 = k\}$. Setting $h(n) = g(j)$ for all $n < \infty$ and $h(\infty) = g(\Delta) = 0$, we see that $g(X_T) = h(T)$ by the definition of X_T. Thus, using the fact that $\{X_n; n \leq T\}$ includes $\{T_1, \ldots, T_m\}$ to apply Theorem (2.2.19), and then applying Corollary (2.2.22), we get

$$\begin{aligned} P_i\{T_{m+1} &- T_m = k \mid T_1, \ldots, T_m\} \\ &= E_i[P_i\{T_{m+1} - T_m = k \mid X_n, n \leq T_m\} \mid T_1, \ldots, T_m] \\ &= E_i[h(T_m) \mid T_1, \ldots, T_m] = h(T_m), \end{aligned}$$

which is as desired. $\square$

Computation of the probability $P_j\{T_1 = k\}$ is easy. More generally, let

(2.3) $$F_k(i, j) = P_i\{T_1 = k\}, \quad i \in E, \, k = 1, 2, \ldots .$$

For $k = 1$,

$$F_k(i, j) = P_i\{T_1 = 1\} = P_i\{X_1 = j\} = P(i, j),$$

and for $k \geq 2$,

$$\begin{aligned} F_k(i, j) &= P_i\{X_1 \neq j, \ldots, X_{k-1} \neq j, X_k = j\} \\ &= \sum_{\substack{b \in E \\ b \neq j}} P_i\{X_1 = b\} P_i\{X_2 \neq j, \ldots, X_{k-1} \neq j, X_k = j \,|\, X_1 = b\} \\ &= \sum_{b \in E - \{j\}} P_i\{X_1 = b\} P_b\{X_1 \neq j, \ldots, X_{k-2} \neq j, X_{k-1} = j\} \end{aligned}$$

by Proposition (1.22). Within the summation, the first factor is $P(i, b)$ and the second is $F_{k-1}(b, j)$. Thus, we have

(2.4) $$F_k(i, j) = \begin{cases} P(i, j) & k = 1, \\ \sum_{b \in E - \{j\}} P(i, b) F_{k-1}(b, j) & k \geq 2. \end{cases}$$

Hence, for $k = 1$, the probabilities $F_1(i, j)$ are the same as the transition probabilities, and for larger k, (2.4) provides a recursive formula. The following illustrates this computation.

(2.5) EXAMPLE. Consider a Markov chain with state space $E = \{1, 2, 3\}$ and the transition matrix

$$P = \begin{bmatrix} 1 & 0 & 0 \\ \frac{1}{2} & \frac{1}{6} & \frac{1}{3} \\ \frac{1}{3} & \frac{3}{5} & \frac{1}{15} \end{bmatrix},$$

and let $j = 3$. Let us define $f_k(i) = F_k(i, j)$ for $i = 1, 2, 3$. Then f_1 is the third column of P, and for $k \geq 2$ we have

$$f_k = Q f_{k-1},$$

where Q is the matrix obtained from P by replacing its jth column by zeros; that is,

$$Q = \begin{bmatrix} 1 & 0 & 0 \\ \frac{1}{2} & \frac{1}{6} & 0 \\ \frac{1}{3} & \frac{3}{5} & 0 \end{bmatrix}.$$

Thus,

$$f_1 = \begin{bmatrix} 0 \\ \frac{1}{3} \\ \frac{1}{15} \end{bmatrix}, \quad f_2 = \begin{bmatrix} 0 \\ \frac{1}{18} \\ \frac{1}{5} \end{bmatrix}, \quad f_3 = \begin{bmatrix} 0 \\ \frac{1}{108} \\ \frac{1}{30} \end{bmatrix}, \quad f_4 = \begin{bmatrix} 0 \\ \frac{1}{648} \\ \frac{1}{180} \end{bmatrix}, \cdots ;$$

that is,

$$F_k(1, 3) = 0 \qquad\qquad k = 1, 2, \ldots ;$$

$$F_k(2, 3) = \frac{1}{3}\left(\frac{1}{6}\right)^{k-1} \qquad k = 1, 2, \ldots ;$$

$$F_k(3, 3) = \begin{cases} \dfrac{1}{15} & k = 1, \\[2mm] \dfrac{3}{5}\left(\dfrac{1}{6}\right)^{k-2}\left(\dfrac{1}{3}\right) & k = 2, 3, \ldots . \end{cases}$$

Thus, starting at 1, the process never visits 3, that is,

$$P_1\{T_1 = +\infty\} = 1.$$

Starting at 2, the first visit to 3 occurs at time k with probability $(\frac{1}{3})(\frac{1}{6})^{k-1}$; hence, the probability that 3 is never visited is, starting at 2,

$$F_2\{T_1 = +\infty\} = 1 - P_2\{T_1 < \infty\}$$
$$= 1 - \sum_{k=1}^{\infty} \frac{1}{3}\left(\frac{1}{6}\right)^{k-1} = \frac{3}{5}.$$

Starting at 3, the probability that the first return to 3 occurs at time k is given as $F_k(3, 3)$; the probability that X never returns to 3 is

$$P_3\{T_1 = +\infty\} = 1 - \sum_{k=1}^{\infty} P_3\{T_1 = k\} = \frac{52}{75}. \qquad\qquad \square$$

For each pair (i, j) we define

(2.6) $$F(i, j) = P_i\{T_1 < \infty\} = \sum_{k=1}^{\infty} F_k(i, j),$$

that is, $F(i, j)$ is the probability that, starting at i, the Markov chain X *ever* visits j. Summing over all k in (2.4), we see that the probabilities $F(i, j)$ satisfy

(2.7) $$F(i, j) = P(i, j) + \sum_{b \in E - \{j\}} P(i, b)F(b, j), \quad i \in E.$$

For each j, this is a system of linear equations in $F(i, j)$ and may be used to solve for the $F(i, j)$. However, very shortly we will have a better method available (cf. Section 1 of Chapter 6).

Consider now the number of visits $N_j(\omega)$ to j by the process X for the realization ω. Note that for any ω, $N_j(\omega) = m$ if and only if $T_1(\omega) < \infty, \ldots,$ $T_m(\omega) < \infty$, and $T_{m+1}(\omega) = +\infty$. By Proposition (2.1), the events $\{T_1 < \infty\}$, $\{T_2 - T_1 < \infty\}, \ldots, \{T_m - T_{m-1} < \infty\}, \{T_{m+1} - T_m = +\infty\}$ are independent and their probabilities are, starting at i, $F(i, j), F(j, j), \ldots, F(j, j)$, $1 - F(j, j)$ respectively. Hence we have the following

(2.8) PROPOSITION. Let N_j be the total number of visits to j. Then

$$(2.9) \qquad P_j\{N_j = m\} = F(j, j)^{m-1}(1 - F(j, j)), \quad m = 1, 2, \ldots;$$

and for $i \neq j$,

$$(2.10) \qquad P_i\{N_j = m\} = \begin{cases} 1 - F(i, j) & m = 0, \\ F(i, j)F(j, j)^{m-1}(1 - F(j, j)) & m = 1, 2, \ldots. \end{cases}$$

By summing over all m in (2.9) we obtain the probability that, starting at j, N_j is finite. If $F(j, j) = 1$, then every term is zero and this sum is zero. If $F(j, j) < 1$, what we have there is the geometric distribution and the sum is unity. We put this next as a

(2.11) COROLLARY. We have

$$P_j\{N_j < \infty\} = \begin{cases} 1 & \text{if } F(j, j) < 1, \\ 0 & \text{if } F(j, j) = 1. \end{cases}$$

If $F(j, j) = 1$, then $N_j = +\infty$ with probability one if the initial state is j; therefore, $E_j[N_j] = +\infty$ in this case. Otherwise, if $F(j, j) < 1$, N_j has the geometric distribution with success probability $p = 1 - F(j, j)$ starting at j; then $E_j[N_j] = 1/p = 1/(1 - F(j, j))$. These and similar computations for the case $i \neq j$ give the following: Let

$$(2.12) \qquad R(i, j) = E_i[N_j].$$

The matrix R whose (i, j)-entry is $R(i, j)$ is called the *potential matrix* of X.

(2.13) COROLLARY. We have

$$R(j, j) = 1/(1 - F(j, j)),$$

and

$$R(i, j) = F(i, j)R(j, j) \quad \text{if } i \neq j$$

(here, $1/0 = \infty$, $0 \cdot \infty = 0$).

Corollary (2.11) is of great significance. It shows that the number $F(j,j)$ determines whether or not a state j will be visited infinitely often. If our computations yielded $F(j,j) = 1$, then for almost any ω, the realization of the process will visit j infinitely often; such a state j will be called *recurrent*. If, on the other hand, our computations show $F(j,j) < 1$, then for almost any ω the realization of X will visit j only finitely many times, so that after a last visit to j, state j will never be entered again; such a state j is called *transient* because it disappears after a while.

The computations of the expected numbers of visits to various states can be made directly, without using Corollary (2.13). In fact, in practice, it is easier to compute the $R(i, j)$ first and then use (2.13) to compute the $F(i, j)$.

Let 1_j be the function whose every entry is zero except the j-entry, which is one; that is, $1_j(k) = 1$ or 0 according as $k = j$ or $k \neq j$ (see also Notation (1.3)). For an $\omega \in \Omega$, $1_j(X_n(\omega))$ is 1 or 0 according as X is in j or not at time n for that ω. Thus, $1_j(X_n(\omega))$ is the number of visits to j at time n, and

$$(2.14) \qquad N_j(\omega) = \sum_{n=0}^{\infty} 1_j(X_n(\omega)), \quad \omega \in \Omega.$$

By the monotone convergence theorem,

$$(2.15) \qquad R(i, j) = E_i \left[\sum_{n=0}^{\infty} 1_j(X_n) \right]$$

$$= \sum_{n=0}^{\infty} E_i[1_j(X_n)]$$

$$= \sum_{n=0}^{\infty} P_i\{X_n = j\} = \sum_{n=0}^{\infty} P^n(i, j).$$

In matrix notation this becomes

$$(2.16) \qquad R = I + P + P^2 + \cdots.$$

If the powers of P are computed beforehand, then (2.16) may be used to compute R. In general, however, it is easier to note first that the potential matrix R satisfies

$$RP = PR = P + P^2 + \cdots = R - I,$$

which yields

$$(2.17) \qquad R(I - P) = (I - P)R = I.$$

In the case where E is finite, (2.17) is a useful computational tool once we figure out which $R(i, j)$ are finite and which are not. In the next section we will develop such criteria which, in many cases, will reduce the task involved to a simple inspection of the matrix P. We will pick up the computational problem itself in the next chapter.

3. Classification of States

Let X be a Markov chain with state space E and transition matrix P. Let T be the time of first visit to state j (this was denoted by T_1 for fixed j in the preceding section), and let N_j be the total number of visits to j as before.

(3.1) DEFINITION. (a) State j is called *recurrent* if

$$P_j\{T < \infty\} = 1;$$

otherwise, if $P_j\{T = +\infty\} > 0$, then j is called *transient*.

(b) A recurrent state j is called *null* if

$$E_j[T] = \infty;$$

otherwise, it is called *non-null*.

(c) A recurrent state j is said to be *periodic with period* δ if $\delta \geq 2$ is the largest integer for which

$$P_j\{T = n\delta \text{ for some } n \geq 1\} = 1;$$

otherwise, if there is no such $\delta \geq 2$, j is called *aperiodic*.   □

If j is recurrent, then by definition $F(j, j) = 1$; in other words, j is recurrent if and only if, starting at j, the probability of returning to j is one. Since $F(j, j) = 1$ implies that $R(j, j) = +\infty$ by Corollary (2.13), the expected number of returns to j is infinite. In fact, as Corollary (2.11) implies, the actual number of returns is infinite almost surely. Thus, j is recurrent if and only if

$$R(j, j) = E_j[N_j] = +\infty,$$

which is also equivalent to

$$P_j\{N_j = +\infty\} = 1.$$

Conversely, if j is transient, then $F(j, j) < 1$ and there is a positive probability $1 - F(j, j)$ of never returning to j. In this case, Corollary (2.11) implies that the total number of returns to j is finite with probability one, and Corollary (2.13) implies that the expected number of returns to j is finite also. Thus, j is transient if and only if

$$R(j, j) = E_j[N_j] < \infty,$$

which is also equivalent to

$$P_j\{N_j < \infty\} = 1.$$

If j is transient, $R(j, j) < \infty$; since $R(i, j) = F(i, j)R(j, j) \leq R(j, j)$ by Corollary (2.13), $R(i, j) < \infty$ for all $i \in E$. By (2.15), $R(i, j)$ is the sum of $P^n(i, j)$ for all n; therefore $R(i, j)$ can be finite only if $P^n(i, j) \rightarrow 0$ as $n \rightarrow \infty$. If j is recurrent, then $R(j, j) = +\infty$, and we are no longer able to deduce easily whether $P^n(j, j)$ will be zero or positive in the limit. It turns out that if j is recurrent null, then the fact that the expected time between two returns to j is infinite implies that $\lim_n P^n(j, j)$ is zero. The following is the basic limit theorem; we are listing it here without proof to enhance further the meanings behind the classification introduced above (see (2.22) in Chapter 9 for a proof).

(3.2) THEOREM.   (a) If j is transient or recurrent null, then for any $i \in E$,

$$\lim_{n \to \infty} P^n(i, j) = 0.$$

(b) If j is recurrent non-null aperiodic, then

$$\pi(j) = \lim_{n \to \infty} P^n(j, j) > 0,$$

and for any $i \in E$,

$$\lim_{n \to \infty} P^n(i, j) = F(i, j)\pi(j).$$

If j is periodic with period δ then a return to j is possible only at steps numbered $\delta, 2\delta, 3\delta, \ldots$. Then the same is true of the time of the second return, and the time of the third return, and so on. Starting at j, the process can be back in state j at time n if and only if n is either the time of the first return or the time of the second return or the time of the third return, etc. Hence, if j is periodic with period δ, then

$$(3.3) \qquad P^n(j, j) = P_j\{X_n = j\} > 0 \quad \text{only if } n \in \{0, \delta, 2\delta, \ldots\}.$$

In other words, we have the following

(3.4) CRITERION.   Let δ be the greatest common divisor of the set of all $n \geq 1$ for which $P^n(j, j) > 0$. If $\delta = 1$, then j is aperiodic. If $\delta \geq 2$, then j is periodic with period δ.

In order to tell whether $P^n(j, j) > 0$ or not, we do not need to compute $P^n(j, j)$. Since, for example,

$$P^3(j, j) = \sum_i \sum_k P(j, i)P(i, k)P(k, j),$$

$P^3(j, j) > 0$ if and only if there are two states i and k such that $P(j, i)$, $P(i, k)$, and $P(k, j)$ are all positive. More generally, $P^n(j, j) > 0$ if and only if there are $i_1, \ldots, i_{n-1} \in E$ such that $P(j, i_1), P(i_1, i_2), \ldots, P(i_{n-1}, j)$ are all positive. This idea turns out to be very useful in the classification of states, and we now formalize it.

We will say that state j *can be reached from* i, and write $i \to j$, if there exists an integer $n \geq 0$ such that $P^n(i, j) > 0$. Thus, for $i \neq j$, $i \to j$ if and only if $F(i, j) > 0$. Since $P^0(j, j) = 1 > 0$, any state j can be reached from itself. In order that $i \to j$, there must be a sequence $i_1, i_2, \ldots, i_n$ of states such that $P(i, i_1) > 0$, $P(i_1, i_2) > 0, \ldots, P(i_n, j) > 0$. For otherwise, if there is no such sequence, $P^n(i, j) = 0$ for all n and therefore $i \nrightarrow j$.

(3.5) DEFINITION. (a) A set of states is said to be *closed* if no state outside it can be reached from any state in it.

(b) A state forming a closed set by itself is called an *absorbing* state.

(c) A closed set is *irreducible* if no proper subset of it is closed.

(d) A Markov chain is called *irreducible* if its only closed set is the set of all states. □

If for some state i, $P(i, j) = 0$ for all $j \neq i$, then no state j can be reached from i. In that case $\{i\}$ is a closed set, that is, i is absorbing. The converse is also easy to see. Thus, absorbing states can be located easily:

(3.6) CRITERION. State j is absorbing if and only if $P(j, j) = 1$.

The following is a simple criterion for irreducibility; its proof follows directly from the definition.

(3.7) CRITERION. A Markov chain is irreducible if and only if all states can be reached from each other.

In general, if C is a closed set, then deleting from the transition matrix P those rows and columns which correspond to states not in C leaves another Markov matrix. If C consists of a single closed set, that is, if C is irreducible, then the chain restricted to C is an irreducible Markov chain with this new Markov matrix as the transition matrix. We put this next as a

(3.8) PROPOSITION. Let $C = \{c_1, c_2, \ldots\} \subset E$ be a closed set and define

$$Q(i, j) = P(c_i, c_j), \quad c_i, c_j \in C.$$

Then Q is a Markov matrix.

In finding the closed sets it is helpful to note that we have

(3.9) PROPOSITION. If $i \longrightarrow j$ and $j \longrightarrow k$, then $i \longrightarrow k$.

Proof. If $i \longrightarrow j$, then there is $r \geq 0$ such that $P^r(i, j) > 0$. Similarly, if $j \longrightarrow k$, there is $s \geq 0$ such that $P^s(j, k) > 0$. Then for $n = r + s$, using (1.8),

$$P^n(i, k) = \sum_v P^r(i, v)P^s(v, k) \geq P^r(i, j)P^s(j, k) > 0$$

so that $i \longrightarrow k$ by definition. $\qquad\square$

To find the closed set C containing a state i we proceed as follows. First, all states j for which $P(i, j) > 0$ must be in C. Now by the preceding proposition, all k for which $P(j, k) > 0$ for some $j \in C$ must be in C too. We proceed in this fashion until we can no longer add any more states to our collection.

(3.10) EXAMPLE. Consider a Markov chain with state space $E = \{a, b, c, d, e\}$ and transition matrix

$$P = \begin{array}{c} \\ a \\ b \\ c \\ d \\ e \end{array} \begin{array}{c} \begin{array}{ccccc} a & b & c & d & e \end{array} \\ \begin{bmatrix} \frac{1}{2} & 0 & \frac{1}{2} & 0 & 0 \\ 0 & \frac{1}{4} & 0 & \frac{3}{4} & 0 \\ 0 & 0 & \frac{1}{3} & 0 & \frac{2}{3} \\ \frac{1}{4} & \frac{1}{2} & 0 & \frac{1}{4} & 0 \\ \frac{1}{3} & 0 & \frac{1}{3} & 0 & \frac{1}{3} \end{bmatrix} \end{array} .$$

To find all the closed sets it is usually helpful to draw a graph with the states as vertices and draw a directed line from i to j if $P(i, j) > 0$. Doing that, we get Figure 5.3.1 below. From this we see that from $\{b, d\}$ the set $\{a, c, e\}$ can be reached but not vice-versa. Thus, once the process leaves the states $\{b, d\}$, neither b nor d can ever be visited again.

The closed sets are $\{a, c, e\}$ and $\{a, b, c, d, e\}$. Since there are two closed

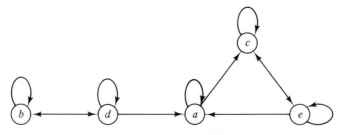

Figure 5.3.1 Transition graph of the Markov chain of Example (3.10): vertices are the states, and a directed line from i to j means $P(i, j) > 0$.

sets, the chain is not irreducible. Deleting the second and the fourth rows and columns, we have

$$Q = \begin{bmatrix} \frac{1}{2} & \frac{1}{2} & 0 \\ 0 & \frac{1}{3} & \frac{2}{3} \\ \frac{1}{3} & \frac{1}{3} & \frac{1}{3} \end{bmatrix},$$

which is the Markov matrix corresponding to the restriction of X to the closed set $\{a, c, e\}$.

If the states were relabeled so that $1 = a$, $2 = c$, $3 = e$, $4 = b$, $5 = d$, then the transition matrix corresponding to this labeling of states becomes

states

$$\bar{P} = \begin{bmatrix} \frac{1}{2} & \frac{1}{2} & 0 & 0 & 0 \\ 0 & \frac{1}{3} & \frac{2}{3} & 0 & 0 \\ \frac{1}{3} & \frac{1}{3} & \frac{1}{3} & 0 & 0 \\ 0 & 0 & 0 & \frac{1}{4} & \frac{3}{4} \\ \frac{1}{4} & 0 & 0 & \frac{1}{2} & \frac{1}{4} \end{bmatrix} \begin{matrix} a \\ c \\ e \\ b \\ d \end{matrix}$$

which is in a form that makes analysis easier. Indeed, the tools we are developing next will enable us to conclude immediately from what has been said that the states a, c, and e are recurrent non-null aperiodic and that the states b and d are transient. □

We start with the following useful

(3.11) LEMMA. *If j is recurrent and $j \rightarrow k$, then $k \rightarrow j$ and $F(k, j) = 1$.*

Proof. If $j \rightarrow k$ then k can be reached from j without returning to j in the process. Let $\alpha > 0$ be the probability of that event. Once k is reached, the probability of never visiting j again is $1 - F(k, j)$. Thus, for the probability $1 - F(j, j)$ of never returning to j we have

$$1 - F(j, j) \geq \alpha(1 - F(k, j)) \geq 0.$$

But since j is recurrent, $1 - F(j, j) = 0$ by Definition (3.1). Since $\alpha > 0$, we must therefore have $1 - F(k, j) = 0$. This shows that $F(k, j) = 1$ and $k \rightarrow j$.

In particular this means that if $j \rightarrow k$ but $k \nrightarrow j$ for some k, then j must be transient. The following solidarity theorem complements the preceding lemma by stating that if j is recurrent and $j \rightarrow k$, then k is also recurrent.

(3.12) THEOREM. *From a recurrent state only recurrent states can be reached.*

Proof. Suppose j is recurrent and $j \rightarrow k$. We need to show that k is then recurrent.

Since $j \rightarrow k$, there is an integer r such that $P^r(j, k) > 0$. Since $j \rightarrow k$ and j is recurrent, $k \rightarrow j$ by Lemma (3.11), and therefore, there is an integer s such that $P^s(k, j) > 0$. Pick the smallest such integers r and s, and put

$$\beta = P^r(j, k)P^s(k, j) > 0.$$

Since $\{X_{s+n+r} = k\} \supset \{X_s = j, X_{s+n} = j, X_{s+n+r} = k\}$, we have

$$
\begin{aligned}
P^{s+n+r}(k, k) &= P_k\{X_{s+n+r} = k\} \\
&\geq P_k\{X_s = j, X_{s+n} = j, X_{s+n+r} = k\} \\
&= P^s(k, j)P^n(j, j)P^r(j, k) = \beta\, P^n(j, j).
\end{aligned}
$$

Hence,

$$
\begin{aligned}
R(k, k) = \sum_{m=0}^{\infty} P^m(k, k) &\geq \sum_{m=r+s}^{\infty} P^m(k, k) = \sum_{n=0}^{\infty} P^{n+r+s}(k, k) \\
&\geq \beta \sum_{n=0}^{\infty} P^n(j, j) = \beta R(j, j).
\end{aligned}
$$

Since j is recurrent, $R(j, j) = \infty$; thus, since $\beta > 0$, this expression implies that $R(k, k) = +\infty$ also, and therefore, k is recurrent. $\square$

From this theorem we know that no transient state can be reached from any recurrent state. Thus, the set of all recurrent states is closed. Within this set, however, there may be more than one closed set.

(3.13) LEMMA. For each recurrent state j there exists an irreducible closed set C which includes j.

Proof. Let j be recurrent and C the set of all states which can be reached from j. This is obviously a closed set. We need to show that if $i, k \in C$ then $i \rightarrow k$.

If $i \in C$, then $j \rightarrow i$. Since j is recurrent, Lemma (3.11) implies that $i \rightarrow j$. Thus, if $k \in C$ is another state, then $j \rightarrow k$, and by Proposition (3.9), $i \rightarrow k$. $\square$

The following is one of the main results of this section. Its proof follows directly from Lemma (3.13) and will be omitted.

(3.14) THEOREM. In a Markov chain the recurrent states can be divided, in a unique manner, into irreducible closed sets $C_1, C_2, \dots$.

In addition to the closed set $C = C_1 \cup C_2 \cup \cdots$, of recurrent states, the chain will in general contain transient states as well. It is possible for the recurrent states to be reached from a transient state (but not vice versa).

On the basis of Proposition (3.8) and Theorem (3.14), we see that if the states are relabeled in an appropriate manner, the transition matrix can be put in the form

$$(3.15) \qquad P = \begin{bmatrix} P_1 & 0 & 0 & \cdot & \cdot & \cdot & 0 \\ 0 & P_2 & 0 & \cdot & \cdot & \cdot & 0 \\ 0 & 0 & P_3 & \cdot & \cdot & \cdot & 0 \\ \cdot & \cdot & \cdot & \cdot & & & \cdot \\ \cdot & \cdot & \cdot & & \cdot & & \cdot \\ \cdot & \cdot & \cdot & & & \cdot & \cdot \\ Q_1 & Q_2 & Q_3 & \cdot & \cdot & \cdot & Q \end{bmatrix},$$

where $P_1, P_2, \ldots$ are the Markov matrices corresponding to the sets C_1, $C_2, \ldots$ of states. If we restrict our attention to one of these sets we obtain an irreducible Markov chain. Thus, the following theorem, together with (3.14), completes the picture.

(3.16) THEOREM. Let X be an irreducible Markov chain. Then either all states are transient, or all are recurrent null, or all are recurrent non-null. Either all states are aperiodic, or else if one is periodic with period δ, then all states are periodic with the same period δ.

Proof. Let j, k be any two states. Since X is irreducible, by Criterion (3.7),

$$(3.17) \qquad j \to k \quad \text{and} \quad k \to j.$$

This in turn implies that there are integers r and s such that $P^r(j, k) > 0$ and $P^s(k, j) > 0$. Pick the smallest such r and s and put

$$\beta = P^r(j, k) P^s(k, j);$$

then $\beta > 0$.

(a) If k is recurrent, then by (3.17) and Theorem (3.12), j is also recurrent. If k is transient, then j must also be transient (otherwise, if j were recurrent, (3.17) and (3.12) would imply that k is recurrent).

(b) Suppose k is recurrent null; then by Theorem (3.2), $P^m(k, k) \to 0$ as $m \to \infty$. Since, as in the proof of (3.12),

$$(3.18) \qquad P^{n+r+s}(k, k) \geq \beta P^n(j, j),$$

we also have that $P^n(j, j) \to 0$ as $n \to \infty$. Since j is recurrent, this can happen only if it is also null, again by (3.2). Interchanging the roles of j and k, we see that, conversely, if j is recurrent null so is k.

(c) Suppose k is periodic with period δ. Since

$$P^{s+r}(k, k) \geq P^s(k, j)P^r(j, k) = \beta > 0,$$

by (3.3), $s + r$ must be a multiple of δ. Therefore, if n is not a multiple of δ, then $n + s + r$ is not a multiple of δ either, and the left-hand side of (3.18) is zero. Since $\beta > 0$, then, $P^n(j, j) = 0$ whenever n is not a multiple of δ. Therefore, j must be periodic with period $\delta' \geq \delta$ by Criterion (3.4).

But if j is periodic with period δ', then reversing the roles of j and k, we must have that k is periodic with period $\delta \geq \delta'$. Thus $\delta = \delta'$, and j and k have the same period. $\qquad\square$

(3.19) COROLLARY. Let C be an irreducible closed set of finitely many states. Then no state in C is recurrent null.

Proof. If one state in C were recurrent null, then all the states in C would be recurrent null by Theorem (3.16). Then by Theorem (3.2),

$$(3.20) \qquad\qquad \lim_{n \to \infty} P^n(i, j) = 0$$

for all $i, j \in C$. But for each $i \in E$ and $n \geq 0$, $\sum_{j \in C} P^n(i, j) = 1$, so that

$$(3.21) \qquad\qquad \lim_{n \to \infty} \sum_{j \in C} P^n(i, j) = 1.$$

But the sum in (3.21) has only finitely many terms, so interchanging the order of summation and limit, we have

$$\lim_{n \to \infty} \sum_{j \in C} P^n(i, j) = \sum_{j \in C} \lim_{n \to \infty} P^n(i, j) = 0$$

by (3.20). This contradicts (3.21). So there can be no null states in C. $\qquad\square$

The same proof goes through for the following

(3.22) COROLLARY. If C is an irreducible closed set with finitely many states, then it has no transient states.

The preceding two corollaries yield the following statement when applied to Markov chains with a finite state space. If E is finite, then no state is recurrent null, and not all states are transient.

(3.23) EXAMPLE. Consider the Markov chain with state space $E = \{1, 2, \ldots, 10\}$ and transition matrix

$$P = \begin{array}{c c} & \begin{array}{c c c c c c c c c c} 1 & 2 & 3 & 4 & 5 & 6 & 7 & 8 & 9 & 10 \end{array} \\ \begin{array}{c} 1 \\ 2 \\ 3 \\ 4 \\ 5 \\ 6 \\ 7 \\ 8 \\ 9 \\ 10 \end{array} & \left[\begin{array}{c c c c c c c c c c} \frac{1}{2} & 0 & \frac{1}{2} & 0 & 0 & 0 & 0 & 0 & 0 & 0 \\ 0 & \frac{1}{3} & 0 & 0 & 0 & 0 & \frac{2}{3} & 0 & 0 & 0 \\ 1 & 0 & 0 & 0 & 0 & 0 & 0 & 0 & 0 & 0 \\ 0 & 0 & 0 & 0 & 1 & 0 & 0 & 0 & 0 & 0 \\ 0 & 0 & 0 & \frac{1}{3} & \frac{1}{3} & 0 & 0 & 0 & \frac{1}{3} & 0 \\ 0 & 0 & 0 & 0 & 0 & 1 & 0 & 0 & 0 & 0 \\ 0 & 0 & 0 & 0 & 0 & 0 & \frac{1}{4} & 0 & \frac{3}{4} & 0 \\ 0 & 0 & \frac{1}{4} & \frac{1}{4} & 0 & 0 & 0 & \frac{1}{4} & 0 & \frac{1}{4} \\ 0 & 1 & 0 & 0 & 0 & 0 & 0 & 0 & 0 & 0 \\ 0 & \frac{1}{3} & 0 & 0 & \frac{1}{3} & 0 & 0 & 0 & 0 & \frac{1}{3} \end{array} \right] \end{array}.$$

Its transition graph is shown in Figure 5.3.2. We see from this graph that $\{1, 3\}$, $\{2, 7, 9\}$, and $\{6\}$ are irreducible closed sets. These closed sets may be reached from the states 4, 5, 8, and 10 (but not vice versa). By Lemma (3.11), the states 4, 5, 8, and 10 are transient. By Corollaries (3.19) and (3.22), none of the states 1, 3, 2, 7, 9, or 6 is transient or recurrent-null. States 1 and 3 form one irreducible closed set of recurrent non-null aperiodic states, states 2, 7, and 9 another, and state 6 another. State 6 is absorbing.

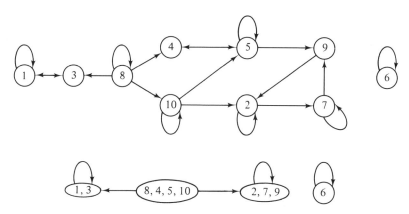

Figure 5.3.2 Transition graph for the Markov chain of Example (3.23) and its reduction corresponding to the groupings of irreducible closed sets.

By relabeling the states we can put P in the form described by (3.15). We have *states*

$$\bar{P} = \begin{bmatrix} 1 & 0 & 0 & 0 & 0 & 0 & 0 & 0 & 0 & 0 \\ 0 & \frac{1}{2} & \frac{1}{2} & 0 & 0 & 0 & 0 & 0 & 0 & 0 \\ 0 & 1 & 0 & 0 & 0 & 0 & 0 & 0 & 0 & 0 \\ 0 & 0 & 0 & \frac{1}{3} & \frac{2}{3} & 0 & 0 & 0 & 0 & 0 \\ 0 & 0 & 0 & 0 & \frac{1}{4} & \frac{3}{4} & 0 & 0 & 0 & 0 \\ 0 & 0 & 0 & 1 & 0 & 0 & 0 & 0 & 0 & 0 \\ 0 & 0 & 0 & 0 & 0 & 0 & 0 & 1 & 0 & 0 \\ 0 & 0 & 0 & 0 & 0 & \frac{1}{3} & \frac{1}{3} & \frac{1}{3} & 0 & 0 \\ 0 & 0 & \frac{1}{4} & 0 & 0 & 0 & \frac{1}{4} & 0 & \frac{1}{4} & \frac{1}{4} \\ 0 & 0 & 0 & \frac{1}{3} & 0 & 0 & 0 & \frac{1}{3} & 0 & \frac{1}{3} \end{bmatrix} \quad \begin{matrix} 6 \\ 1 \\ 3 \\ 2 \\ 7 \\ 9 \\ 4 \\ 5 \\ 8 \\ 10 \end{matrix}$$

(3.24) EXAMPLE. Consider the Markov chain $\{N_n\}$ where N_n is the number of successes in the first n trials of a Bernoulli process (Example (1.11)). For any j, we have $j \rightarrow j + 1$ but not vice versa. Thus, by Lemma (3.11), j cannot be recurrent. Hence, all states of this Markov chain are transient. □

(3.25) EXAMPLE. Consider Example (1.17); suppose $p_1, p_2, \dots$ are all positive. Then $0 \rightarrow j$ for any j in one step, and we have $j \rightarrow j - 1 \rightarrow j - 2 \rightarrow \cdots \rightarrow 2 \rightarrow 1 \rightarrow 0$. Thus, all states can be reached from each other, and by Criterion (3.7) the Markov chain is irreducible. Since $P^1(0, 0) = p_1 > 0$, state 0 is aperiodic. Then by Theorem (3.16), all states are aperiodic.

Return to state 0 occurs if the lifetime of the equipment is finite. The probability that the lifetime is finite is $\sum_j p_j = 1$. Thus, $F(0, 0) = \sum_j p_j = 1$, and state 0 is recurrent. By Theorem (3.16) this implies that all states are recurrent. If the expected lifetime $\sum_j jp_j = +\infty$, then 0 is null and therefore all states are recurrent null. If $\sum_j jp_j < \infty$, then 0 is non-null and all states are recurrent non-null aperiodic. □

In the case where there are only finitely many states, we have all the tools we need to classify the states. In summary: we first identify the irreducible closed sets; then by using Theorem (3.16) and Corollaries (3.19) and (3.22) we conclude that all states belonging to an irreducible closed set are recurrent non-null; the remaining states (if there are any) are transient; periodicity is determined by applying Criterion (3.4) to each irreducible closed set and then using Theorem (3.16).

In the case where there are infinitely many states, as Examples (3.24) and (3.25) show, it is possible to have irreducible closed sets with infinitely many states all of which are transient or recurrent null. Then, applying Theorem

(3.26), we can determine whether they are all recurrent non-null or not. In the latter case, if they are not recurrent non-null, applying Theorem (3.29) below we can determine whether they are transient or not. (Of course, if the states are not recurrent non-null and are not transient, then they must all be recurrent null.) The following two theorems, then, complete our account of the classification of states for the present. The proofs will be provided later in Chapter 6; see Theorem (6.2.1) and Proposition (6.4.5) ff.

(3.26) THEOREM. Let X be an irreducible Markov chain, and consider the system of linear equations

$$(3.27) \qquad v(j) = \sum_{i \in E} v(i)P(i,j), \quad j \in E.$$

Then all states are recurrent non-null if and only if there exists a solution v with

$$(3.28) \qquad \sum_{j \in E} v(j) = 1.$$

If (3.27) and (3.28) have a solution v, then $v(j) > 0$ for every $j \in E$, and there are no other solutions.

(3.29) THEOREM. Let X be an irreducible Markov chain with transition matrix P, and let Q be the matrix obtained from P by deleting the k-row and the k-column for some $k \in E$. Then all states are recurrent if and only if the only solution of

$$(3.30) \qquad h(i) = \sum_{j \in E_0} Q(i,j)h(j), \quad 0 \le h(i) \le 1; \quad i \in E_0;$$

is $h(i) = 0$ for all $i \in E_0$; here $E_0 = E - \{k\}$.

(3.31) EXAMPLE. *Random walks.* Let X be a Markov chain with state space $E = \{0, 1, 2, \ldots\}$ and the transition matrix

$$P = \begin{bmatrix} 0 & 1 & & & \\ q & 0 & p & & 0 \\ & q & 0 & p & \\ & & q & 0 & p & \\ 0 & & & \cdot & \cdot & \cdot \\ & & & & \cdot & \cdot & \cdot \\ & & & & & \cdot & \cdot \end{bmatrix}$$

where $0 < p < 1, q = 1 - p$. This chain is called a random walk, and is used to describe the walk of a sufficiently intoxicated person: if he is at position i after step n, his next step leads him to either $i + 1$ or $i - 1$ with respective probabilities p and q except that at $i = 0$ there is a barrier and when he hits it he is sure to step back to 1.

All states can be reached from each other, and by Criterion (3.7), the chain is irreducible. Starting from 0, to come back to 0 the process must take as many steps to the left as to the right. Hence, the return to 0 can occur only at steps numbered 2, 4, 6, 8, . . . ; that is, state 0 is periodic with period $\delta = 2$. Since X is irreducible, by Theorem (3.16), all states are periodic with period 2. By the same theorem, either all states are transient, or all are recurrent null, or all are recurrent non-null.

To find out whether the states are recurrent non-null we use Theorem (3.26). The equation $v = vP$, that is, Equation (3.27), becomes (we write $v_i = v(i)$)

(3.32)
$$v_0 = qv_1,$$
$$v_1 = v_0 + qv_2,$$
$$v_2 = pv_1 + qv_3,$$
$$v_3 = pv_2 + qv_4,$$

etc. Solving the first equation for v_1, the second for v_2, and so on, we obtain

$$v_1 = \frac{1}{q}v_0,$$
$$v_2 = \frac{1}{q}\left(\frac{1}{q}v_0 - v_0\right) = \frac{p}{q^2}v_0,$$
$$v_3 = \frac{1}{q}\left(\frac{p}{q^2} - \frac{p}{q}\right)v_0 = \frac{p^2}{q^3}v_0,$$

etc. That is, any solution of (3.27) is of the form

(3.33)
$$v_j = \frac{1}{q}\left(\frac{p}{q}\right)^{j-1}v_0, \quad j = 1, 2, \ldots,$$

for some constant $v_0 \geq 0$.

If $p < q$, then $p/q < 1$ and

$$\sum_{j=0}^{\infty} v_j = \left(1 + \frac{1}{q}\sum_{j=1}^{\infty}\left(\frac{p}{q}\right)^{j-1}\right)v_0 = \left(1 + \frac{1}{q}\cdot\frac{1}{1-p/q}\right)v_0 = \frac{2q}{q-p}v_0,$$

and by choosing

$$v_0 = \frac{q-p}{2q} = \frac{1}{2}\left(1 - \frac{p}{q}\right),$$

we can make the sum $\sum v_j = 1$. Hence, if $p < q$,

(3.34)
$$v(j) = \begin{cases} \frac{1}{2}\left(1 - \frac{p}{q}\right) & \text{if } j = 0, \\ \frac{1}{2q}\left(1 - \frac{p}{q}\right)\left(\frac{p}{q}\right)^{j-1} & \text{if } j \geq 1 \end{cases}$$

is a solution for (3.27) and (3.28), and in this case all states are recurrent non-null.

On the other hand, if $p \geq q$, then the solution (3.33) of (3.27) is such that $\sum v_j$ is either 0 (this is true if $v_0 = 0$) or infinite (this is true if $v_0 \neq 0$). Hence there is no way of satisfying (3.28) and (3.27) together, and the states are not recurrent non-null.

If $p \geq q$, therefore, either all states are recurrent null or all states are transient. To distinguish between the two cases we use Theorem (3.29). Excluding the row and the column corresponding to state 0, we obtain

$$Q = \begin{bmatrix} 0 & p & & & \\ q & 0 & p & & 0 \\ & q & 0 & p & \\ & & q & 0 & p \\ 0 & & & \cdot & \cdot & \cdot \\ & & & & \cdot & \cdot & \cdot \\ & & & & & \cdot & \cdot & \cdot \end{bmatrix}.$$

Now the equation $h = Qh$ gives, writing $h_i = h(i)$,

$$h_1 = ph_2,$$
$$h_2 = qh_1 + ph_3,$$
$$h_3 = qh_2 + ph_4, \ldots.$$

The typical equation is $h_i = qh_{i-1} + ph_{i+1}$. Writing $h_i = ph_i + qh_i$, passing the term ph_i to the right and the term qh_{i-1} to the left, we get

(3.35)
$$p(h_{i+1} - h_i) = q(h_i - h_{i-1}); \quad i = 2, 3, \ldots;$$

and directly from the first equation,

(3.36)
$$p(h_2 - h_1) = qh_1.$$

Iterating on i in (3.35), we obtain

$$h_{i+1} - h_i = \left(\frac{q}{p}\right)^{i-1}(h_2 - h_1) = \left(\frac{q}{p}\right)^i h_1, \quad i = 1, 2, \ldots,$$

from which we have, for any $i \geq 1$,

$$h_{i+1} = (h_{i+1} - h_i) + (h_i - h_{i-1}) + \cdots + (h_2 - h_1) + h_1$$
$$= \left[\left(\frac{q}{p} \right)^i + \left(\frac{q}{p} \right)^{i-1} + \cdots + \frac{q}{p} + 1 \right] h_1.$$

We have thus shown that any solution of the equation $h = Qh$ is of the form

$$(3.37) \qquad h_i = c \left[1 + \frac{q}{p} + \cdots + \left(\frac{q}{p} \right)^{i-1} \right], \quad i = 1, 2, \ldots,$$

where c is some constant.

If $p = q$, then (3.37) shows that $h_i = ic$ for all $i \geq 1$, and the only way to have $0 \leq h_i \leq 1$ for all i is by choosing $c = 0$. That is, if $p = q$, the only solution of (3.30) is $h = 0$, which implies by Theorem (3.29) that all states are recurrent. (Since we know they are not recurrent non-null, they must be recurrent null.)

If $p > q$, then choosing $c = 1 - (q/p)$, we get

$$(3.38) \qquad\qquad h_i = 1 - \left(\frac{q}{p} \right)^i, \quad i = 1, 2, \ldots,$$

which satisfies $0 \leq h_i \leq 1$ for all i. Hence, in this case all states are transient.

To summarize: if $p < q$, then all states are recurrent non-null periodic with period 2; if $p = q$, then all states are recurrent null periodic with period 2; if $p > q$, then all states are transient. $\qquad\square$

4. Exercises

(4.1) Let X be a Markov chain with state space $E = \{a, b, c\}$, transition matrix

$$P = \begin{bmatrix} 0 & \frac{1}{3} & \frac{2}{3} \\ \frac{1}{4} & \frac{3}{4} & 0 \\ \frac{2}{5} & 0 & \frac{3}{5} \end{bmatrix},$$

and the initial distribution $\pi = (\frac{2}{5}, \frac{1}{5}, \frac{2}{5})$. Compute the following (recall the notation $P_i(\cdot) = P\{\cdot \mid X_0 = i\}$):

    (a) $P_a\{X_1 = b, X_2 = b, X_3 = b, X_4 = a, X_5 = c\}$,
    (b) $P_c\{X_1 = a, X_2 = c, X_3 = c, X_4 = a, X_5 = b\}$,
    (c) $P_a\{X_1 = b, X_3 = a, X_4 = c, X_6 = b\}$,
    (d) $P\{X_1 = b, X_2 = b, X_3 = a\}$,
    (e) $P\{X_2 = b, X_5 = b, X_6 = b\}$.

(4.2) Let X be a Markov chain with state space $E = \{r, w, b, y\}$ and transition matrix

$$P = \begin{bmatrix} 0 & 0 & 1 & 0 \\ 0 & 0.4 & 0.6 & 0 \\ 0.8 & 0 & 0.2 & 0 \\ 0.2 & 0.3 & 0 & 0.5 \end{bmatrix}.$$

(a) Compute $P\{X_5 = b, X_6 = r, X_7 = b, X_8 = b \,|\, X_4 = w\}$.

(b) Compute

$$E[f(X_5)f(X_6)\,|\,X_4 = y]$$

for the function f with values 2, 4, 7, and 3 at r, w, b, and y respectively.

(c) Compute the distribution $\{F_k(i, j); k = 1, 2, \ldots\}$ of the first passage time from state i to state j for $(i, j) = (w, b)$ by direct inspection. Do the same for $(i, j) = (b, b)$.

(4.3) Classify the states of the Markov chain of Example (1.9).

(4.4) Classify the states of the Markov chains with the following transition matrices:

(a) $\begin{bmatrix} 0.8 & 0 & 0.2 & 0 \\ 0 & 0 & 1 & 0 \\ 1 & 0 & 0 & 0 \\ 0.3 & 0.4 & 0 & 0.3 \end{bmatrix}$ (b) $\begin{bmatrix} 0 & 0 & 0.4 & 0.6 & 0 \\ 0 & 0.2 & 0 & 0.5 & 0.3 \\ 0.5 & 0 & 0.5 & 0 & 0 \\ 0 & 0 & 1 & 0 & 0 \\ 0.3 & 0 & 0.5 & 0 & 0.2 \end{bmatrix}$

(c) $\begin{bmatrix} 0.5 & 0 & 0 & 0.5 & 0 \\ 0 & 0.6 & 0 & 0 & 0.4 \\ 0.3 & 0 & 0.7 & 0 & 0 \\ 0 & 0 & 1 & 0 & 0 \\ 0 & 1 & 0 & 0 & 0 \end{bmatrix}$

(d) $\begin{bmatrix} 0.8 & 0 & 0 & 0 & 0 & 0.2 & 0 \\ 0 & 0 & 0 & 0 & 1 & 0 & 0 \\ 0.1 & 0 & 0.9 & 0 & 0 & 0 & 0 \\ 0 & 0 & 0 & 0.5 & 0 & 0 & 0.5 \\ 0 & 0.3 & 0 & 0 & 0.7 & 0 & 0 \\ 0 & 0 & 1 & 0 & 0 & 0 & 0 \\ 0 & 0.5 & 0 & 0 & 0 & 0.5 & 0 \end{bmatrix}$

(e)
$$\begin{bmatrix} 0 & 0 & 1 & 0 & 0 & 0 & 0 \\ 0 & 0.2 & 0 & 0 & 0.4 & 0.4 & 0 \\ 0.8 & 0 & 0 & 0 & 0.2 & 0 & 0 \\ 0 & 0 & 0 & 0 & 0 & 1 & 0 \\ 0 & 0 & 1 & 0 & 0 & 0 & 0 \\ 0 & 0 & 0 & 0.7 & 0 & 0.3 & 0 \\ 0 & 0 & 0 & 0 & 0 & 0 & 1 \end{bmatrix}$$

(f)
$$\begin{bmatrix} 0 & 0 & 1 & 0 & 0 & 0 & 0 & 0 \\ 0 & 0 & 1 & 0 & 0 & 0 & 0 & 0 \\ 0 & 0 & 0 & 0.5 & 0 & 0.5 & 0 & 0 \\ 1 & 0 & 0 & 0 & 0 & 0 & 0 & 0 \\ 0 & 1 & 0 & 0 & 0 & 0 & 0 & 0 \\ 0.3 & 0.7 & 0 & 0 & 0 & 0 & 0 & 0 \\ 0.2 & 0.4 & 0 & 0 & 0.1 & 0 & 0.1 & 0.2 \\ 0 & 0 & 0.3 & 0 & 0 & 0.4 & 0 & 0.3 \end{bmatrix}$$

(4.5) In Example (1.17), suppose $p_1 = p_3 = p_5 = \cdots = 0$, $p_2 = \frac{1}{2}$, $p_4 = \frac{1}{4}$, $p_6 = 0$, $p_8 = \frac{1}{4}$, $p_{10} = p_{12} = \cdots = 0$. Classify all the states.

(4.6) Consider an irreducible Markov chain with transition matrix P. Show that, if $P(j, j) > 0$ for some j, then all states are aperiodic.

(4.7) In Example (1.16) suppose the reviews are weekly, so that $t_0 = 0$, $t_1 = 1$, $t_2 = 2, \ldots$; and suppose the critical values are set at $s = 3$ and $S = 8$. Suppose that the demand process is Poisson with parameter $\lambda = 4$ per week, and that the demand during a week is independent of the inventory on hand.

(a) Show that X is a Markov chain with state space $E = \{0, 1, \ldots, 8\}$ and transition matrix

$$P = \begin{bmatrix} r_7 & p_7 & p_6 & p_5 & p_4 & p_3 & p_2 & p_1 & p_0 \\ r_7 & p_7 & p_6 & p_5 & p_4 & p_3 & p_2 & p_1 & p_0 \\ r_7 & p_7 & p_6 & p_5 & p_4 & p_3 & p_2 & p_1 & p_0 \\ r_7 & p_7 & p_6 & p_5 & p_4 & p_3 & p_2 & p_1 & p_0 \\ r_3 & p_3 & p_2 & p_1 & p_0 & 0 & 0 & 0 & 0 \\ r_4 & p_4 & p_3 & p_2 & p_1 & p_0 & 0 & 0 & 0 \\ r_5 & p_5 & p_4 & p_3 & p_2 & p_1 & p_0 & 0 & 0 \\ r_6 & p_6 & p_5 & p_4 & p_3 & p_2 & p_1 & p_0 & 0 \\ r_7 & p_7 & p_6 & p_5 & p_4 & p_3 & p_2 & p_1 & p_0 \end{bmatrix}$$

where

$$p_n = \frac{e^{-4}4^n}{n!}, \qquad r_n = \sum_{k=n+1}^{\infty} \frac{e^{-4}4^k}{k!}, \qquad n = 0, 1, \ldots.$$

(b)   Starting with the maximum amount possible, what is the probability that a shortage occurs before the period ends?

(c)   Starting with the maximum possible, find the probability that no replenishment will be necessary at t_1 and t_2.

(d)   Starting with $i = 4$ units, what is the distribution of the time until the first replenishment?

(4.8)   Let X_n denote the capital of a gambler at the end of the nth play. His strategy is as follows. If his capital is 4 dollars or more, then he bets 2 dollars which earn him 4, 3, or 0 dollars with respective probabilities 0.25, 0.30, 0.45. If his capital is 1, 2, or 3 dollars, then he plays more conservatively, bets 1 dollar, and this earns him either 2 or 0 dollars with respective probabilities 0.45 and 0.55. When his capital becomes 0, he stops.

(a)   Let Y_{n+1} be the net earnings at the $(n + 1)$th play, that is, $X_{n+1} = X_n + Y_{n+1}$. Compute

$$P\{Y_{n+1} = k \,|\, X_n = i\}, \qquad i = 0, 1, \ldots; k = -2, -1, 0, 1, \ldots.$$

(b)   Show that $X = \{X_n; n \in \mathbb{N}\}$ is a Markov chain.

(c)   Compute the transition probabilities for X.

(d)   Classify the states.

(4.9)   Classify the states of the Markov chain of Example (1.17) by using Theorems (3.26) and (3.29).

(4.10)   *Age process.* In Example (1.17) again, instead of considering the remaining lifetime at time n, let us consider the age Y_n of the equipment in use at time n. For any n, $Y_{n+1} = 0$ if the equipment failed during the period $n + 1$, and $Y_{n+1} = Y_n + 1$ if it did not fail during the period $n + 1$.

(a)   Show that

$$q_i = P\{Y_{n+1} = i + 1 \,|\, Y_n = i\} = \begin{cases} 1 - p_1 & \text{if } i = 0, \\ \dfrac{1 - p_1 - \cdots - p_{i+1}}{1 - p_1 - \cdots - p_i} & \text{if } i \geq 1. \end{cases}$$

(b)   Show that $\{Y_n; n \in \mathbb{N}\}$ is a Markov chain with state space $E = \{0, 1, \ldots\}$ and the transition matrix

$$P = \begin{bmatrix} 1 - q_0 & q_0 & & & & \\ 1 - q_1 & 0 & q_1 & & 0 & \\ 1 - q_2 & 0 & 0 & q_2 & & \\ \cdot & \cdot & \cdot & \cdot & \cdot & \\ \cdot & \cdot & \cdot & \cdot & & \cdot \\ \cdot & \cdot & \cdot & \cdot & & \cdot \end{bmatrix}.$$

(c)   Classify the states of this Markov chain.

(4.11)   *Bernoulli–Laplace model of diffusion.* Consider two urns each of which contains m balls; b of these $2m$ balls are black, and the remaining $2m - b$ are white.

We say that the system is in state i if the first urn contains i black balls (then it also contains $m - i$ white balls, and the second urn contains $b - i$ black and $m - b + i$ white balls). Each trial consists of choosing one ball at random from each urn and interchanging the two.

Show that the successive states of this system form a Markov chain, and compute the transition probabilities.

This is a probabilistic model of the flow of two incompressible fluids between two containers.

(4.12) *Space–time processes.* Suppose $X = \{X_n; n \in \mathbb{N}\}$ is a Markov chain with state space E and transition probabilities $P(i, j)$. Let N_0 be a random variable taking values in $\mathbb{N}$ which is independent of X. For each $n \in \mathbb{N}$, let $N_n = N_0 + n$, and define Y_n to be the pair (X_n, N_n).

    (a)   Show that $Y = \{Y_n; n \in \mathbb{N}\}$ is a Markov chain with the countable set $F = E \times \mathbb{N}$ as its state space.

    (b)   Compute the transition probabilities $P((i, i'), (j, j'))$ for this Markov chain.

    (c)   Show that the same results hold if, instead of N_0 being independent of X, it is only assumed that N_0 is conditionally independent of $\{X_1, X_2, \ldots\}$ given X_0.

Such a Markov chain Y is called a *space–time* process.

(4.13) *Non-time-homogeneous processes.* Suppose $X = \{X_n; n \in \mathbb{N}\}$ is a Markov chain in the sense of Definition (1.1) but is not necessarily time-homogeneous; that is,

$$P_n(i, j) = P\{X_{n+1} = j \mid X_n = i\}, \qquad i, j \in E,$$

may also depend on n.

    Let N_0 be a random variable taking values in $\mathbb{N}$ which is conditionally independent of $\{X_1, X_2, \ldots\}$ given X_0. Define $N_n = N_0 + n$, $Y_n = (X_n, N_n)$ for every $n \in \mathbb{N}$.

    (a)   Show that $Y = \{Y_n; n \in \mathbb{N}\}$ is a *time-homogeneous* Markov chain with a countable state space.

    (b)   Compute the transition probabilities $P((i, i'), (j, j'))$ for this Markov chain.

(4.14) *K-dependent processes.* Let $X = \{X_n; n \in \mathbb{N}\}$ be a stochastic process taking values in a countable state space E. Suppose there exists an integer $K \in \mathbb{N}$ such that

$$P\{X_n = j \mid X_0, \ldots, X_{n-1}\} = P\{X_n = j \mid X_{n-K}, \ldots, X_{n-1}\}$$

for every $j \in E$ and $n \in \mathbb{N}$, $n \geq K$. In other words, given all the past, the future depends only on the last K values. Such a process X is called a *K-dependent* chain. For $K = 1$, we have the ordinary Markov chains. Their theory can, however, be reduced to that of the ordinary Markov chains by the following device.

For each $n \in \mathbb{N}$, let

$$Y_n = (X_n, X_{n+1}, \ldots, X_{n+K-1}).$$

Then $Y = \{ Y_n; n \in \mathbb{N} \}$ is a stochastic process taking values in the countable set $F = E^K = E \times \cdots \times E$. Show that Y is an ordinary Markov chain.

$$* \qquad * \qquad *$$

Our subject matter is named after A. A. MARKOV, *who laid the foundations of the theory in a series of papers starting in 1907. He considered the finite state space case and showed that limits of the transition probabilities $P^n(i, j)$ exist in the case of an aperiodic chain with only one recurrent class.* MARKOV's *first application was to give a rigorous solution of the urn problem of Exercise (4.11), which was first posed by* DANIEL BERNOULLI *in 1769 and later analyzed by* LAPLACE *in 1812. Another interesting application was an investigation of the way the vowels and consonants alternate in Russian literature.* MARKOV *carried out such a study on Pushkin's* Eugene Onegin *and Aksakov's "The Childhood Years of Bagrov's Grandson."*

The early work on Markov chains was restricted to the finite state space case, and matrix theory played an important role. For a fairly detailed account of the work of this period we refer the reader to M. FRÉCHET [1].

The infinite state space case was introduced by A. N. KOLMOGOROV [2]. *The present probabilistic approach to studying Markov chains was first championed by* FELLER [2]. *Much of the present terminology originates from the book by* FELLER [3].

CHAPTER 6

Limiting Behavior
and Applications
of Markov Chains

This chapter is to complete the elementary treatment of Markov chains that we started in the preceding chapter. We will first consider the computation of the expected number $R(i, j)$ of visits to j and the probability $F(i, j)$ of ever reaching j, both starting at i. Considering the recurrent states we will show how to compute the probability $P^n(i, j)$ of being in state j at time n, starting at i, for large n. Studying the periodic states is easily reduced to the aperiodic case, and we will show how to do that. Then we will take up the computation of the probability of remaining in a set of transient states forever. Finally, we will give a brief treatment of two important models in queueing theory and an introduction to branching processes.

Throughout, Ω will be a fixed sample space and P a fixed probability measure on it. For a Markov chain X defined on it we will write, as we did before, $P_i(A)$ for the conditional probability $P\{A \mid X_0 = i\}$ and, similarly, write $E_i[Z]$ for the conditional expectation $E[Z \mid X_0 = i]$.

1. Computation of R and F

Let X be a Markov chain with state space E and transition matrix P. Let $R(i, j)$ be the expected number of visits to state j starting at i, and let $F(i, j)$ be the probability of ever reaching j starting from i; these were defined by (5.2.12) and (5.2.6). We first take up the computation of the potential matrix R.

Suppose j is a recurrent state. Then by Definition (5.3.1), $F(j, j) = 1$; and by Corollary (5.2.13) this implies that $R(j, j) = +\infty$. Furthermore, by the same corollary, if j can be reached from i, then $F(i, j) > 0$ and $R(i, j) = \infty$ again. If, on the other hand, j cannot be reached from i, then $F(i, j) = 0$ and $R(i, j) = 0$. Hence, for j recurrent,

(1.1)
$$R(i, j) = \begin{cases} 0 & \text{if } F(i, j) = 0, \\ +\infty & \text{if } F(i, j) > 0. \end{cases}$$

If j is transient and i recurrent, then by Theorem (5.3.12) j cannot be reached from i; therefore, in that case $F(i, j) = 0$ and

(1.2)
$$R(i, j) = 0.$$

Thus, the only remaining case which is not trivial is when both i and j are transient.

Let D denote the set of all transient states, and let Q and S be the matrices obtained from P and R, respectively, by deleting all the rows and columns corresponding to the recurrent states; that is,

(1.3)
$$Q(i, j) = P(i, j), \quad S(i, j) = R(i, j), \quad i, j \in D.$$

If the states are labeled so that the recurrent states precede the transient ones, then as in (5.3.15), we can write P in the form

(1.4)
$$P = \begin{bmatrix} K & 0 \\ L & Q \end{bmatrix}$$

by defining K and L suitably. Then for any $m \in \mathbb{N}$,

(1.5)
$$P^m = \begin{bmatrix} K^m & 0 \\ L_m & Q^m \end{bmatrix},$$

where Q^m is the mth power of Q (similarly for K^m but not for L_m). Thus, from (5.2.15),

$$R = \sum_{m=0}^{\infty} P^m = \begin{bmatrix} \sum K^m & 0 \\ \sum L_m & \sum Q^m \end{bmatrix},$$

which shows that

(1.6)
$$S = \sum_{m=0}^{\infty} Q^m = I + Q + Q^2 + \cdots.$$

Now the computation of S is not difficult: noting that

$$SQ = QS = Q + Q^2 + \cdots = S - I,$$

we have

(1.7) $$(I - Q)S = I, \qquad S(I - Q) = I;$$

that is, S satisfies the two systems of linear equations (1.7), and we may use either one to solve for S. In particular, if D is finite, (1.7) shows that the finite-dimensional matrix S is the inverse of the matrix $I - Q$; that is,

(1.8) PROPOSITION. If there are only finitely many transient states, then

$$S = (I - Q)^{-1}.$$

On the other hand, when the set D of transient states is infinite, it is possible to have more than one solution to the system (1.7). In that case, the next theorem shows that S is the smallest possible solution:

(1.9) THEOREM. S is the minimal solution of

(1.10) $$(I - Q)Y = I, \quad Y \geq 0.$$

Proof. We know by (1.7) that S satisfies (1.10). To show that it is the minimal solution, let Y be another solution. Then we have

$$Y = I + QY.$$

Replacing Y on the right by $I + QY$, we get

$$Y = I + Q(I + QY) = I + Q + Q^2 Y,$$

and repeating thus,

(1.11) $$Y = I + Q + \cdots + Q^n + Q^{n+1} Y \geq \sum_{m=0}^{n} Q^m$$

for any $n \in \mathbb{N}$, since $Q^{n+1} \geq 0$ and $Y \geq 0$. Now taking limits in (1.11) as $n \to \infty$, we get $Y \geq S$ as claimed. $\qquad \square$

If Y is another solution, then $Y = I + QY$ and $S = I + QS$ together imply that $H = Y - S \geq 0$ satisfies $H = QH$. Then every column of H satisfies $h = Qh$, $h \geq 0$. Moreover, every column of S is bounded (since $S(i, j) = F(i, j)S(j, j)$ for $i \neq j$). It follows that S is the unique solution of (1.10) if and only if the only bounded solution of $h = Qh$ is $h = 0$, or equivalently, if and only if

$$h = Qh, \quad 0 \leq h \leq 1$$

implies $h = 0$. For further discussion of this point we refer the reader to Section 4 (see also Section 2 of Chapter 7).

(1.12) EXAMPLE. Let X be a Markov chain with state space $E = \{1, 2, 3, 4, 5, 6, 7, 8\}$ and transition matrix

$$P = \begin{bmatrix} 0.4 & 0.3 & 0.3 & & & & & \\ 0 & 0.6 & 0.4 & & 0 & & 0 & \\ 0.5 & 0.5 & 0 & & & & & \\ & & & 0 & 1 & & & \\ & 0 & & 0.8 & 0.2 & & 0 & \\ 0 & 0 & 0 & 0 & 0 & 0.4 & 0.6 & 0 \\ 0.4 & 0.4 & 0 & 0 & 0 & 0 & 0 & 0.2 \\ 0.1 & 0 & 0.3 & 0 & 0 & 0.6 & 0 & 0 \end{bmatrix}.$$

States 1, 2, and 3 are recurrent non-null aperiodic and form an irreducible class. States 4 and 5 form another irreducible class of recurrent non-null aperiodic states. States 6, 7, and 8 are transient; from these states only 1, 2, and 3 can be reached.

Here

$$Q = \begin{bmatrix} 0.4 & 0.6 & 0 \\ 0 & 0 & 0.2 \\ 0.6 & 0 & 0 \end{bmatrix};$$

so using Proposition (1.8) above, we get

$$S = (I - Q)^{-1} = \begin{bmatrix} 0.6 & -0.6 & 0 \\ 0 & 1 & -0.2 \\ -0.6 & 0 & 1 \end{bmatrix}^{-1} = \tfrac{1}{66} \begin{bmatrix} 125 & 75 & 15 \\ 15 & 75 & 15 \\ 75 & 45 & 75 \end{bmatrix}.$$

Thus, the potential matrix R for this Markov chain is

$$R = \begin{bmatrix} \infty & \infty & \infty & & & & & \\ \infty & \infty & \infty & & 0 & & 0 & \\ \infty & \infty & \infty & & & & & \\ & & & \infty & \infty & & & \\ & 0 & & \infty & \infty & & 0 & \\ \infty & \infty & \infty & 0 & 0 & \frac{125}{66} & \frac{75}{66} & \frac{15}{66} \\ \infty & \infty & \infty & 0 & 0 & \frac{15}{66} & \frac{75}{66} & \frac{15}{66} \\ \infty & \infty & \infty & 0 & 0 & \frac{75}{66} & \frac{45}{66} & \frac{75}{66} \end{bmatrix}.$$

□

Next we consider the computation of the probability $F(i, j)$ of ever reaching j from i. If i and j are both recurrent and belong to the same irreducible closed set, then by Lemma (5.3.11),

(1.13) $F(i, j) = 1.$

If i is recurrent and j transient, or if i and j are recurrent but belong to different irreducible sets, then

(1.14) $F(i, j) = 0.$

If i and j are both transient, then $R(i, j) < \infty$, and by Corollary (5.2.13) we have

(1.15) $F(j, j) = 1 - \dfrac{1}{R(j, j)}, \quad F(i, j) = \dfrac{R(i, j)}{R(j, j)}$

for $i \neq j$, and the $R(i, j)$ may be obtained by using (1.8) or (1.9) above.

The only remaining case is when i is transient and j recurrent. The computations are simplified somewhat by the following.

(1.16) LEMMA.   Let C be an irreducible closed set of recurrent states. Then for any transient state i,

$$F(i, j) = F(i, k)$$

for all $j, k \in C$.

Proof. For $j, k \in C$, by Lemma (5.3.11), $F(j, k) = F(k, j) = 1$. Thus, once the chain reaches any one of the states of C, it also visits all the other states. Hence, $F(i, j) = F(i, k)$ is the probability of ever entering the set C from i.   □

In view of this lemma we need only talk about the probability of reaching an irreducible recurrent set. Let $C_1, C_2, \ldots$ be the irreducible recurrent classes, and let D be the set of all transient states. Suppose the states are labeled so that the states in C_1 precede those in C_2, and so on, and the transient states are after all the recurrent ones. Then the transition matrix P is in the form (5.3.15). Since we are interested only in ever reaching C_j, we may lump all the states of C_j together to make one absorbing state. The matrix of transition probabilities then becomes

(1.17) $\hat{P} = \begin{bmatrix} 1 & 0 & \cdot & \cdot & \cdot & 0 & 0 \\ 0 & 1 & \cdot & \cdot & \cdot & 0 & 0 \\ \cdot & \cdot & & \cdot & & \cdot & \cdot \\ \cdot & \cdot & & & \cdot & \cdot & \cdot \\ \cdot & \cdot & & \cdot & & \cdot & \cdot \\ 0 & 0 & \cdot & \cdot & \cdot & 1 & 0 \\ b_1 & b_2 & \cdot & \cdot & \cdot & b_m & Q \end{bmatrix},$

where

$$b_j(i) = \sum_{k \in C_j} P(i, k), \quad i \in D.$$

The probability of ever reaching the absorbing state j from the transient state i by the chain with the transition matrix $\hat{P}$ is the same as that of ever reaching C_j from the transient state i by the original chain with the transition matrix P.

We now rewrite $\hat{P}$ as

(1.18)
$$\hat{P} = \begin{bmatrix} I & 0 \\ B & Q \end{bmatrix}$$

by defining B as the matrix with columns $b_1, b_2, \ldots$; that is,

(1.19)
$$B(i, j) = \sum_{k \in C_j} P(i, k), \quad i \in D, j = 1, 2, \ldots.$$

Then, for any $n \in \mathbb{N}$, the nth power of $\hat{P}$ is

$$\hat{P}^n = \begin{bmatrix} I & 0 \\ B_n & Q^n \end{bmatrix}$$

where

(1.20)
$$B_n = (I + Q + \cdots + Q^{n-1})B.$$

Evidently, $B_n(i, j)$ is the probability that starting from i, the chain enters the recurrent class C_j at or before the nth step. Hence, if we put

(1.21)
$$G = \lim_{n \to \infty} B_n = \left(\sum_{k=0}^{\infty} Q^k \right) B = SB,$$

then $G(i, j)$ is the probability of ever reaching the set C_j from the transient state i; that is, we have proved the following

(1.22) PROPOSITION. Let Q be the matrix obtained from P by deleting all the rows and columns corresponding to the recurrent states, and let B be defined as in (1.19) for each transient i and recurrent class C_j. Compute S by using either (1.8) or (1.9), and put

$$G = SB.$$

Then for each transient state i and recurrent class C_j,

$$G(i, j) = F(i, k)$$

for all $k \in C_j$.

If there is only one recurrent class and if there are only finitely many transient states, then the computations involved disappear. For in that case, the matrix B becomes a column vector, and since $P1 = 1$, we have $B + Q1 = 1$, or $B = 1 - Q1$. Putting this in (1.20), we obtain $B_n = 1 - Q^n 1$, so in the limit we have $G = 1 - \lim_n Q^n 1$. Since $\sum Q^n < \infty$, $\lim Q^n = 0$, and since there are only finitely many states, Q is a finite-dimensional matrix. Thus $\lim_n (Q^n 1) = (\lim Q^n)1 = 0$, and $G = 1$. Hence we have also proved the following.

(1.23) COROLLARY.   Let i be a transient state from which only finitely many transient states can be reached. Furthermore, suppose that there is only one recurrent class C which can be reached from i. Then $F(i, j) = 1$ for all $j \in C$.

(1.24) EXAMPLE.   In Example (1.12) above, from the set of transient states $\{6, 7, 8\}$ only the recurrent irreducible set $\{1, 2, 3\}$ can be reached. By using R computed above along with (1.15), we obtain the $F(i, j)$ for $i, j \in \{6, 7, 8\}$. The $F(i, j)$ for $i, j \in \{1, 2, 3, 4, 5\}$ follow from (1.13) and (1.14). Finally, for $i \in \{6, 7, 8\}$ and $j \in \{1, 2, 3\}$ we can apply the preceding corollary. The resulting matrix F is

$$
F = \begin{bmatrix}
1 & 1 & 1 & & & & & \\
1 & 1 & 1 & & 0 & & 0 & \\
1 & 1 & 1 & & & & & \\
& & & 1 & 1 & & & \\
0 & & & 1 & 1 & & 0 & \\
1 & 1 & 1 & 0 & 0 & 0.472 & 1 & 0.20 \\
1 & 1 & 1 & 0 & 0 & 0.12 & 0.12 & 0.20 \\
1 & 1 & 1 & 0 & 0 & 0.60 & 0.60 & 0.12
\end{bmatrix}.
$$

$\square$

(1.25) EXAMPLE.   Let X be a Markov chain with state space $E = \{1, 2, \ldots, 7\}$ and transition matrix

$$
P = \begin{bmatrix}
0.5 & 0.5 & & & & & \\
0.8 & 0.2 & & 0 & & & 0 \\
& & 0 & 0.4 & 0.6 & & \\
0 & & 1 & 0 & 0 & & 0 \\
& & 1 & 0 & 0 & & \\
0.1 & 0 & 0.2 & 0.2 & 0.1 & 0.3 & 0.1 \\
0.1 & 0.1 & 0.1 & 0 & 0.1 & 0.2 & 0.4
\end{bmatrix}.
$$

There are two recurrent classes: $C_1 = \{1, 2\}$ and $C_2 = \{3, 4, 5\}$. Lumping the states of C_1 and C_2 into two absorbing states, we get a new transition matrix

$$\begin{bmatrix} 1 & 0 & 0 & 0 \\ 0 & 1 & 0 & 0 \\ 0.1 & 0.5 & 0.3 & 0.1 \\ 0.2 & 0.2 & 0.2 & 0.4 \end{bmatrix}.$$

Thus,

$$S = (I - Q)^{-1} = \begin{bmatrix} 0.7 & -0.1 \\ -0.2 & 0.6 \end{bmatrix}^{-1} = \begin{bmatrix} 1.50 & 0.25 \\ 0.50 & 1.75 \end{bmatrix},$$

$$G = SB = \begin{bmatrix} 1.50 & 0.25 \\ 0.50 & 1.75 \end{bmatrix}\begin{bmatrix} 0.1 & 0.5 \\ 0.2 & 0.2 \end{bmatrix} = \begin{bmatrix} 0.2 & 0.8 \\ 0.4 & 0.6 \end{bmatrix}.$$

Hence the matrix of the probabilities $F(i, j)$ is

$$F = \left[\begin{array}{cc|ccc|cc} 1 & 1 & & & & & \\ 1 & 1 & & 0 & & & 0 \\ \hline & & 1 & 1 & 1 & & \\ 0 & & 1 & 1 & 1 & & 0 \\ & & 1 & 1 & 1 & & \\ \hline 0.2 & 0.2 & 0.8 & 0.8 & 0.8 & \frac{1}{3} & \frac{1}{7} \\ 0.4 & 0.4 & 0.6 & 0.6 & 0.6 & \frac{1}{3} & \frac{3}{7} \end{array}\right].$$ □

(1.26) EXAMPLE. Let X_n be the total assets (measured in dollars) of a certain firm at the end of the nth year of its existence. If $X_n = 0$ at some time n, then the firm is said to be bankrupt, and we put $X_{n+1} = X_{n+2} = \cdots = 0$. From any other state it is possible to reach (of course, not necessarily in one year) the state 0. Suppose $X = \{X_n; n \geq 0\}$ is a Markov chain. By our hypothesis, state 0 is absorbing and all other states are transient, and the transition matrix is in the form

$$P = \begin{bmatrix} 1 & 0 \\ b & Q \end{bmatrix}.$$

Though we have only one recurrent class, the probability of ever reaching it is not necessarily equal to one (as it would have been if the state space were finite). Putting $S = I + Q + \cdots$, and $g = Sb$, we have that

$$F(i, 0) = g(i), \quad i = 1, 2, \ldots.$$

This is not necessarily one, and the probability that the firm never goes bankrupt is (if it was started with an initial capital of i dollars)

$$1 - F(i, 0) = 1 - g(i). \qquad \square$$

2. Recurrent States and the Limiting Probabilities

It follows from Theorem (5.3.14) that the recurrent states can be divided into irreducible closed sets in a unique manner. Therefore, it is sufficient to consider one irreducible recurrent set of states. By Proposition (5.3.8) the restriction of the chain to a closed set is also a Markov chain. Hence, we may restrict ourselves without loss of generality to irreducible recurrent Markov chains.

Let X be a Markov chain with state space E and transition matrix P. Part of the following theorem was already listed as Theorem (5.3.26) to aid in classifying the states of an irreducible Markov chain.

(2.1) Theorem. Suppose X is irreducible and aperiodic. Then all states are recurrent non-null if and only if the system of linear equations

$$(2.2) \qquad \pi(j) = \sum_{i \in E} \pi(i) P(i, j), \quad j \in E,$$

$$(2.3) \qquad \sum_{j \in E} \pi(j) = 1$$

has a solution π. If there exists a solution π, then it is strictly positive, there are no other solutions, and we have

$$(2.4) \qquad \pi(j) = \lim_{n \to \infty} P^n(i, j)$$

for all $i, j \in E$.

Proof. Suppose all the states are recurrent non-null. Then by Lemma (5.3.11), $F(i, j) = 1$ for all $i, j \in E$; and therefore, by Theorem (5.3.2), the limits

$$(2.5) \qquad \lim_{n \to \infty} P^n(i, j) = \gamma(j), \quad i, j \in E,$$

exist, and furthermore,

$$(2.6) \qquad \gamma(j) > 0, \quad \sum_{j \in E} \gamma(j) = 1.$$

For any finite subset A of E,

$$P^{n+1}(i, j) = \sum_{k \in E} P^n(i, k) P(k, j) \geq \sum_{k \in A} P^n(i, k) P(k, j),$$

and in the limit, as $n \to \infty$, this implies by (2.5) that

$$\gamma(j) \geq \sum_{k \in A} \gamma(k) P(k, j), \quad j \in E.$$

Since this is true for any finite subset A, we see by taking a sequence of such subsets A increasing to E that

$$(2.7) \qquad \gamma(j) \geq \sum_{k \in E} \gamma(k) P(k, j), \quad j \in E.$$

If the strict inequality were to hold in (2.7) for some j, then summing both sides over j and noting that $\sum_j P(k, j) = 1$, we would have that $\sum_j \gamma(j) > \sum_k \gamma(k)$. This being absurd, we must have

$$(2.8) \qquad \gamma(j) = \sum_{k \in E} \gamma(k) P(k, j), \quad j \in E.$$

Now (2.8) and (2.6) show that γ is a solution to (2.2) and (2.3).

To show that there is only one solution, let π be a solution to (2.2) and (2.3). Then in matrix notation, $\pi = \pi P$, which by iteration gives $\pi = \pi P = \pi P^2 = \cdots = \pi P^n$; that is,

$$(2.9) \qquad \pi(j) = \sum_{k \in E} \pi(k) P^n(k, j).$$

Now taking limits in (2.9), by the bounded convergence theorem (Theorem (2.1.35) applied to the sequence $Y_n(i) = P^n(i, j)$ of "random variables" defined on the probability space (E, π)), we get

$$(2.10) \qquad \pi(j) = \sum_{k \in E} \pi(k) \lim_{n \to \infty} P^n(k, j) = \sum_k \pi(k) \gamma(j) = \gamma(j);$$

that is, there is only one solution π and (2.4) holds.

It remains to show that the existence of a solution π to (2.2) and (2.3) implies that all the states are recurrent non-null. Supposing otherwise, if the states were not recurrent non-null, then (2.9) will still hold, and (2.10) will become

$$\pi(j) = \sum_{k \in E} \pi(k) \lim_{n \to \infty} P^n(k, j) = \sum_{k \in E} \pi(k) \cdot 0 = 0,$$

which contradicts $\sum \pi(j) = 1$. $\qquad \square$

By Corollaries (5.3.19) and (5.3.22), an irreducible chain X with finitely many states has no null states and no transient states. Hence, we have the following (we are using the matrix notation explained in (5.1.3))

(2.11) COROLLARY. If X is an irreducible aperiodic Markov chain with finitely many states, then

$$\pi P = \pi, \quad \pi 1 = 1$$

has a unique solution. The solution π is strictly positive, and

$$\pi(j) = \lim_{n \to \infty} P^n(i, j)$$

for all $i, j \in E$.

A probability distribution π which satisfies (2.2), namely $\pi = \pi P$, is called an *invariant distribution for X* (or, sometimes, for P).

Theorem (2.1) states that if an irreducible aperiodic Markov chain has an invariant distribution, then all its states are recurrent non-null. Otherwise, if it has no invariant distribution, then all its states are either transient or recurrent null.

The justification for the term "invariant" comes from the fact that $\pi = \pi P = \pi P^2 = \cdots = \pi P^n$ for any n; therefore, if π is the initial distribution of X, that is, if

$$P\{X_0 = j\} = \pi(j), \quad j \in E,$$

then also

(2.12) $$P\{X_n = j\} = \sum_i \pi(i) P^n(i, j) = \pi(j), \quad j \in E$$

for any $n \in \mathbb{N}$.

The preceding two propositions show how to obtain the limiting probabilities $\lim_{n \to \infty} P^n(j, j)$ for recurrent, non-null, aperiodic states j. Namely, consider the irreducible closed set containing j, and solve for the invariant distribution for the restriction of X to that set. Furthermore, we know by Theorem (5.3.2) that

(2.13) $$\lim_{n \to \infty} P^n(i, j) = F(i, j) \lim_{n \to \infty} P^n(j, j)$$

for any i and recurrent aperiodic j. Once the $F(i, j)$ are computed as in the preceding section, this completes the specifications of the limits of $P^n(i, j)$ for all such j. In the next section we will show a simple device to handle the periodic states as well.

(2.14) COMPUTATIONAL HINT. If π is a solution of $\pi = \pi P$, then any constant c times π is again a solution. Since the equations $\pi P = \pi$ and $\pi 1 = 1$ have at most one solution, any solution of $\pi P = \pi$ can differ from it by at most a constant term of multiplication. In solving $\pi P = \pi$ and $\pi 1 = 1$, therefore, it is best to solve $\pi P = \pi$ first and then normalize the resulting solution (by dividing each term by the sum of all the terms) to satisfy the remaining equation $\pi 1 = 1$. Furthermore, in the finite state space case, the equations of $\pi P = \pi$ are not linearly independent; therefore, one can throw one of the equations out of consideration.

(2.15) EXAMPLE. Let X be a Markov chain with state space $E = \{1, 2, 3\}$, and transition matrix

$$P = \begin{bmatrix} 0.3 & 0.5 & 0.2 \\ 0.6 & 0 & 0.4 \\ 0 & 0.4 & 0.6 \end{bmatrix}.$$

Clearly, all states are recurrent non-null aperiodic. The invariant distribution is obtained by solving for π in $\pi P = \pi$, $\pi 1 = 1$. We first solve for π in $\pi = \pi P$, which is, in open form,

$$\pi(1) = \pi(1)0.3 + \pi(2)0.6$$
$$\pi(2) = \pi(1)0.5 \qquad\qquad + \pi(3)0.4$$
$$\pi(3) = \pi(1)0.2 + \pi(2)0.4 + \pi(3)0.6.$$

We can delete one of these equations; the worst-looking one is the third, so we do not consider it. To obtain a solution to the first two, we take a wholesome number for $\pi(1)$ to start out with, say $\pi(1) = 60$. Then the first equation gives $\pi(2) = (60 - 60 \times 0.3)/0.6 = 70$, and the second gives $\pi(3) = (70 - 60 \times 0.5)/0.4 = 100$. Hence $(60, 70, 100)$ is a solution for $\pi = \pi P$; any constant times that is again a solution; and the particular solution whose terms add up to one is the constant $1/(60 + 70 + 100)$ times $(60, 70, 100)$. That is, the unique solution of $\pi = \pi P$ and $\pi 1 = 1$ is

$$\pi = (\tfrac{6}{23}, \tfrac{7}{23}, \tfrac{10}{23}).$$

By Theorem (2.1) (or Corollary (2.11), which is also applicable),

$$P^\infty = \lim_{n \to \infty} P^n = \begin{bmatrix} \tfrac{6}{23} & \tfrac{7}{23} & \tfrac{10}{23} \\ \tfrac{6}{23} & \tfrac{7}{23} & \tfrac{10}{23} \\ \tfrac{6}{23} & \tfrac{7}{23} & \tfrac{10}{23} \end{bmatrix}.$$

$\square$

(2.16) EXAMPLE. Let X be a Markov chain with state space $E = \{1, 2, 3, 4, 5, 6, 7\}$ and transition matrix

$$P = \begin{bmatrix} 0.2 & 0.8 & 0 & 0 & 0 & 0 & 0 \\ 0.7 & 0.3 & 0 & 0 & 0 & 0 & 0 \\ 0 & 0 & 0.3 & 0.5 & 0.2 & 0 & 0 \\ 0 & 0 & 0.6 & 0 & 0.4 & 0 & 0 \\ 0 & 0 & 0 & 0.4 & 0.6 & 0 & 0 \\ 0 & 0.1 & 0.1 & 0.2 & 0.2 & 0.3 & 0.1 \\ 0.1 & 0.1 & 0.1 & 0 & 0.1 & 0.2 & 0.4 \end{bmatrix}.$$

Then the states 1 and 2 form one irreducible closed set and the states 3, 4, and 5 another. The remaining states are transient.

Theorem (2.1) applies to each irreducible closed set separately. Thus, for

$$P_1 = \begin{bmatrix} 0.2 & 0.8 \\ 0.7 & 0.3 \end{bmatrix},$$

the invariant distribution is found by solving $\pi = \pi P_1$, $\pi 1 = 1$; doing that we obtain $(\frac{7}{15}, \frac{8}{15})$. Similarly, for

$$P_2 = \begin{bmatrix} 0.3 & 0.5 & 0.2 \\ 0.6 & 0 & 0.4 \\ 0 & 0.4 & 0.6 \end{bmatrix},$$

the invariant distribution was found to be $(\frac{6}{23}, \frac{7}{23}, \frac{10}{23})$ in Example (2.15).

Using the methods of Section 1 to compute the probabilities $F(i, j)$, we obtain (see also Example (1.25))

$$\begin{bmatrix} F(6, 1) & \cdots & F(6, 5) \\ F(7, 1) & \cdots & F(7, 5) \end{bmatrix} = \begin{bmatrix} 0.2 & 0.2 & 0.8 & 0.8 & 0.8 \\ 0.4 & 0.4 & 0.6 & 0.6 & 0.6 \end{bmatrix}$$

(note that these are the only needed values). Hence, it follows from (2.13) and (2.1) that

$$P^\infty = \lim_{n \to \infty} P^n = \begin{bmatrix} \frac{7}{15} & \frac{8}{15} & 0 & 0 & 0 & 0 & 0 \\ \frac{7}{15} & \frac{8}{15} & 0 & 0 & 0 & 0 & 0 \\ 0 & 0 & \frac{6}{23} & \frac{7}{23} & \frac{10}{23} & 0 & 0 \\ 0 & 0 & \frac{6}{23} & \frac{7}{23} & \frac{10}{23} & 0 & 0 \\ 0 & 0 & \frac{6}{23} & \frac{7}{23} & \frac{10}{23} & 0 & 0 \\ \frac{1.4}{15} & \frac{1.6}{15} & \frac{4.8}{23} & \frac{5.6}{23} & \frac{8}{23} & 0 & 0 \\ \frac{2.8}{15} & \frac{3.2}{15} & \frac{3.6}{23} & \frac{4.2}{23} & \frac{6}{23} & 0 & 0 \end{bmatrix}.$$

$\square$

(2.17) EXAMPLE. *Random Walks.* Consider the Markov chain X with state space $E = \{0, 1, \ldots\}$ and transition matrix

$$P = \begin{bmatrix} q & p & & & \\ q & 0 & p & & 0 \\ & q & 0 & p & \\ & & \cdot & \cdot & \cdot \\ 0 & & & \cdot & \cdot & \cdot \\ & & & & \cdot \end{bmatrix}.$$

It is clear that X is irreducible and that all states are aperiodic (since 0 is obviously so). See (5.3.31) for a similar chain.

In open form $\pi = \pi P$ gives (we write π_i for $\pi(i)$),

$$\pi_0 = \pi_0 q + \pi_1 q,$$
$$\pi_1 = \pi_0 p + \pi_2 q,$$
$$\pi_2 = \pi_1 p + \pi_3 q, \ldots.$$

Starting with $\pi_0 = 1$, the first equation yields $\pi_1 = p/q$, the second yields $\pi_2 = (p/q - p)/q = p^2/q^2$, the third yields $\pi_3 = (p^2/q^2 - p \cdot p/q)/q = p^3/q^3$, and so on. Hence,

$$\pi = \left(1, \frac{p}{q}, \frac{p^2}{q^2}, \frac{p^3}{q^3}, \cdots \right)$$

is a solution for $\pi = \pi P$; and any other solution is of the form $(c, cp/q, cp^2/q^2, \ldots)$ for some constant c.

If $p \geq q$, then the sum of the terms of any solution is either not finite (if we take $c \neq 0$) or is equal to 0 (if we take $c = 0$). In either case, the sum cannot be 1, and hence $\pi P = \pi$, $\pi 1 = 1$ has no solution in the case $p \geq q$.

On the other hand, if $p < q$, then by taking $c = 1 - p/q$ we obtain a solution to $\pi P = \pi$, $\pi 1 = 1$. Thus, in this case, all the states are recurrent non-null aperiodic, and

$$\lim_{n \to \infty} P^n(i, j) = \left(1 - \frac{p}{q} \right) \left(\frac{p}{q} \right)^j, \quad j = 0, 1, \ldots. \qquad \square$$

(2.18) EXAMPLE. Consider the remaining lifetime problem of Example (5.1.17). In open form, the equation $\pi = \pi P$ becomes (we write π_i for $\pi(i)$)

$$\pi_0 = \pi_0 p_1 + \pi_1,$$
$$\pi_1 = \pi_0 p_2 \qquad + \pi_2,$$
$$\pi_2 = \pi_0 p_3 \qquad\qquad + \pi_3, \ldots.$$

Taking $\pi_0 = 1$ and solving these equations successively, we obtain a solution v:

$$v_0 = 1,$$
$$v_1 = 1 - p_1,$$
$$v_2 = 1 - p_1 - p_2, \ldots.$$

It is possible to normalize this to obtain a solution to $\pi P = \pi$, $\pi 1 = 1$ if and only if

$$\sum_0^\infty v_j = (p_1 + p_2 + p_3 + \cdots) + (p_2 + p_3 + \cdots) + (p_3 + \cdots) + \cdots$$
$$= p_1 + 2p_2 + 3p_3 + \cdots = m$$

is finite. Note that m is the expected lifetime of one of the components. So if the expected lifetime m is infinite, all states are recurrent null and $\lim_n P^n(i, j) = 0$. If, on the other hand, the expected lifetime m is finite, then all states are recurrent non-null, and

$$\pi(j) = \lim_{n\to\infty} P^n(i, j) = \frac{1}{m}(1 - p_1 - \cdots - p_j), \quad i, j \in E. \qquad \square$$

The preceding example yields an interesting interpretation of the limiting probabilities. By Proposition (5.2.1), for any Markov chain X, the times between visits to a fixed state j are independent and identically distributed. Therefore, we may think of entrances to state j as the times of failure in a system like that in Example (5.1.17). In other words, if we define Y_n as the time from n until the next visit to state j, then $\{Y_n; n \geq 0\}$ is a Markov chain as in Example (5.1.17); and the chain Y is in state 0 if and only if the chain X is in state j. Then the expected time $m(j)$ between two successive visits of X to state j becomes the expected lifetime. Thus, by Example (2.18) above, the probability $P_j\{Y_n = 0\}$ has the limit $1/m(j)$. But the event $\{Y_n = 0\}$ is the same as the event $\{X_n = j\}$. Hence, we have proved the following.

(2.19) PROPOSITION. Let j be an aperiodic recurrent non-null state, and let $m(j)$ be the expected time between two returns to j. Then

$$(2.20) \qquad\qquad \pi(j) = \lim_{n\to\infty} P^n(j, j) = \frac{1}{m(j)}.$$

In other words, the limiting probability $\pi(j)$ of being in state j is equal to the *rate* at which j is visited. So if the mean time between visits to j is $m(j) = 7.2$, for example, then j is being visited once every 7.2 steps on the average, and therefore the limiting probability of being in state j is $\pi(j) = 1/7.2$. In fact, a somewhat more precise statement holds:

(2.21) PROPOSITION. Let j be a recurrent non-null aperiodic state, and let $\pi(j)$ be as in (2.20). Then for almost all $\omega \in \Omega$,

$$\lim_{n\to\infty} \frac{1}{n+1} \sum_{m=0}^{n} 1_j(X_m(\omega)) = \pi(j).$$

In other words, if we were to observe the process X, the average number of visits to j during the first n steps is equal to $\pi(j)$ for large n. Of course, this is of practical value in estimating the probabilities $\pi(j)$ from the data.

If f is a bounded function defined on E, then

$$\sum_{m=0}^{n} f(X_m) = \sum_{j\in E} f(j) \sum_{m=0}^{n} 1_j(X_m);$$

therefore, applying the preceding proposition to each term separately, we see that the following is plausible.

(2.22) COROLLARY. Suppose X is an irreducible recurrent Markov chain having a limiting distribution π. Then for any bounded function f defined on E,

$$\lim_{n\to\infty} \frac{1}{n+1} \sum_{m=0}^{n} f(X_m) = \pi f$$

almost surely.

Of course, π being a row vector and f a column vector, we have

$$\pi f = \sum_{j\in E} \pi(j)f(j).$$

A similar statement also holds for the expectations:

(2.23) COROLLARY. Suppose X is an irreducible recurrent Markov chain with a limiting distribution π. Then for any bounded function f on E,

$$\lim_{n\to\infty} \frac{1}{n+1} \sum_{m=0}^{n} E_i[f(X_m)] = \pi f$$

independent of i.

In other words, if $f(j)$ is the reward received whenever X is in j, then both the expected average reward in the long run and the actual average reward in the long run converge to the constant πf.

Next, consider the ratio of the total reward received during the steps $0, 1, \ldots, n$ by using the reward function f to the total received during the same period by using an alternative reward function g. If the hypotheses of (2.22) and (2.23) hold, then it is easy to see that

$$(2.24) \qquad \lim_{n\to\infty} \frac{\displaystyle\sum_{m=0}^{n} f(X_m)}{\displaystyle\sum_{m=0}^{n} g(X_m)} = \frac{\pi f}{\pi g}$$

almost surely; and a similar statement holds for the ratio of the expectations. Of course, (2.24) becomes (2.22) upon taking $g = 1$.

The formula (2.24) also holds under less restrictive conditions; that is, when X is only recurrent (and can be null or periodic or both). The following are two theorems which summarize such results. We omit the proofs.

(2.25) THEOREM. Let X be an irreducible recurrent chain with transition matrix P. Then

$$v = vP$$

has a strictly positive solution; any other solution is a constant multiple of that one.

(2.26) THEOREM.   Suppose X is irreducible recurrent, and let v be a solution of $v = vP$. Then for any two functions f and g on E for which the two sums

$$vf = \sum_{i \in E} v(i)f(i), \qquad vg = \sum_{i \in E} v(i)g(i)$$

converge absolutely and at least one is not zero, we have

$$\lim_{n \to \infty} \frac{\sum\limits_{m=0}^{n} E_i[f(X_m)]}{\sum\limits_{m=0}^{n} E_j[g(X_m)]} = \frac{vf}{vg}$$

independent of $i, j \in E$. Moreover, we also have

$$\lim_{n \to \infty} \frac{\sum\limits_{m=0}^{n} f(X_m(\omega))}{\sum\limits_{m=0}^{n} g(X_m(\omega))} = \frac{vf}{vg}$$

for almost all $\omega \in \Omega$.

Any non-negative solution of $v = vP$ is called an *invariant measure* for X. Theorem (2.25) above states that any irreducible recurrent chain X has an invariant measure, and this is unique up to multiplication by a constant. Furthermore, if X is also non-null, then $v1 = \sum v(j)$ is finite, and v is a constant multiple of the limiting distribution π satisfying $\pi P = \pi$ and $\pi 1 = 1$. On the other hand, it is worth noting that the existence of an invariant measure v for X does not imply that X is recurrent (for example, the Markov chain of (2.17) with $p > q$ is transient and has an invariant measure).

By using Theorem (2.26), we can attach an intuitive meaning to an invariant measure v of an irreducible recurrent chain X. In (2.26), taking $f = 1_k$ and $g = 1_j$ we see that for $i = j$,

$$\lim_{n \to \infty} \frac{E_j\left[\sum\limits_{m=0}^{n} 1_k(X_m)\right]}{E_j\left[\sum\limits_{m=0}^{n} 1_j(X_m)\right]} = \frac{v(k)}{v(j)}.$$

In other words, $v(k)/v(j)$ is equal to the ratio of the expected number of visits to k during the first n steps to the expected number of returns to j during the same period, in the limit, as the number of steps n approaches infinity. It can further be shown that $v(k)/v(j)$ is also equal to the expected number of visits to k between two visits to j.

3.   Periodic States

In the discussion of limiting probabilities we had to leave out the periodic states momentarily. This is mainly due to minor complications which make otherwise simple results appear hard. In this context, we need worry only about the recurrent periodic states. By Theorem (5.3.14), the recurrent states can be divided into disjoint closed sets $C_1, C_2, \ldots$. Each one of these, taken by itself, defines an irreducible Markov chain with recurrent states. Thus, it is sufficient for us to consider only an irreducible Markov chain with periodic recurrent states. By Theorem (5.3.16), then, all states have the same period. The following is the main result.

(3.1) LEMMA. Let X be an irreducible Markov chain with recurrent periodic states with period δ. Then the states can be divided into δ disjoint sets B_1, $B_2, \ldots, B_\delta$ such that $P(i, j) = 0$ unless $i \in B_1$ and $j \in B_2$, or $i \in B_2$ and $j \in B_3, \ldots$, or $i \in B_\delta$ and $j \in B_1$.

Proof. Consider a fixed state 0 and a state j. Since the chain is irreducible, states 0 and j can be reached from each other; that is, there exist integers r and s such that $P^r(0, j) > 0$, $\beta = P^s(j, 0) > 0$. We also have, from the Chapman–Kolmogorov equation, that

$$(3.2) \qquad P^{m+s}(0, 0) \geq P^m(0, j)P^s(j, 0) = \beta P^m(0, j).$$

If $P^m(0, j) > 0$, then $P^{m+s}(0, 0) > 0$. Since state 0 has period δ, $P^{m+s}(0, 0) > 0$ implies that $m + s = k\delta$ for some integer k. Hence, $P^m(0, j) > 0$ implies that $m = k\delta - s$. Since s is fixed, this means that $P^m(0, j)$ is positive only if $m = \alpha$, $\alpha + \delta, \alpha + 2\delta, \ldots$ and the number α is a characteristic of the state j. Then we put j in the set B_α.

Thus B_α is the set of all states which can be reached from 0 only in steps numbered $\alpha, \alpha + \delta, \alpha + 2\delta, \ldots$; and $B_{\alpha+1}$ is the set of all states that can be reached from the state 0 in steps numbered $\alpha + 1, \alpha + \delta + 1$, $\alpha + 2\delta + 1, \ldots$. Hence $P(i, j) = 0$ unless $i \in B_\alpha$ and $j \in B_{\alpha+1}$ for some $\alpha = 1, \ldots, \delta$ (if $\alpha = \delta$ we take $B_{\alpha+1} = B_1$). $\qquad \square$

(3.3) EXAMPLE. Suppose the transition matrix P of an irreducible Markov chain is

$$(3.4) \qquad P = \begin{bmatrix} 0 & 0 & \frac{1}{2} & \frac{1}{4} & \frac{1}{4} & 0 & 0 \\ 0 & 0 & \frac{1}{3} & 0 & \frac{2}{3} & 0 & 0 \\ 0 & 0 & 0 & 0 & 0 & \frac{1}{3} & \frac{2}{3} \\ 0 & 0 & 0 & 0 & 0 & \frac{1}{2} & \frac{1}{2} \\ 0 & 0 & 0 & 0 & 0 & \frac{3}{4} & \frac{1}{4} \\ \frac{1}{2} & \frac{1}{2} & 0 & 0 & 0 & 0 & 0 \\ \frac{1}{4} & \frac{3}{4} & 0 & 0 & 0 & 0 & 0 \end{bmatrix}$$

(suppose the states are numbered 1, 2, 3, 4, 5, 6, and 7). Note that in one step, from states 1 and 2, only the states 3, 4, and 5 can be reached. From states

3, 4, and 5 only 6 and 7 can be reached in one step; and from 6 and 7 a step leads to either 1 or 2. Hence, we have that all states are periodic with period 3. Then, in the proof of the preceding lemma, if we take state 0 to be state 7 of this one, we get $B_1 = \{1, 2\}$, $B_2 = \{3, 4, 5\}$, $B_3 = \{6, 7\}$, and the chain travels cyclically through B_1, B_2, and B_3.

Obviously, from B_1, in one step the Markov chain reaches B_2, in two steps B_3, and in three steps B_1. The powers of P will reflect this. For example,

$$(3.5) \qquad P^2 = \left[\begin{array}{cc|ccc|cc} & & & & & \frac{23}{48} & \frac{25}{48} \\ \multicolumn{2}{c|}{0} & \multicolumn{3}{c|}{0} & \frac{11}{18} & \frac{7}{18} \\ \hline \frac{1}{3} & \frac{2}{3} & & & & & \\ \frac{3}{8} & \frac{5}{8} & \multicolumn{3}{c|}{0} & \multicolumn{2}{c}{0} \\ \frac{7}{16} & \frac{9}{16} & & & & & \\ \hline & & \frac{5}{12} & \frac{1}{8} & \frac{11}{24} & & \\ \multicolumn{2}{c|}{0} & \frac{3}{8} & \frac{1}{16} & \frac{9}{16} & \multicolumn{2}{c}{0} \end{array}\right],$$

and

$$(3.6) \qquad P^3 = \left[\begin{array}{cc|ccc|cc} \frac{71}{192} & \frac{121}{192} & & & & & \\ \frac{29}{72} & \frac{43}{72} & \multicolumn{3}{c|}{0} & \multicolumn{2}{c}{0} \\ \hline & & \frac{14}{36} & \frac{3}{36} & \frac{19}{36} & & \\ \multicolumn{2}{c|}{0} & \frac{19}{48} & \frac{3}{32} & \frac{49}{96} & \multicolumn{2}{c}{0} \\ & & \frac{13}{32} & \frac{7}{64} & \frac{31}{64} & & \\ \hline & & & & & \frac{157}{288} & \frac{131}{288} \\ \multicolumn{2}{c|}{0} & \multicolumn{3}{c|}{0} & \frac{111}{192} & \frac{81}{192} \end{array}\right].$$

It is important to note that the matrix $\bar{P} = P^3$ is of the form

$$\bar{P} = \begin{bmatrix} P_1 & 0 & 0 \\ 0 & P_2 & 0 \\ 0 & 0 & P_3 \end{bmatrix}.$$

Thus, the chain corresponding to $\bar{P}$ has three closed sets: B_1, B_2, and B_3, and each one of these is irreducible, recurrent, and aperiodic. Then the theory of Section 2 applies to compute $\lim_m P_1^m$, $\lim_m P_2^m$, and $\lim_m P_3^m$ separately.   □

Next is a theorem which reduces the study of periodic states to that of aperiodic case by formalizing what is said in the preceding example.

(3.7) THEOREM.   Let P be the transition matrix of an irreducible Markov chain with recurrent periodic states of period δ, and let $B_1, \ldots, B_\delta$ be as defined in Lemma (3.1). Then in the Markov chain with transition matrix $\bar{P} = P^\delta$, the classes $B_1, B_2, \ldots, B_\delta$ are irreducible closed sets of aperiodic states.

Proof. In δ steps, the Markov chain defined by P moves from B_1 to B_1, from B_2 to $B_2, \ldots,$ from B_δ to B_δ. Hence, $\bar{P}$ is of the form

(3.8)
$$\bar{P} = \begin{bmatrix} P_1 & & & 0 \\ & P_2 & & \\ & & \ddots & \\ 0 & & & P_\delta \end{bmatrix}$$

where $P_\alpha(i, j) = P^\delta(i, j)$ for $i, j \in B_\alpha$. This shows that $B_1, \ldots, B_\delta$ are closed sets for the new chain. Since δ steps of the original chain correspond to one step in the new chain with matrix $\bar{P}$, states of the $\bar{P}$ chain are aperiodic. $\square$

It follows from Lemma (3.1) that if $i \in B_\alpha$, then

(3.9)
$$P_i\{X_m \in B_\beta\} = 1, \quad \beta = \alpha + m \;(\text{mod }\delta).$$

Thus, $P^n(i, j)$ does not have a limit as $n \longrightarrow \infty$ except when all the states are null (in which case $P^n(i, j) \longrightarrow 0$ for all i, j as $n \longrightarrow \infty$). However, the limits of $P^{n\delta+m}(i, j)$ exist as $n \longrightarrow \infty$ but are dependent on the initial state i.

(3.10) THEOREM. Let P and the B_α be as in (3.7) and suppose the chain is non-null. Then for any $m \in \{0, 1, \ldots, \delta - 1\}$,

(3.11)
$$\lim_{n \to \infty} P^{n\delta+m}(i, j) = \begin{cases} \pi(j) & \text{if } i \in B_\alpha, j \in B_\beta, \; \beta = \alpha + m \;(\text{mod }\delta); \\ 0 & \text{otherwise.} \end{cases}$$

The probabilities $\pi(j), j \in E$, form the unique solution of

(3.12)
$$\pi(j) = \sum_{i \in E} \pi(i)P(i, j), \quad \sum_{i \in E} \pi(i) = \delta, \quad j \in E.$$

Proof. It follows from Theorem (3.7) that the chain corresponding to $P^\delta = \bar{P}$ has δ irreducible classes each one of which is recurrent, non-null, and aperiodic. Let $\pi_1, \ldots, \pi_\delta$ be the limiting distributions corresponding to $B_1, \ldots, B_\delta$, and let $\pi(j) = \pi_\alpha(j)$ if $j \in B_\alpha$.

Let $m \in \{0, 1, \ldots, \delta - 1\}$ be fixed, suppose $i \in B_\alpha$, and let $\beta = \alpha + m$ (mod δ). Now,

(3.13)
$$P^{n\delta+m}(i, j) = \sum_k P^m(i, k)P^{n\delta}(k, j)$$

$$= \begin{cases} \sum_{k \in B_\beta} P^m(i, k)P^{n\delta}(k, j) & \text{if } j \in B_\beta, \\ 0 & \text{otherwise,} \end{cases}$$

since $P^m(i, k) = 0$ unless $k \in B_\beta$, and for $k \in B_\beta$, $P^{n\delta}(k, j) = 0$ unless $j \in B_\beta$

also. For $k, j \in B_\beta$, $\lim_n P^{n\delta}(k, j) = \pi(j)$ independent of k by the preceding paragraph. Putting this in (3.13), by the bounded convergence theorem, we obtain (3.11).

There remains to show that π is the unique solution of (3.12). The uniqueness is immediate: by Theorem (2.25), $\pi = \pi P$ has only one solution up to multiplication by a constant, and the normalization in (3.12) fixes that constant. To show that π satisfies $\pi = \pi P$, first suppose P is in the canonical form

$$(3.14) \qquad P = \begin{bmatrix} 0 & K_1 & & & & & 0 \\ 0 & 0 & K_2 & & & & \\ \cdot & \cdot & \cdot & \cdot & & & \\ \cdot & \cdot & \cdot & & \cdot & & \\ \cdot & \cdot & \cdot & & & \cdot & \\ 0 & 0 & 0 & \cdot & \cdot & \cdot & K_{\delta-1} \\ K_\delta & 0 & 0 & \cdot & \cdot & \cdot & 0 \end{bmatrix},$$

and let $P^\delta = \bar{P}$ be as in (3.8). Since $\pi_\alpha = \pi_\alpha P_\alpha$ and $\pi = (\pi_1, \ldots, \pi_\delta)$, (3.8) shows that π satisfies $\pi = \pi P^\delta$. Now $\pi P = (\pi P^\delta)P = (\pi P)P^\delta$ implies, by the uniqueness of the solution of $v_\alpha = v_\alpha P_\alpha$ up to multiplication by a constant, that

$$(3.15) \qquad \pi P = (c_1 \pi_1, \ldots, c_\delta \pi_\delta)$$

where $c_1, \ldots, c_\delta$ are positive constants. On the other hand, multiplying π with P in the form (3.14) directly, we obtain

$$\pi P = (\pi_\delta K_\delta, \pi_1 K_1, \ldots, \pi_{\delta-1} K_{\delta-1}).$$

From this and (3.15) we have $c_1 \pi_1 = \pi_\delta K_\delta$, which implies $c_1 \pi_1 1 = \pi_\delta K_\delta 1$, which yields $c_1 = 1$, since $\pi_1 1 = 1$ and $\pi_\delta K_\delta 1 = \pi_\delta 1 = 1$. Similarly, $c_2 = \cdots = c_\delta = 1$. In view of (3.15), this completes the proof. $\square$

For ratio limit theorems we refer the reader to Theorems (2.25) and (2.26). In particular, if X is recurrent non-null periodic with period δ, then taking $f = 1_j$ and $g = 1$ in (2.26) yields

$$\lim_{n \to \infty} \frac{1}{n+1} \sum_{m=0}^n P^m(i, j) = \frac{1}{\delta} \pi(j).$$

(3.16) EXAMPLE. Let X be a Markov chain with state space $E = \{1, 2, 3, 4, 5\}$ and transition matrix

$$P = \begin{bmatrix} 0 & 0 & 0 & 0.5 & 0.5 \\ 0 & 0 & 0 & 0.4 & 0.6 \\ 0 & 0 & 0 & 0 & 1 \\ 0.8 & 0 & 0.2 & 0 & 0 \\ 0 & 1 & 0 & 0 & 0 \end{bmatrix}.$$

This chain is irreducible, recurrent, non-null, and periodic with period $\delta = 2$. The cyclic classes are $B_1 = \{1, 2, 3\}$ and $B_2 = \{4, 5\}$. An invariant measure v (satisfying $v = vP$) is $v = (32, 60, 8, 40, 60)$. Hence,

$$\lim_{n \to \infty} P^{2n} = \begin{bmatrix} 0.32 & 0.60 & 0.08 & 0 & 0 \\ 0.32 & 0.60 & 0.\dot{0}8 & 0 & 0 \\ 0.32 & 0.60 & 0.08 & 0 & 0 \\ 0 & 0 & 0 & 0.4 & 0.6 \\ 0 & 0 & 0 & 0.4 & 0.6 \end{bmatrix},$$

and

$$\lim_{n \to \infty} P^{2n+1} = \begin{bmatrix} 0 & 0 & 0 & 0.4 & 0.6 \\ 0 & 0 & 0 & 0.4 & 0.6 \\ 0 & 0 & 0 & 0.4 & 0.6 \\ 0.32 & 0.60 & 0.08 & 0 & 0 \\ 0.32 & 0.60 & 0.08 & 0 & 0 \end{bmatrix}. \qquad \square$$

(3.17) EXAMPLE. Consider the Markov chain of Example (5.3.31) for $p < q$. We had shown there that all states are recurrent, non-null, and periodic with period $\delta = 2$. The cyclic classes are $B_1 = \{0, 2, 4, \ldots\}$ and $B_2 = \{1, 3, 5, \ldots\}$. We had obtained a solution to $v = vP$ with

$$v_0 = 1, \quad v_2 = \frac{p}{q^2}, \quad v_4 = \frac{p^3}{q^4}, \ldots$$

$$v_1 = \frac{1}{q}, \quad v_3 = \frac{p^2}{q^3}, \quad v_5 = \frac{p^4}{q^5}, \ldots.$$

To normalize, since

$$\sum v_i = 1 + \frac{1}{q}\left[1 + \frac{p}{q} + \left(\frac{p}{q}\right)^2 + \cdots\right]$$

$$= 1 + \frac{1}{q}\frac{1}{1 - \dfrac{p}{q}} = \frac{2}{1 - \dfrac{p}{q}},$$

we multiply each term by $1 - p/q$. Thus, the invariant distributions corresponding to the classes B_1 and B_2 are

$$(\pi_0, \pi_2, \pi_4, \ldots) = \left(1 - \frac{p}{q}\right)\left(1, \frac{p}{q^2}, \frac{p^3}{q^4}, \ldots\right)$$

and

$$(\pi_1, \pi_3, \pi_5, \ldots) = \left(1 - \frac{p}{q}\right)\left(\frac{1}{q}, \frac{p^2}{q^3}, \frac{p^4}{q^5}, \ldots\right),$$

respectively. Hence,

$$\lim_{n \to \infty} P^{2n} = \left(1 - \frac{p}{q}\right) \begin{bmatrix} 1 & 0 & \frac{p}{q^2} & 0 & \frac{p^3}{q^4} & 0 & \cdots \\ 0 & \frac{1}{q} & 0 & \frac{p^2}{q^3} & 0 & \frac{p^4}{q^5} & \cdots \\ 1 & 0 & \frac{p}{q^2} & 0 & \frac{p^3}{q^4} & 0 & \cdots \\ 0 & \frac{1}{q} & 0 & \frac{p^2}{q^3} & 0 & \frac{p^4}{q^5} & \cdots \\ \cdot & \cdot & \cdot & \cdot & \cdot & \cdot & \cdots \\ \cdot & \cdot & \cdot & \cdot & \cdot & \cdot & \cdots \\ \cdot & \cdot & \cdot & \cdot & \cdot & \cdot & \cdots \end{bmatrix},$$

and

$$\lim_{n \to \infty} P^{2n+1} = \left(1 - \frac{p}{q}\right) \begin{bmatrix} 0 & \frac{1}{q} & 0 & \frac{p^2}{q^3} & 0 & \frac{p^4}{q^5} & \cdots \\ 1 & 0 & \frac{p}{q^2} & 0 & \frac{p^3}{q^4} & 0 & \cdots \\ 0 & \frac{1}{q} & 0 & \frac{p^2}{q^3} & 0 & \frac{p^4}{q^5} & \cdots \\ 1 & 0 & \frac{p}{q^2} & 0 & \frac{p^3}{q^4} & 0 & \cdots \\ \cdot & \cdot & \cdot & \cdot & \cdot & \cdot & \cdots \\ \cdot & \cdot & \cdot & \cdot & \cdot & \cdot & \cdots \\ \cdot & \cdot & \cdot & \cdot & \cdot & \cdot & \cdots \end{bmatrix}.$$

$\square$

4. Transient States

If a Markov chain X has only finitely many transient states, then it eventually will leave the set of transient states never to return. On the other hand, if there are infinitely many transient states, it is possible for the chain to remain in the set of transient states forever.

(4.1) EXAMPLE. Consider Example (5.1.14) and suppose the trivial case $p_0 = 1$ is excluded. Then all states are transient, and the chain travels from one transient state to another. If the initial state is i, then the chain stays forever in the set $\{i, i + 1, i + 2, \ldots\}$. It is clear in this case that in the limit, as $n \to \infty$, $X_n(\omega) \to \infty$ for almost all ω. $\square$

Let A be a subset of the state space E, and let Q be the matrix obtained from P by deleting all the rows and columns corresponding to states which are *not* in A. Then for $i, j \in A$, the (i, j)-entry of the nth power of Q is

(4.2)
$$Q^n(i, j) = \sum_{i_1 \in A} \cdots \sum_{i_{n-1} \in A} Q(i, i_1) Q(i_1, i_2) \cdots Q(i_{n-1}, j)$$
$$= P_i\{X_1 \in A, \ldots, X_{n-1} \in A, X_n = j\},$$

and in particular,

(4.3)
$$\sum_{j \in A} Q^n(i, j) = P_i\{X_1 \in A, \ldots, X_n \in A\}.$$

It is clear that the event $\{X_1 \in A, \ldots, X_n \in A\}$ is implied by the event $\{X_1 \in A, \ldots, X_{n+1} \in A\}$. Hence, the number above in (4.3) decreases as n increases. Let

(4.4)
$$f(i) = \lim_{n \to \infty} \sum_{j \in A} Q^n(i, j), \qquad i \in A;$$

clearly, $f(i)$ is the probability that starting at $i \in A$, the chain stays in the set A forever. For its computation we have the following.

(4.5) PROPOSITION.   The function f is the maximal solution of the system of linear equations

(4.6)
$$h = Qh, \quad 0 \le h \le 1.$$

Either $f = 0$ or $\sup_{i \in A} f(i) = 1$.

Proof. Let $f_n(i) = \sum_A Q^n(i, j)$, then we had already noted that $f_0 \ge f_1 \ge f_2 \ge \cdots$ and $f = \lim_n f_n$. Since $Q^{n+1} = QQ^n$, we have

$$f_{n+1}(i) = \sum_{j \in A} Q(i, j) f_n(j);$$

so in the limit we have

$$f(i) = \sum_{j \in A} Q(i, j) f(j)$$

by the bounded convergence theorem. This shows that f satisfies (4.6). Let h be another solution; then $h \le 1$ implies that $Q^n h \le Q^n 1 = f_n$. Since $Q^n h = h$, we see that $h \le f_n$ and therefore $h \le f$.

To prove the last statement, suppose $f \ne 0$; then $c = \sup_{i \in A} f(i) > 0$, and clearly $f \le c1$. Thus, $f = Q^n f \le cQ^n 1 = cf_n$ for all n, which in the limit gives $f \le cf$. This is possible only if $c = 1$. $\square$

An application of this to the classification of states was given as Theorem (5.3.29). We now are in a position to prove that theorem.

Proof of (5.3.29).   Fix the particular state chosen, and name it 0. Since X

is irreducible, it is possible to go from 0 to some $i \in A = E - \{0\}$. If the probability $f(i)$ of remaining in A forever is $f(i) = 0$ for all $i \in A$, then with probability one, the chain will leave A and enter 0 again. Hence, if the only solution to (5.3.30) is $h = 0$, then state 0 is recurrent, and that in turn implies that all states are recurrent.

Conversely, if all states are recurrent, then the probability of remaining in the set A forever must be zero, since 0 will be reached with probability one from any state $i \in A$ (by Lemma (5.3.11)). $\qquad\square$

(4.7) EXAMPLE. Consider the Markov chain X of Example (5.3.31) again. We had shown that if $p > q$, all states are transient. We had obtained, as a solution to $Qh = h$, $0 \le h \le 1$,

$$f(i) = 1 - \left(\frac{q}{p}\right)^i, \quad i = 1, 2, \dots ;$$

since $\sup_i f(i) = 1$, this is the maximal solution. We now have the interpretation that starting at state 7, for example, the probability of staying forever within the set $\{1, 2, 3, \dots\}$ is equal to $1 - (q/p)^7$. Note that if the initial state i is greater, then the probability of remaining in $\{1, 2, \dots\}$ is also greater.

Further, it is clear from the shape of the matrix P that the restriction of P to the set $\{k, k + 1, \dots\}$ is the same as the matrix Q on page 137. Hence, for any $k \in \{1, 2, \dots\}$,

$$P_{k+i}\{X_1 \ge k, X_2 \ge k, \dots\} = 1 - \left(\frac{q}{p}\right)^{i+1}$$

for all $i \ge 0$. $\qquad\square$

For any subset A of E, let $f_A(i)$ denote the probability of remaining forever in A given the initial state $i \in A$. If A is an irreducible recurrent class, then $f_A = 1$. If A is any proper subset of an irreducible recurrent class, then $f_A = 0$. If A is a finite set of transient states, then $f_A = 0$ again. The only non-trivial case, then, is when A is an infinite set of transient states. Even then, it is possible that $f_A = 0$. On the other hand, if $f_A \ne 0$, then by (4.5) the set

$$A_1 = \{i \in A : f_A(i) > 1 - \varepsilon\}$$

is not empty (here $\varepsilon > 0$), and furthermore, $f_{A_1} \ne 0$. By repeating this we may obtain a sequence $A_1, A_2, A_3, \dots$ of subsets of A such that

$$A \supset A_1 \supset A_2 \supset A_3 \supset \cdots$$

and

$$\lim_n A_n = \bigcap_n A_n = \varnothing,$$

for which $f_{A_1} \neq 0$, $f_{A_2} \neq 0$, $f_{A_3} \neq 0$, ... Typically, the Markov chain X travels through such a sequence of sets on its way toward "infinity."

5. Applications to Queueing Theory: M/G/1 Queue

Consider a system where "customers" arrive requiring "service," and where there are some restrictions on the service that can be provided. Usually, arrivals form a stochastic process, and the service times of customers are random variables. If the number of "servers" is limited, then some customers will experience delays before they can be served. Queueing theory studies the general characteristics of such systems.

The terms customer, server, service time, and waiting time may acquire different meanings in different applications. For example, customers may be patients arriving at a doctor's office; then the doctor is the server, and the service times are the examination times. Or customers may be calls arriving at a telephone exchange; then the servers are the trunk lines, and the service times are the lengths of conversations. Or customers may be aircraft arriving at an airport requiring to land; then the runways are the servers, and a waiting time is the time spent by an aircraft circling around the airport.

In this section and the next we present a streamlined introduction to queueing theory by considering the queue-size process in two single-server queueing systems. Our objective is to give two examples of complex stochastic processes which can be studied by identifying a Markovian structure imbedded in them.

In the remainder of this section we are considering a single-server queueing system subject to a Poisson process of arrivals and an arbitrary distribution for service times, which are assumed to be independent of each other and of the arrival process. The shorthand notation M/G/1 refers to such a queue.

For a possible outcome ω of such a system, we will denote by $N_t(\omega)$ the number of arrivals occurring during the time interval $[0, t]$; by $Z_1(\omega)$, $Z_2(\omega)$, . . . the service times of the customers who depart first, second, . . . ; and by $Y_t(\omega)$ the number of customers in the system (waiting or being served) at time t. We are assuming that the arrival process $N = \{N_t; t \geq 0\}$ is a Poisson process with rate a and that the service times $Z_1, Z_2, \ldots$ are independent of each other and of the process N and have the same distribution function φ.

We are interested in the queue-size process $Y = \{Y_t; t \geq 0\}$. However, this process is hard to deal with directly because the future of the process from t onward depends not only on the present value Y_t but also on the amount of time already spent serving the customer being served at that time t. This difficulty disappears if we consider the future of Y from a time T of departure onward. Accordingly, we define X_n to be the number of customers in the system just after the instant of the nth departure, and assume (for reasons of simplicity) that the time origin is chosen to be an instant of departure. The following provides the reason for our interest in the process $X = \{X_n; n = 0, 1, \ldots\}$.

(5.1) THEOREM. X is a Markov chain with the transition matrix

$$(5.2) \qquad P = \begin{bmatrix} q_0 & q_1 & q_2 & q_3 & \cdot & \cdot & \cdot \\ q_0 & q_1 & q_2 & q_3 & \cdot & \cdot & \cdot \\ & q_0 & q_1 & q_2 & \cdot & \cdot & \cdot\cdot \\ & & q_0 & q_1 & \cdot & \cdot & \cdot \\ & & & \cdot & \cdot & \\ \mathbf{0} & & & & \cdot & \cdot \\ & & & & & \cdot & \cdot \end{bmatrix}$$

where

$$(5.3) \qquad q_k = \int_0^\infty \frac{e^{-at}(at)^k}{k!} \, d\varphi(t), \quad k = 0, 1, \ldots .$$

Proof. We need to show that

$$(5.4) \qquad P\{X_{n+1} = j \mid X_0, \ldots, X_n\} = P\{X_{n+1} = j \mid X_n\}$$

and that

$$(5.5) \qquad P\{X_{n+1} = j \mid X_n = i\} = \begin{cases} q_j & \text{if } i = 0, j \geq 0, \\ q_{j+1-i} & \text{if } i > 0, j \geq i - 1, \\ 0 & \text{otherwise.} \end{cases}$$

Let T be the time of the nth departure, and let $Z = Z_{n+1}$ (the service time of the $(n + 1)$th customer). If $X_n > 0$, the service time of the $(n + 1)$th customer starts at T and lasts until $T + Z$; during this time, there will be $N_{T+Z} - N_T$ arrivals, and therefore the queue size just after the $(n + 1)$th departure will be $X_n + (N_{T+Z} - N_T) - 1$. On the other hand, if $X_n = 0$, then the server will remain idle until the $(n + 1)$th customer arrives, say at S; then his service will start and last until $S + Z$; during this time there will be $N_{S+Z} - N_S$ arrivals occurring, and these arrivals will be precisely the ones left behind by the $(n + 1)$th departure. Hence,

$$(5.6) \qquad X_{n+1} = \begin{cases} X_n + (N_{T+Z} - N_T) - 1 & \text{if } X_n > 0, \\ N_{S+Z} - N_S & \text{if } X_n = 0. \end{cases}$$

From the independence of the arrival process from the service times and the fact that in a Poisson process the number of arrivals during an interval of length Z has a distribution which depends only on Z, it follows that

$$(5.7) \qquad P\{N_{T+Z} - N_T = k \,|\, X_0, \dots, X_n; T\} = P\{N_Z = k\},$$

and similarly when T is replaced by S. This together with (5.6) shows that (5.4) holds.

To show (5.5), first we note that

$$(5.8) \qquad q_k = P\{N_Z = k\} = E[P\{N_Z = k \,|\, Z\}]$$
$$= E\left[\frac{e^{-aZ}(aZ)^k}{k!}\right] = \int_0^\infty \frac{e^{-at}(at)^k}{k!}\, d\varphi(t).$$

Now if $i = 0$, from (5.6) we have

$$(5.9) \qquad P\{X_{n+1} = j \,|\, X_n = 0\} = P\{N_{S+Z} - N_S = j\} = P\{N_Z = j\} = q_j.$$

If $i > 0$, then from (5.6) again,

$$(5.10) \qquad \begin{aligned} P\{X_{n+1} = j \,|\, X_n = i\} &= P\{N_{T+Z} - N_T = j + 1 - i\} \\ &= P\{N_Z = j + 1 - i\} = \begin{cases} q_{j+1-i} & \text{if } j \geq i - 1, \\ 0 & \text{if } j < i - 1. \end{cases} \end{aligned}$$

This completes the proof. $\qquad\qquad\qquad\qquad\qquad\qquad\qquad\qquad$ $\square$

It is clear from an inspection of (5.2) that the Markov chain X is irreducible and aperiodic. By Theorem (5.3.16), either all states are recurrent nonnull, or all are recurrent null, or all are transient. It turns out that depending on whether the parameter

$$(5.11) \qquad\qquad r \equiv E[N_Z] = aE[Z] \equiv ab$$

is less than one, equal to one, or greater than one, all three cases are possible. The parameter r is called the *traffic intensity*. Being the expected number of arrivals during a service time, it is a good measure of the long-run behavior of the queue size: if $r > 1$, it is clear that the server will be unable to keep up with the arrivals, and the queue size will increase to infinity (in which case all the states are transient). If $r \leq 1$, the server will be able to clear the work load presented to him, so we expect the queue to be empty again and again; in other words, the state 0 will be recurrent and therefore all states will be recurrent. However, if $r = 1$, our analysis will show that the times between returns to zero will have infinite expectation so that all the states are then recurrent null. Following is a precise development leading up to the results heuristically described in these remarks.

We introduce the notation

$$(5.12) \qquad\qquad r_k = 1 - q_0 - \cdots - q_k,$$

and note that

$$(5.13) \qquad\qquad r = r_0 + r_1 + \cdots.$$

(5.14) PROPOSITION. The chain X is recurrent non-null aperiodic if and only if $r < 1$.

Proof. In view of Theorem (2.1), we need to show that

$$(5.15) \qquad\qquad \pi = \pi P,$$
$$(5.16) \qquad\qquad \pi 1 = 1$$

has a solution if and only if $r < 1$. Writing π_j for $\pi(j)$, the equations (5.15) are

$$\pi_0 = \pi_0 q_0 + \pi_1 q_0,$$
$$\pi_1 = \pi_0 q_1 + \pi_1 q_1 + \pi_2 q_0,$$
$$\pi_2 = \pi_0 q_2 + \pi_1 q_2 + \pi_2 q_1 + \pi_3 q_0, \dots.$$

For each j, add the equations for $\pi_0, \dots, \pi_j$ side by side and solve for $\pi_{j+1} q_0$. This gives

$$(5.17) \qquad\qquad \pi_1 q_0 = \pi_0 r_0,$$
$$\pi_2 q_0 = \pi_0 r_1 + \pi_1 r_1,$$
$$\pi_3 q_0 = \pi_0 r_2 + \pi_1 r_2 + \pi_2 r_1, \dots.$$

It is clear that for any fixed $\pi_0 \geq 0$, this system of equations has a unique non-negative solution. Hence (5.15) has a non-negative solution which is unique up to multiplication by a constant.

Next we test for (5.16). Summing all the equations (5.17) side by side, noting that $q_0 = 1 - r_0$ and that $r_0 + r_1 + \cdots = r$, we have

$$(5.18) \qquad\qquad (1 - r_0) \sum_{j=1}^{\infty} \pi_j = \pi_0 r + (r - r_0) \sum_{j=1}^{\infty} \pi_j.$$

If $r < 1$, (5.18) implies

$$\sum_{j=1}^{\infty} \pi_j = \frac{r}{1 - r} \pi_0 < \infty,$$

so (5.16) is satisfied by taking

$$(5.19) \qquad\qquad \pi_0 = 1 - r.$$

Hence, if $r < 1$, all states are recurrent non-null aperiodic, and the unique invariant distribution is obtained from (5.19) and (5.17).

On the other hand, if $r \geq 1$, the equality (5.18) can hold only if either $\pi_0 = 0$ (which implies $\pi_1 = \pi_2 = \cdots = 0$ by (5.17)) or $\pi_0 > 0$ and $\sum_{j=1}^{\infty} \pi_j = \infty$. In either case, we see that (5.16) cannot hold; consequently, if $r \geq 1$, X is not recurrent non-null. $\qquad\square$

The preceding proof gave, as a by-product, the set of equations which must be solved in order to obtain the limiting distribution in the recurrent non-null case. The following completes the picture in this respect.

(5.20) THEOREM. The limits $\pi(j) = \lim_{n \to \infty} P^n(i, j)$ exist for all $j \in E$ and are independent of the initial state i. If $r \geq 1$, then $\pi(j) = 0$ for all j. If $r < 1$, then

$$(5.21) \qquad\qquad \pi(0) = 1 - r,$$

$$(5.22) \qquad\qquad \pi(1) = (1 - r)\frac{r_0}{q_0},$$

and, for any $j \in \{1, 2, \ldots\}$,

$$(5.23) \qquad \pi(j + 1) = (1 - r) \sum_{k=1}^{j} \left(\frac{1}{q_0}\right)^{k+1} \sum_{\alpha \in S_{jk}} r_{\alpha_1} r_{\alpha_2} \cdots r_{\alpha_k}$$

where S_{jk} is the set of all k-tuples $\boldsymbol{\alpha} = (\alpha_1, \ldots, \alpha_k)$ of integers $\alpha_i \geq 1$ with $\alpha_1 + \cdots + \alpha_k = j$.

Proof. If $r \geq 1$, the preceding proposition shows that all states are recurrent null or transient. In either case, Theorem (5.3.2) implies that the limits $\pi(j)$ are all zero.

If $r < 1$, all states are recurrent non-null aperiodic, and by Theorem (2.1) the limiting distribution exists independent of the initial state and is the unique solution of (2.2) and (2.3). From the proof of (5.14) above, it follows that (5.19) and (5.17) specify this distribution. It is easy to see that (5.21) and (5.22) hold; and directly from the second equation in (5.17) we obtain $\pi(2) = (1 - r)r_1/q_0^2$, which is the same as that given by (5.23) for $j = 1$. To complete the proof, we make the generalized induction hypothesis that (5.23) is true for all $j = 1, \ldots, n$ and show that (5.23) is then also true for $j = n + 1$.

Let S_{jk} be as defined above, and define

$$(5.24) \qquad\qquad q_{jk} = \sum_{\alpha \in S_{jk}} r_{\alpha_1} \cdots r_{\alpha_k}$$

for all $j \geq k \geq 1$. First observe that separating the sum over $\beta = \alpha_{k+1}$ from the rest, we can write

$$(5.25) \qquad q_{j+1,k+1} = \sum_{\beta=1}^{j+1-k} \left(\sum_{\alpha \in S_{j+1-\beta,k}} r_{\alpha_1} \cdots r_{\alpha_k} \right) r_\beta$$

$$= r_1 q_{jk} + r_2 q_{j-1,k} + \cdots + r_{j+1-k} q_{kk}.$$

Now, for $j = n+1$, (5.17) gives

$$q_0 \pi_{n+2} = r_{n+1}(\pi_0 + \pi_1) + r_n \pi_2 + r_{n-1}\pi_3 + \cdots + r_1 \pi_{n+1}.$$

Replacing $\pi_0 + \pi_1$ by its value $(1 - r)/q_0$ and using the induction hypothesis for $\pi_2, \pi_3, \ldots, \pi_{n+1}$, we obtain

$$\frac{q_0 \pi_{n+2}}{1-r} = \frac{r_{n+1}}{q_0} + r_n \frac{q_{11}}{q_0^2}$$

$$+ r_{n-1}\left[\frac{q_{21}}{q_0^2} + \frac{q_{22}}{q_0^3} \right]$$

$$\vdots$$

$$+ r_1 \left[\frac{q_{n1}}{q_0^2} + \frac{q_{n2}}{q_0^3} + \cdots + \frac{q_{nn}}{q_0^{n+1}} \right]$$

$$= \frac{1}{q_0} r_{n+1} + \left(\frac{1}{q_0}\right)^2 (r_n q_{11} + r_{n-1}q_{21} + \cdots + r_1 q_{n1})$$

$$+ \left(\frac{1}{q_0}\right)^3 (r_{n-1}q_{22} + r_{n-2}q_{32} + \cdots + r_1 q_{n2})$$

$$\vdots$$

$$+ \left(\frac{1}{q_0}\right)^{n+1} r_1 q_{nn}$$

$$= \frac{1}{q_0} q_{n+1,1} + \left(\frac{1}{q_0}\right)^2 q_{n+1,2} + \left(\frac{1}{q_0}\right)^3 q_{n+1,3} + \cdots + \left(\frac{1}{q_0}\right)^{n+1} q_{n+1,n+1}$$

by (5.25). But this is precisely (5.23) for $j = n+1$. $\qquad \square$

    Before moving on to a consideration of the transient case, we now discuss a few consequences of the recurrent non-null case. We had already computed the limiting distribution in this case. With that known we can, in principle, compute $E[X_n]$, $\mathrm{Var}(X_n)$, etc., in the limit as $n \longrightarrow \infty$. However, it is easier to use formula (5.6) for this purpose: we rewrite it as

$$(5.26) \qquad\qquad X_{n+1} = X_n + M_n - U_n$$

by defining $U_n = 1 - 1_0(X_n)$ and letting M_n denote the number of arrivals during the $(n + 1)$th service. Note that in the recurrent non-null case,

$$(5.27) \qquad \lim_{n\to\infty} E[U_n] = 1 - \lim_{n\to\infty} P\{X_n = 0\} = 1 - \pi(0) = r = ab$$

and that $E[M_n] = r = ab$ as in (5.11); and using (4.1.14) for the expected value of the square of a Poisson-distributed random variable,

$$(5.28) \qquad E[M_n^2] = E[E[N_Z^2 | Z]]$$
$$= E[aZ + a^2 Z^2] = ab + a^2 c^2$$

where

$$c^2 = E[Z^2] = \int_0^\infty t^2 d\varphi(t).$$

Now taking the square of both sides in (5.26) and noting that $X_n U_n = X_n$ and $U_n^2 = U_n$, we obtain

$$X_{n+1}^2 = X_n^2 + M_n^2 + U_n + 2X_n M_n - 2X_n - 2M_n U_n.$$

Taking expectations on both sides, observing that M_n is independent of both X_n and U_n, and letting $n \to \infty$, we obtain by using (5.27) and (5.28) that

$$(5.29) \qquad 0 = ab + a^2 c^2 + ab + 2qab - 2q - 2a^2 b^2$$

where we put $q = \lim_n E[X_n]$. Solving for q in (5.29), we obtain

$$(5.30) \qquad \lim_{n\to\infty} E[X_n] = ab + a^2 c^2 / (2 - 2ab).$$

These results are useful in discussing the characteristics of the system at large. The following is a useful device in order to deduce from these some measure of the plight of the individual customers. Let W_n be the time spent waiting by the nth customer, and let $V_n = W_n + Z_n$ (that is, V_n is the time spent in the system). Observe that if the service is on a first-come, first-served basis, then the customers that will be left behind by the nth customer will be exactly those who arrived during the time interval between the instant S_n of arrival and the instant $S_n + V_n$ of departure of that nth customer. Thus,

$$(5.31) \qquad X_n = N_{S_n + V_n} - N_{S_n}.$$

The distribution of the random variable on the right depends only on V_n, and therefore, all quantities of interest concerning V_n (and W_n) can be obtained from this relation and the known results concerning X_n in the limit. We leave these as exercises for the reader; cf. Exercises (8.19) and (8.20).

We return to the classification of states in the case $r \geq 1$ and to the computation of probabilities of the queue never becoming empty. First consider the probability $f_k(j)$ that starting from the state $k + j$, the Markov chain X never enters the set $\{0, 1, \ldots, k\}$. From Proposition (4.5) it follows that the function f_k is the maximal solution of the system of equations $Qh = h$, $0 \leq h \leq 1$ where Q is the matrix obtained from P by deleting all rows and columns corresponding to the states $0, \ldots, k$. Doing this, we obtain

(5.32)
$$Q = \begin{bmatrix} q_1 & q_2 & q_3 & \cdot & \cdot & \cdot \\ q_0 & q_1 & q_2 & \cdot & \cdot & \cdot \\ & q_0 & q_1 & \cdot & \cdot & \cdot \\ & & & \cdot & \cdot & \\ & 0 & & & \cdot & \cdot \\ & & & & & \cdot & \cdot \end{bmatrix}.$$

Observe that Q does not depend on k; therefore, $f_k(j) = f_0(j)$ for all j and k. We put this useful observation next.

(5.33) LEMMA.   The probability that X never enters $\{0, 1, \ldots, k\}$ starting from $k + j$ is the same as the probability $f(j)$ that X never enters 0 starting from j.

Next we compute this probability $f(j)$ of the queue never becoming empty if it started with j customers. Of course, this will be zero if X is recurrent and therefore, when the next theorem is put together with Proposition (5.14), we conclude that X is recurrent non-null if $r < 1$, recurrent null if $r = 1$, and transient if $r > 1$.

(5.34) THEOREM.   Let $f(j)$ be the probability that the queue, starting with j customers, never becomes empty. Then

(5.35)
$$f(j) = 1 - \beta^j, \quad j = 1, 2, \ldots,$$

where β is the smallest number in $[0, 1]$ satisfying

(5.36)
$$\beta = q_0 + q_1\beta + q_2\beta^2 + \cdots.$$

This number β is strictly less than one if and only if the traffic intensity $r > 1$. Therefore, X is transient if and only if $r > 1$.

Proof. We observe from the shape (5.2) of the transition matrix P that starting from $j + 1$, in order for the queue to become empty, it must first reach j. Therefore, starting from $j + 1$, the queue can avoid becoming empty either by never reaching j or, if it does reach j, by never becoming empty starting from j. Since the probability of never reaching j starting with $j + 1$ customers is $f(1)$ by Lemma (5.33), this argument implies that

$$f(j + 1) = f(1) + (1 - f(1))f(j), \quad j = 1, 2, \ldots.$$

Putting $f(1) = 1 - \beta$ and solving this recursively, we obtain (5.35). There remains to compute β.

By Proposition (4.5), f is the maximal solution of $h = Qh$, $0 \le h \le 1$ with Q as given by (5.32). From the first equation of $h = Qh$ we see that f must satisfy

$$f(1) = q_1 f(1) + q_2 f(2) + \cdots,$$

which, upon putting the values (5.35) in, implies that β satisfies the equation (5.36). Since f is the maximal such solution, β must be the smallest possible solution of (5.36).

Finally, we need to show that $\beta < 1$ (and therefore that $f \neq 0$, which in turn implies that X is transient) if and only if $r > 1$. Consider the function

$$G(\beta) = q_0 + q_1 \beta + q_2 \beta^2 + \cdots, \quad 0 \le \beta \le 1.$$

This is non-negative, increasing, and has an increasing derivative (cf. Figure 6.5.1). This implies that the graph of G can intersect the line $H: H(\beta) = \beta$ at most once in $(0, 1)$ (of course they always intersect at $\beta = 1$). Consider the derivative $G'(1)$ at 1. If $G'(1) > 1$, the point of intersection β is strictly less than one; otherwise, if $G'(1) \le 1$, then G and H have no point of intersection in $(0, 1)$ and the smallest number β desired is 1. Finally, note that $G'(1) = r$.

$\square$

It is possible to use the foregoing results about the imbedded Markov chain X to study the actual queue-size process $\{Y_t; t \ge 0\}$ in continuous time. It can be shown, for example, that the limiting distribution $\lim_{t \to \infty} P\{Y_t = j\}$

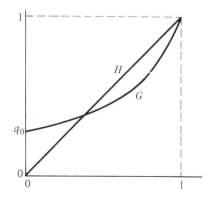

Figure 6.5.1 Since $G(0) > 0$, $G(1) = 1$, and $G'(1) = r > 1$, the curve G intersects the line H at least once. Since G is convex, it cannot intersect H twice in $(0, 1)$.

$(j = 0, 1, \ldots)$ is exactly the same as the π we computed in Theorem (5.20) above. For results of this nature we refer the reader to Chapter 10.

6.   Queueing System G/M/1

Consider a single-server queueing system with exponentially distributed service times which are independent of each other and of the arrival process, and with an arrival process where the successive interarrival times are independent and identically distributed. We will let a denote the parameter of the service time distribution and let φ denote the interarrival time distribution. Then q_n defined by (5.3) becomes the probability that the server completes exactly n services during an interarrival time provided that there are that many customers available to be served. We define $r_n = q_{n+1} + q_{n+2} + \cdots$ as before and let

$$r = \sum_{n=1}^{\infty} n q_n = r_0 + r_1 + \cdots,$$

so that r is the expected number of services which the server is capable of completing during an interarrival time.

If $r \geq 1$, then the server can keep up with arrivals, and we expect the queue size to be recurrent; if $r < 1$, then the queue size is likely to increase to infinity, and the process is transient. Note that this queueing system is the dual of the system M/G/1 studied in the preceding section. This duality comes about by the reversed roles of the interarrival times and service times. Therefore, with r fixed, when one system is transient the other is recurrent and vice versa. Later we will make these heuristic remarks more precise. The duality mentioned extends somewhat further and enables us to obtain the limiting distribution of the queue size for the G/M/1 system as a function of the probabilities of never becoming empty in the M/G/1 system. There is a similar functional relationship between the limiting distribution of the queue size for the M/G/1 system and the probabilities of never becoming empty for the G/M/1 system.

In this section we denote by X_n^* the number of customers present in the system just before the time T_n of the nth arrival.

(6.1) THEOREM. $X^* = \{X_n^*; n \in \mathbb{N}\}$ is a Markov chain with state space $E = \{0, 1, \ldots\}$ and transition matrix

(6.2)

$$P^* = \begin{bmatrix} r_0 & q_0 & & & & 0 \\ r_1 & q_1 & q_0 & & & \\ r_2 & q_2 & q_1 & q_0 & & \\ \cdot & \cdot & \cdot & \cdot & \cdot & \\ \cdot & \cdot & \cdot & & \cdot & \cdot \\ \cdot & \cdot & \cdot & & & \cdot \end{bmatrix}.$$

Proof. Let M_{n+1} be the number of services completed during the $(n + 1)$th interarrival time $[T_n, T_{n+1})$. Then

(6.3)
$$X_{n+1}^* = X_n^* + 1 - M_{n+1};$$

that is, X_{n+1}^* is the number of customers that were there before the nth arrival plus the nth customer less those whose services were completed during the next interarrival time. It follows from (6.3) that to show that X^* is a Markov chain, it is sufficient to show that M_{n+1} is conditionally independent of the past history before T_n given the present number X_n^*. But this follows from the independence of the service times and interarrival times and the fact that by the memorylessness of the exponential distribution, the remaining service time of the customer who was being served at time T_n (if any) has the same exponential distribution as any service time.

Furthermore, the above reasoning shows that as long as there are customers to be served, the conditional distribution of the number of services completed will be Poisson: with $Z = T_{n+1} - T_n$,

$$P\{M_{n+1} = k \mid X_n^*, Z\} = \begin{cases} \dfrac{e^{-aZ}(aZ)^k}{k!} & \text{if } X_n^* + 1 > k, \\[2mm] \displaystyle\sum_{m=k}^{\infty} \dfrac{e^{-aZ}(aZ)^m}{m!} & \text{if } X_n^* + 1 = k, \\[2mm] 0 & \text{otherwise.} \end{cases}$$

Taking expectations with respect to Z, which is independent of X_n^*, we obtain

(6.4)
$$P\{M_{n+1} = k \mid X_n^* = i\} = \begin{cases} q_k & \text{if } k \le i, \\ r_{k-1} & \text{if } k = i + 1, \\ 0 & \text{otherwise.} \end{cases}$$

Putting (6.3) and (6.4) together, we see that the transition probabilities $P^*(i, j)$ are as claimed. ☐

It is clear from an inspection of (6.2) that the Markov chain X^* is irreducible aperiodic. Therefore all the states are of the same type. The following theorem shows that they are all recurrent non-null if and only if $r > 1$.

(6.5) THEOREM. X^* is recurrent non-null if and only if $r > 1$. If $r > 1$, the limiting distribution

(6.6)
$$\pi^*(j) = \lim_{n \to \infty} P\{X_n^* = j \mid X_0^* = i\}$$

is given by

(6.7)
$$\pi^*(j) = (1 - \beta)\beta^j, \quad j = 0, 1, 2, \ldots,$$

where β is the unique number satisfying

(6.8) $$\beta = q_0 + q_1\beta + q_2\beta^2 + \cdots, \quad 0 < \beta < 1.$$

If $r \leq 1$, then $\pi^*(j) = 0$ for all j.

Proof. By Theorem (2.1), X^* is recurrent non-null if and only if $\nu = \nu P^*$, $\nu 1 = 1$ has a solution; and if there is a solution, then $\pi^* = \nu$. Writing out the equations for $\nu = \nu P^*$, we obtain (observe that the initial equation uses $r_n = q_{n+1} + q_{n+2} + \cdots$)

$$\begin{aligned}
\nu_0 = {}& q_1\nu_0 + q_2\nu_0 + q_3\nu_0 + \cdots \\
& \qquad\quad\; + q_2\nu_1 + q_3\nu_1 + \cdots \\
& \qquad\qquad\qquad\; + q_3\nu_2 + \cdots, \\
\nu_1 = {}& q_0\nu_0 + q_1\nu_1 + q_2\nu_2 + q_3\nu_3 + \cdots, \\
\nu_2 = {}& q_0\nu_1 + q_1\nu_2 + q_2\nu_3 + q_3\nu_4 + \cdots, \dots.
\end{aligned}$$

Let a function f be defined by

(6.9) $$f(j) = \nu_0 + \cdots + \nu_{j-1}, \quad j = 1, 2, \dots.$$

Now the equation for ν_0 gives an equation for $f(1)$, summing the equations for ν_0 and ν_1 yields an equation for $f(2)$, and so on. We obtain

$$\begin{aligned}
f(1) &= q_1 f(1) + q_2 f(2) + q_3 f(3) + \cdots, \\
f(2) &= q_0 f(1) + q_1 f(2) + q_2 f(3) + \cdots, \\
f(3) &= \qquad\quad\; q_0 f(2) + q_1 f(3) + \cdots, \dots.
\end{aligned}$$

In other words, f satisfies

$$f = Qf$$

with Q as defined by (5.32), and we are interested in a solution of this satisfying

$$\lim_{j \to \infty} f(j) = \sum_{j=0}^{\infty} \nu_j = 1.$$

Remembering that Q was obtained from P by deleting the 0th row and column, we conclude by Proposition (4.5) that such a solution f exists if and only if the chain X is transient. By Theorem (5.34), this is true if and only if $r > 1$. If $r > 1$, the solution for f is given by (5.35), where β satisfies (5.36), which is the same relation as (6.8).

Solving for ν, by using (6.9), out of the expression (5.35) for f, we obtain

$$\nu_0 = f(1) = 1 - \beta, \quad \nu_1 = f(2) - f(1) = (1 - \beta)\beta, \quad \dots,$$

which is the same as the claimed limiting distribution π^*. This completes the proof, since the last statement, that $r \leq 1$ implies $\pi^* = 0$, follows directly from Theorem (5.3.2). $\square$

The limiting distribution obtained in the recurrent non-null case, then, is a geometric distribution. It is easy to work with, and in particular, we see that

$$(6.10) \qquad \lim_{n \to \infty} E[X_n^*] = \beta/(1 - \beta).$$

It is worth mentioning that this last result may be used, in practice, to estimate β directly; so after ascertaining that the model fits the system of practical interest, all one needs is an estimate of the average number of customers in the system just before an arrival. Once this number $\bar{q}$ is obtained, (6.10) may be used to estimate $\bar{\beta}$ by setting $\bar{q} = \bar{\beta}(1 - \bar{\beta})$. In other words, if one is interested only in the queue-size process, one does not need to estimate the probabilities q_n or try to solve the equation (6.8) for β.

It is easy to relate these results to the waiting times of the individual customers. Let W_n be that of the nth customer. Then, if the service is on a first-come, first-served basis, the nth customer's waiting time will be equal to the sum of the service times of those customers whom he found there upon his arrival. That is, W_n is equal to the sum of X_n^* random variables each of which has the exponential distribution with parameter a. From the interpretation of the geometric distribution in its relation to the time of first success in a sequence of Bernoulli trials (cf. Chapter 3 for details), we may think of W_n for large n as the time of first success with trials being performed at the times of arrivals in a Poisson process with rate a (note that there is a trial at time 0 in this setup). From the results on decomposition of Poisson processes (cf. Chapter 4) we see that W_n will have an exponential distribution, for large n, with parameter equal to a times the probability of success $1 - \beta$, except that at $t = 0$ there is a probability of immediate success equal to $1 - \beta$. Hence,

$$(6.11) \qquad \lim_{n \to \infty} P\{W_n \leq t\} = 1 - \beta + \beta \cdot (1 - e^{-a(1-\beta)t})$$

$$= 1 - \beta e^{-a(1-\beta)t}, \quad t \geq 0.$$

Some further quantities of interest can be obtained from this by straightforward methods. We leave some of these as exercises.

We return to the queue size process X^* and consider it when it is not recurrent non-null, namely, when $r \leq 1$. The next theorem shows that when $r < 1$, X^* is transient and when $r = 1$, X^* is recurrent null. In the transient case we compute the probability that the queue never becomes empty given the number of initial customers.

(6.12) THEOREM. X^* is transient if and only if $r < 1$. If $r < 1$, the probability $f^*(j)$ that the queue starting with j customers never becomes empty is given by

(6.13) $$f^*(j) = \pi(0) + \pi(1) + \cdots + \pi(j), \quad j = 1, 2, \ldots,$$

where the $\pi(j)$ are given by (5.21), (5.22), and (5.23).

Proof. In view of Proposition (4.5), f^* is the solution of $h = Q^*h$, $0 \le h \le 1$, where Q^* is obtained from P^* by deleting the 0th row and column. Then the equations for $h = Q^*h$ become (write $f^*(j) = h_j$)

(6.14) $$h_1 = q_0 h_2 + q_1 h_1,$$
$$h_2 = q_0 h_3 + q_1 h_2 + q_2 h_1,$$
$$h_3 = q_0 h_4 + q_1 h_3 + q_2 h_2 + q_3 h_1, \ldots.$$

Define $\pi_0 = q_0 h_1$, $\pi_1 = (1 - q_0)h_1$, and let $\pi_j = h_j - h_{j-1}$ for $j = 2, 3, \ldots$. Then the first equation of (6.14), along with $\pi_0 = q_0 h_1$, implies the two equations

$$\pi_0 = q_0 \pi_0 + q_0 \pi_1,$$
$$\pi_1 = q_1 \pi_0 + q_1 \pi_1 + q_0 \pi_2,$$

and subtracting the equation for h_{j-1} from the one for h_j for $j = 2, 3, \ldots$ yields

$$\pi_2 = q_2 \pi_0 + q_2 \pi_1 + q_1 \pi_2 + q_0 \pi_3,$$
$$\pi_3 = q_3 \pi_0 + q_3 \pi_1 + q_2 \pi_2 + q_1 \pi_3 + q_0 \pi_4, \ldots.$$

In other words, π satisfies $\pi = \pi P$ with P as given in (5.2), and we are interested in the solution of $\pi = \pi P$ with $\sum \pi_j = \lim_j h_j = 1$. From Theorem (5.14), such a solution exists if and only if $r < 1$. When $r < 1$, the solution is given by (5.21), (5.22), and (5.23), and the solution π is connected to h by the relation $h_j = \pi_0 + \cdots + \pi_j$. This completes the proof.   □

A special case of some interest is the single-server queueing system with Poisson arrivals and exponential service times (usually denoted by M/M/1 queue). One may consider it either as a special M/G/1 queue or as a special G/M/1 queue. It is easier, from the computational point of view, to take it as a special case of the G/M/1 queue with the interarrival distribution

$$\varphi(t) = 1 - e^{-\lambda t}, \quad t \ge 0$$

(and, of course, exponential service times with parameter a).

To compute the limiting distribution of the queue size X_n^* just before the

nth arrival, we use Theorem (6.5). Now, to solve for β in (6.8), first note that

$$\sum_{k=0}^{\infty} q_k \beta^k = \sum_{k=0}^{\infty} \beta^k \int_0^{\infty} \frac{e^{-at}(at)^k}{k!} \lambda e^{-\lambda t}\, dt$$

$$= \int_0^{\infty} e^{-at(1-\beta)} \lambda e^{-\lambda t}\, dt = \frac{\lambda}{\lambda + a - a\beta}.$$

Thus, equation (6.8) becomes

$$\beta = \frac{\lambda}{\lambda + a - a\beta} \quad \text{or} \quad (1 - \beta)(\lambda - a\beta) = 0.$$

We know that 1 is a solution. To obtain the other, we see that when $r = a/\lambda > 1$ we have, as the smallest solution,

$$\beta = \frac{\lambda}{a}.$$

So we have

(6.15) $$\lim_{n \to \infty} P\{X_n^* = j\} = \left(1 - \frac{\lambda}{a}\right)\left(\frac{\lambda}{a}\right)^j, \quad j = 0, 1, \ldots.$$

It turns out, though it is not at all apparent from anything which precedes this, that we also have

(6.16) $$\lim_{t \to \infty} P\{Y_t = j\} = \left(1 - \frac{\lambda}{a}\right)\left(\frac{\lambda}{a}\right)^j, \quad j = 0, 1, \ldots$$

for the queue size Y_t at time t, and similarly

(6.17) $$\lim_{n \to \infty} P\{X_n = j\} = \left(1 - \frac{\lambda}{a}\right)\left(\frac{\lambda}{a}\right)^j, \quad j = 0, 1, \ldots$$

for the queue size X_n just after the nth departure (cf. Exercise (8.21) for this).

For a different formulation of this problem we refer the reader to Section 6 of Chapter 8. For the general $G/M/1$ queue and its time-dependent behavior, see Section 7 of Chapter 10.

7. Branching Processes

Consider some objects which can generate additional objects of the same kind. An initial set of objects, which we call the initial generation, have "children" that make up the first generation, and they in turn have children which form the second generation, and so on. We are interested in the sizes of successive generations, the possibility of extinction, and the total size of all the generations.

Let $N(n, i)$ be the number of children born to the ith individual of the nth generation (supposing there is such an individual). Our basic assumption is that the random variables $N(n, i)$, ($n \in \mathbb{N}, i = 1, 2, \ldots,$) are independent and identically distributed. This implies that the number of children born to an individual does not depend on the past history or on how many other individuals are present, or on how many children others have. We let

$$(7.1) \qquad p_k = P\{N(n, i) = k\}, \quad k = 0, 1, \ldots,$$

be the common distribution, and define

$$(7.2) \qquad G(\alpha) = \sum_{k=0}^{\infty} \alpha^k p_k, \quad 0 \le \alpha \le 1.$$

Let X_n be the size of the nth generation. If $X_n = 0$, the population is said to be extinct by the nth generation; then $X_{n+1} = 0$ also. Otherwise, if $X_n = i > 0$, then X_{n+1} is equal to $N(n, 1) + \cdots + N(n, i)$. That is, for any $n \in \mathbb{N}$,

$$(7.3) \qquad X_{n+1} = \begin{cases} 0 & \text{on } \{X_n = 0\}, \\ N(n, 1) + \cdots + N(n, X_n) & \text{on } \{X_n > 0\}. \end{cases}$$

It follows that $X = \{X_n; n = 0, 1, \ldots\}$ is a Markov chain with state space $E = \{0, 1, \ldots\}$. X is called a *branching process*.

(7.4) EXAMPLE. *Electron multipliers.* The device consists of a series of plates; each electron, as it strikes the first of these plates, gives rise to a random number of electrons which in turn strike the second plate, and so on. Here X_n is the number of electrons produced at the nth plate, and we are interested in its distribution for various n. □

(7.5) EXAMPLE. *Extinction of families.* Let $p_0, p_1, p_2, \ldots$ be the respective probabilities that a man has $0, 1, 2, \ldots$ sons, let each son have the same probabilities for sons of his own, and so on. Furthermore, suppose that the number of sons a man has is not influenced by either the family history or the numbers of sons his brothers and cousins may have. Then the numbers X_n of descendants in the male line in the nth generation form a branching process. We are interested in the distribution of X_n and in the probability that the male line eventually becomes extinct. □

As usual, we write P_i for the conditional probability $P\{\cdot \mid X_0 = i\}$ and write E_i for the corresponding expectation.

(7.6) PROPOSITION. The transition probabilities $P^n(i, j)$ satisfy

(7.7) $$\sum_{j \in E} P^n(i, j) \alpha^j = [G_n(\alpha)]^i, \quad i \in E,$$

where $G_0(\alpha) = \alpha$ and

$$G_{n+1}(\alpha) = G(G_n(\alpha)), \quad n \in \mathbb{N}.$$

Proof will be left as an exercise. Evidently, $G_n(\alpha)$ is $E_1[\alpha^{X_n}]$, and (7.7) can be explained as follows. If the initial generation consists of i individuals, then the population is composed of i subpopulations, each initiated by one individual, progressing simultaneously, independent of each other, and according to the same laws.

It follows from the definition (7.3) that state 0 is absorbing; let T be the time of absorption; that is, let

(7.8) $$T = \inf\{n : X_n = 0\}.$$

Then T is called the time of extinction. If $T < \infty$, the population becomes extinct after some finite number of generations, and we are interested in the probability of the event $\{T < \infty\}$.

If $p_0 = 0$, then $X_0 \leq X_1 \leq \cdots$ and $T = +\infty$ almost surely, starting with $i \geq 1$ individuals. If $p_0 > 0$ and $p_0 + p_1 = 1$, then $X_0 \geq X_1 \geq \cdots$ and $T < \infty$ almost surely (in fact, $P_i\{T \leq n\} = (1 - p_1^n)^i$, $i \in E$, $n \in \mathbb{N}$). We exclude these trivial cases and assume that

(7.9) $$p_0 > 0; \quad p_0 + p_1 < 1; \quad m = \sum_k k p_k < \infty.$$

The following is the main result of this section.

(7.10) THEOREM. Suppose (7.9) holds. Then for any $i \in E$,

(7.11) $$P_i\{T < \infty\} = \eta^i.$$

If $m \leq 1$, then $\eta = 1$. If $m > 1$, then η is the unique number satisfying

(7.12) $$\alpha = G(\alpha), \quad 0 < \alpha < 1.$$

Proof. Let $T_1, T_2, \ldots$ be the times of extinction of the sub-processes initiated by the first, second, ... individuals of the initial generation. Then $T_1, T_2, \ldots$ are independent and $P\{T_i < \infty\} = \eta$ for some number η independent of i. If $X_0 = i$, then T is the maximum of $T_1, \ldots, T_i$, and therefore $T < \infty$ if and only if $T_1 < \infty, \ldots, T_i < \infty$. These yield (7.11).

To compute the number η, let $f(i) = P_i\{T = +\infty\} = 1 - \eta^i$; then $f(i)$ is the probability that starting at $i \geq 1$, the chain X remains forever in $\{1, 2, \ldots\}$. By Proposition (4.5), therefore, f is the maximal solution of $h = Qh$, $0 \leq h \leq 1$, where Q is obtained from P by deleting the 0th row and column. In particular, then,

$$1 - \eta = f(1) = \sum_{j=1}^{\infty} Q(1, j) f(j) = \sum_{j=1}^{\infty} p_j(1 - \eta^j) = 1 - \sum_{j=0}^{\infty} p_j \eta^j;$$

that is, the number η satisfies $\alpha = G(\alpha)$. Moreover, by the maximality of f, η is the minimal solution of $\alpha = G(\alpha)$.

Since $p_0 > 0$, $G(0) = p_0 > 0$; since $p_0 + p_1 < 1$, G is convex; and obviously, $G(1) = 1$. Thus the graph of G can intersect the line $H(\alpha) = \alpha$ at most twice in $[0, 1]$ (see Figure 6.5.1). If $G'(1) = m \leq 1$, then G and H intersect only at $\alpha = 1$ and therefore $\eta = 1$. If $G'(1) = m > 1$, then they intersect twice: at $\alpha = 1$ and at $\alpha = \eta < 1$. $\quad\square$

Under the assumptions (7.9), all states $i \geq 1$ are transient and can be reached from each other. Hence, for almost all ω, $\lim_n X_n(\omega)$ is either 0 or $+\infty$. If $T(\omega) < \infty$, then $X_n(\omega) = 0$ for some finite n and therefore for all large n; that is, $T(\omega) < \infty$ if and only if $\lim_n X_n(\omega) = 0$. Thus we have the following limiting behavior.

(7.13) PROPOSITION.   For almost all ω,

$$(7.14) \qquad \lim_{n \to \infty} X_n(\omega) = \begin{cases} 0 & \text{if } T(\omega) < \infty, \\ +\infty & \text{if } T(\omega) = \infty. \end{cases}$$

Therefore, for any $i \in E$,

$$(7.15) \qquad P_i\{\lim_n X_n = 0\} = 1 - P_i\{\lim_n X_n = +\infty\} = \eta^i,$$

where η is the number defined in Theorem (7.10).

To summarize, if the mean number m of children born to an individual is less than or equal to one, then the population becomes extinct with probability one after only finitely many generations. If $m > 1$, however, there are two possibilities: the population may become extinct after finitely many generations (this has probability η if the initial size is one), or the population never becomes extinct and instead its size grows to infinity (this has probability $1 - \eta$).

(7.16) EXAMPLE.   In Example (7.5), suppose $p_k = bc^{k-1}$ for $k = 1, 2, \ldots$ with b and $c > 0$ and $b + c < 1$ (then $p_0 = a = 1 - b/(1 - c)$). Then

$G(\alpha) = a + b\alpha/(1 - c\alpha)$, so the equation $G(\alpha) = \alpha$ is equivalent to $(\alpha - 1)$ $(c\alpha - a) = 0$, whose two roots are $\alpha = 1$ and $\alpha = a/c$. If $b \leq (1 - c)^2$, then the smallest root is $\eta = 1$. Otherwise, if $b > (1 - c)^2$, the smallest root is $\eta = a/c < 1$. Thus, a family for which $m = b/(1 - c)^2 \leq 1$ will become extinct (on the male side) with probability one. If $m = b/(1 - c)^2 > 1$, the family will become extinct with probability $\eta = a/c$ and will grow to become a very large clan with probability $1 - \eta = 1 - a/c$. $\qquad \square$

Finally, we consider the cumulative size of all generations, that is,

$$(7.17) \qquad Y = \sum_{n=0}^{\infty} X_n = \sum_{n=0}^{T-1} X_n.$$

If $T(\omega)$ is finite, then $Y(\omega)$ is the sum of a finite number of finite quantities; therefore, $T(\omega) < \infty$ implies $Y(\omega) < \infty$. If $T(\omega)$ is infinite, then $Y(\omega)$ is the sum of infinitely many terms each of which is equal to one or more; therefore, $T(\omega) = \infty$ implies $Y(\omega) = \infty$. We list this observation next along with the probability distribution of Y.

(7.18) PROPOSITION. For any ω, $Y(\omega) < \infty$ if and only if $T(\omega) < \infty$. Therefore,

$$(7.19) \qquad P_i\{Y < \infty\} = \eta^i, \quad i \in E.$$

Moreover, for any $\beta \in [0, 1)$,

$$(7.20) \qquad E_i[\beta^Y] = F(\beta)^i, \quad i \in E,$$

where the number $F(\beta)$ is the unique solution of

$$(7.21) \qquad \alpha = \beta G(\alpha), \quad 0 < \alpha \leq \eta.$$

Proof. We have already shown the first statement; (7.19) follows from it and the expression (7.11). To show (7.20), suppose $X_0 = i$; then Y is the sum of the numbers $Y_1, Y_2, \ldots, Y_i$ of total descendants of the first, second, $\ldots$, ith individual of the initial generation. Now (7.20) is immediate from the fact that $Y_1, Y_2, \ldots$ are independent and identically distributed. For fixed $\beta \in (0, 1)$,

$$(7.22) \qquad F(\beta) = E_1[\beta^Y] = \beta E_1[\beta^{X_1 + X_2 + \cdots}] = \beta E_1[E_1[\beta^{X_1 + X_2 + \cdots} \,|\, X_1]].$$

Now, $X_1 + X_2 + \cdots$ is the total size of the population whose 0th generation is the first one of the original population. Then, by Proposition (5.1.22) and the already shown result (7.20),

$$(7.23) \qquad E[\beta^{X_1 + X_2 + \cdots} \,|\, X_1 = j] = E_j[\beta^Y] = F(\beta)^j.$$

Putting (7.23) into (7.22), we obtain the result that

$$F(\beta) = \beta \sum p_j F(\beta)^j = \beta G(F(\beta)).$$

Thus the number $F(\beta)$ satisfies $\alpha = \beta G(\alpha)$, and clearly $F(\beta) \leq P_1\{Y < \infty\}$ $= \eta$. For $\beta < 1$ the graph of $\alpha \to \beta G(\alpha)$ takes the value $\beta p_0 > 0$ at $\alpha = 0$, is convex in $(0, 1)$, and is equal to $\beta G(1) = \beta < 1$ at $\alpha = 1$. Therefore, that graph can intersect the bisector $H(\alpha) = \alpha$ only once. ☐

Considering the expected numbers of individuals, note that

(7.24) $$E[X_{n+1} \,|\, X_n] = mX_n,$$

so that $E_i[X_n] = im^n$ and

(7.25) $$E_i[Y] = i(1 + m + m^2 + \cdots) = \begin{cases} +\infty & \text{if } m \geq 1, \\ i/(1 - m) & \text{if } m < 1. \end{cases}$$

That is, if $m < 1$, the population size is finite and has the finite expectation $i/(1 - m)$. If $m = 1$, the population size is finite with probability one, but its expected value is infinite. If $m > 1$, the population size is finite with a probability less than one and, of course, its expectation is infinite.

(7.26) EXAMPLE. Consider an $M/G/1$ queueing system (see Section 5 for the description). Customers who are initially there form the 0th generation, those arriving during their service times form the first generation, and so on. Let $N(n, i)$ be the number of customers arriving during the service time of the ith individual of the nth generation. These are independent and identically distributed, and therefore describe a branching process. The number Y defined by (7.17) is the total number of customers served during the first busy period. If $Y < \infty$, then the busy period lasts only a finite length of time. Thus, the queue-size process is recurrent non-null if, in addition, $E_i[Y] < \infty$. Propositions (7.13) and (7.18) contain information on these points; the results are the same as those obtained in Section 5. Furthermore, (7.20) and (7.21) yield the distribution of the number of customers served during a busy period starting with i customers. ☐

8. Exercises

(8.1) Compute the potential matrices R corresponding to the following transition matrices:

(a) $$\begin{bmatrix} 0.3 & 0.7 & 0 \\ 0.4 & 0.4 & 0.2 \\ 0 & 0 & 1 \end{bmatrix}$$

(b) $$\begin{bmatrix} 0.2 & 0.8 & 0 & 0 \\ 0.6 & 0.4 & 0 & 0 \\ 0 & 0.2 & 0.3 & 0.5 \\ 0 & 0 & 0.5 & 0.5 \end{bmatrix}$$

(c)
$$\begin{bmatrix} 0.3 & 0.7 & 0 & 0 & 0 \\ 0.4 & 0.4 & 0.2 & 0 & 0 \\ 0 & 0 & 0.8 & 0.2 & 0 \\ 0 & 0 & 0 & 0.5 & 0.5 \\ 0 & 0 & 0.5 & 0.4 & 0.1 \end{bmatrix}$$
(d)
$$\begin{bmatrix} 0.3 & 0.7 & 0 & 0 & 0 \\ 0.5 & 0.5 & 0 & 0 & 0 \\ 0 & 0.2 & 0.4 & 0.4 & 0 \\ 0 & 0 & 0 & 1 & 0 \\ 0.2 & 0.3 & 0 & 0 & 0.5 \end{bmatrix}$$

(8.2) Compute the probabilities $F(i, j)$ of ever reaching j from i for all i, j for the Markov chains with transition matrices given above in (8.1a), (8.1b), (8.1c), and (8.1d).

(8.3) Considering the Markov chain of Example (5.3.10), compute
(a) the potential matrix R,
(b) the limiting distribution.

(8.4) Compute the limiting distributions for the Markov chains of
(a) Example (5.1.9),
(b) Example (5.1.10).

(8.5) Consider the Markov chain of Example (5.3.23) and
(a) compute, for all $i, j \in E$, the probability $F(i, j)$ of ever reaching j from i;
(b) compute the limiting distribution $\lim_n P_i\{X_n = j\}$, $j \in E$, for every initial state $i \in E$;
(c) compute the expected value of the recurrence time of each state.

(8.6) Compute the limiting distribution for the Markov chain of Example (5.1.15) (see also the next exercise).

(8.7) Let X be an irreducible aperiodic Markov chain with $m < \infty$ states, and suppose its transition matrix P is doubly Markov. Show that then

$$\pi(i) = \frac{1}{m}, \quad i \in E,$$

is the limiting distribution.

(8.8) Compute the limiting distribution of the age Y_n of the machine in use at time n as $n \to \infty$ in Exercise (5.4.10).

(8.9) Compute the limiting distribution of the number of black balls in the first container in Exercise (5.4.11).

(8.10) Let X be an irreducible Markov chain having the invariant distribution $\pi = (0.2, 0.1, 0.3, 0.2, 0.2)$.
(a) Compute the limit of the expected average reward

$$\frac{1}{n+1} \sum_{m=0}^{n} E_i[f(X_m)]$$

as $n \to \infty$ for the function $f = (3, -1, 2, -5, 3)$.

(b)   Compute the limit of the ratio

$$\frac{\sum\limits_{m=0}^{n} f(X_m)}{\sum\limits_{m=0}^{n} g(X_m)}$$

as $n \longrightarrow \infty$ with f as given in part (a) and $g = (2, -1, 2, -3, 0)$.

(8.11)   Consider the Markov chain of Example (5.3.10) (see also (8.3) above), and let $f = (2, 8, -3, -5, 6)$ be defined on its state space. Compute the limiting value of the average actual reward

$$\frac{1}{n+1} \sum\limits_{m=0}^{n} f(X_m)$$

as $n \longrightarrow \infty$.

(8.12)   Consider the Markov chain of Example (1.12) (see also Example (1.24) later).

(a)   Compute the limit of P^n as $n \longrightarrow \infty$.

(b)   Find the expected value of the recurrence time of each state.

(c)   Compute the limit of the expected average reward

$$\frac{1}{n+1} \sum\limits_{m=0}^{n} E_i[f(X_m)]$$

as $n \longrightarrow \infty$ for $f = (0, -1, 2, 4, -3, 4, 6, 3)$ for each initial state i.

(8.13)   Consider the Markov chain of Example (1.25), and let g be defined as

$$g(i) = \lim_{n\to\infty} \frac{1}{n+1} \sum\limits_{m=0}^{n} E_i[f(X_m)], \quad i \in E.$$

(a)   Compute g for $f = (2, 3, 0, 0, 0, 2, 1)$.

(b)   Compute g for $f = (0, 0, 2, 1, -1, 0, 0)$.

(c)   Compute g for $f = (2, 3, 2, 1, -1, 2, 1)$.

(d)   Compute g for arbitrary $f = (f(1), \ldots, f(7))$.

(8.14)   Let X be a Markov chain with state space $E = \{a, b\}$ and transition matrix

$$P = \begin{bmatrix} 0.4 & 0.6 \\ 1 & 0 \end{bmatrix},$$

and suppose that a reward of $g(i, j)$ units is received for every jump from i to j where

$$g = \begin{bmatrix} 3 & 2 \\ -1 & 1 \end{bmatrix}.$$

(a)   Considering the reward received at the mth jump, show that

$$E_i[g(X_m, X_{m+1})] = E_i[f(X_m)]$$

if f is defined by

$$f(j) = \sum_{k \in E} P(j, k)g(j, k) = \begin{cases} 2.4 & \text{if } j = a, \\ -1 & \text{if } j = b. \end{cases}$$

(b) Compute the average expected reward

$$\frac{1}{n+1} \sum_{m=0}^{n} E_i[g(X_m, X_{m+1})]$$

in the limit as $n \longrightarrow \infty$.

(8.15) Let X be a Markov chain with state space E and transition matrix P. Suppose a reward of $g(i, j)$ units is received for every jump from i to j. Suppose X is irreducible and has an invariant distribution π.

  (a) Show that for any initial state i,

$$\lim_{n \to \infty} \frac{1}{n+1} \sum_{m=0}^{n} E_i[g(X_m, X_{m+1})] = \sum_{j \in E} \pi(j) \sum_{k \in E} P(j, k)g(j, k).$$

  (b) Show that

$$\lim_{n \to \infty} \frac{1}{n+1} \sum_{m=0}^{n} g(X_m, X_{m+1}) = \sum_{j \in E} \pi(j) \sum_{k \in E} P(j, k)g(j, k)$$

almost surely. (*Hint:* the sequence $\{Y_n = (X_n, X_{n+1}); n = 0, 1, \ldots\}$ is a Markov chain.)

(8.16) Consider the periodic Markov chain of Example (3.3).

  (a) Show that there is an invariant distribution, although it is not the limiting distribution (which does not exist).

  (b) Compute P^{3n}, P^{3n+1}, P^{3n+2} in the limit as $n \longrightarrow \infty$.

  (c) Compute the limit

$$\lim_{n \to \infty} \frac{1}{n+1} \sum_{m=0}^{n} f(X_m)$$

for $f = (2, 5, 4, 2, -1, -4, 6)$.

(8.17) Consider two gamblers whose capitals sum to 7 dollars, so that as soon as one has all seven dollars the other is ruined and the game stops. Plays form independent trials with even chances for winning and losing. Let X_n be the capital of the first gambler at the end of the nth play.

  (a) Show that X is a Markov chain with state space $E = \{0, \ldots, 7\}$ and transition probabilities $P(i, i - 1) = P(i, i + 1) = 1/2$ for $i = 1, \ldots, 6$ and $P(0, 0) = P(7, 7) = 1$.

  (b) Compute the probability $f(i) = F(i, 0)$ of eventual ruin for the first player for each possible initial capital i.

(8.18) Consider an M/G/1 queueing system. Show that, for a given arrival rate a and mean service time b, the expected queue size in the long run is minimized when the service times are deterministic constants equal to b. Compute, explicitly, π_0, π_1, and π_2 in this case of deterministic service times.

(8.19)   *Waiting times in an M/G/1 queue.* In the notation of Section 5, use the relation (5.31) to show that

$$E[X_n] = aE[V_n].$$

Use this to compute $\lim_{n\to\infty} E[V_n]$ and $\lim_{n\to\infty} E[W_n]$.

(8.20)   *Continuation.* Use (5.31) to show that

$$\pi(k) = \int_0^\infty \frac{e^{-at}(at)^k}{k!}\,d\psi(t), \quad k = 0, 1, \ldots$$

holds for the limiting distribution π of the queue size and the limiting distribution

$$\psi(t) = \lim_{n\to\infty} P\{V_n \leq t\}$$

of the time spent in the system by a customer. This formula determines the distribution ψ by identifying its Laplace transform:

$$\int_0^\infty e^{-\alpha t}d\psi(t) = \sum_{k=0}^\infty \pi(k)\left(1 - \frac{\alpha}{a}\right)^k; \quad 0 \leq \alpha < a.$$

(8.21)   Consider the limiting distribution π given in Theorem (5.20), and suppose that the service distribution is exponential with parameter μ.
(a)   Show that $q_n = (\mu/a)(a/(a + \mu))^{n+1}$ by using the results of Chapter 4 on the Laplace transforms of times of arrivals in a Poisson process.
(b)   Show that $r_n = (a/(a + \mu))^{n+1}$, $n = 0, 1, \ldots$ and that

$$q(j, k) = \binom{j-1}{k-1}\left(\frac{a}{a+\mu}\right)^{j+k}$$

for $j \geq k \geq 1$.
(c)   Show that when $r = a/\mu < 1$,

$$\pi(j) = (1 - r)r^j, \quad j = 0, 1, \ldots,$$

by putting the results obtained in (a) and (b) above into the formulas of (5.20). Compare this result with those given in (6.15), (6.16), and (6.17).

(8.22)   It is known, from direct statistical studies, that the expected queue size in the long run for a G/M/1 queueing system is $q = 4.8$ customers. Find the limiting distribution of the queue size and compute the variance.

(8.23)   Consider an M/G/1 queue (single server, Poisson process of arrivals, independent and identically distributed service times) with a finite waiting room of capacity m. Then all customers who arrive to find the waiting room full (namely, $m + 1$ customers in the system) leave the system never to return. Let X_n be the number of customers in the system just after the nth departure.
(a)   Show that $X = \{X_n; n \in \mathbb{N}\}$ is a Markov chain with state space

$E = \{0, 1, \ldots, m\}$ and transition matrix

$$P = \begin{bmatrix} q_0 & q_1 & q_2 & \cdot & \cdot & \cdot & q_{m-1} & r_{m-1} \\ q_0 & q_1 & q_2 & \cdot & \cdot & \cdot & q_{m-1} & r_{m-1} \\ & q_0 & q_1 & \cdot & \cdot & \cdot & q_{m-2} & r_{m-2} \\ & & \cdot & \cdot & & & \cdot & \cdot \\ & & & \cdot & \cdot & & \cdot & \cdot \\ & & & & \cdot & \cdot & \cdot & \cdot \\ & & & & & q_0 & q_1 & r_1 \\ & \mathbf{0} & & & & & q_0 & r_0 \end{bmatrix},$$

where q_n is the probability that exactly n arrivals occur during a service time and where $r_n = q_{n+1} + q_{n+2} + \cdots$.

(b) Compute the limiting distribution π explicitly for $m = 3$, and then for arbitrary m.

(8.24) Consider a G/M/1 queue with a waiting room of capacity $m - 1$. A customer finding m customers upon his arrival cannot join the system and leaves never to return. Let X_n^* be the queue size just before the nth arrival.

(a) Show that $X^* = \{X_n^*; n \in \mathbb{N}\}$ is a Markov chain with state space $E = \{0, 1, \ldots, m\}$ and transition matrix

$$P^* = \begin{bmatrix} r_0 & q_0 & & & & & & \mathbf{0} \\ r_1 & q_1 & q_0 & & & & & \\ r_2 & q_2 & q_1 & q_0 & & & & \\ \cdot & \cdot & \cdot & \cdot & \cdot & & & \\ \cdot & \cdot & \cdot & & \cdot & \cdot & & \\ \cdot & \cdot & \cdot & & & \cdot & \cdot & \\ r_{m-1} & q_{m-1} & q_{m-2} & \cdot & \cdot & \cdot & q_1 & q_0 \\ r_{m-1} & q_{m-1} & q_{m-2} & \cdot & \cdot & \cdot & q_1 & q_0 \end{bmatrix}.$$

(b) Note that if the states were relabeled so that $\bar{0} = m$, $\bar{1} = m - 1, \ldots, \bar{m} = 0$, then the transition matrix $\bar{P}^*$ corresponding to this labeling would be

$$\bar{P}^* = P$$

where P is the transition matrix of Exercise (8.23).

(c) Conclude from this that the limiting distribution π^* of the Markov chain X^* is related to the limiting distribution π of X through

$$\pi(j) = \pi^*(m - j), \quad j = 0, \ldots, m.$$

(8.25) Prove Proposition (7.6).

(8.26) *Extinction of families.* For the distribution given in Example (7.16), show that for each $n \in \mathbb{N}$, the distribution of X_n has the form

$$P_1\{X_n = k\} = b_n c_n^{k-1}, \quad k = 1, 2, \ldots$$

again. (Hint: use induction to show that $G_n(\alpha) = a_n + b_n\alpha/(1 - c_n\alpha)$.)

(8.27)   *Continuation.* In (7.16), let $b = 0.4$ and $c = 0.2$ (then $a = 0.5$ and $\eta = 1$).
Compute the generating function F of the total size Y of the population.

(8.28)   *Nuclear chain reactions.* Here the individuals are certain nuclear particles.
Each particle has a certain probability p of scoring a hit, in which case it produces
3 particles. With probability $q = 1 - p$ the particle moves out of the system
without a hit. If the number of particles grows to infinity, then there occurs an
explosion; otherwise, the process dies out. Compute the probability of explosion
when $p = 0.4$, starting with one particle.

*With the exception of Section 4, this chapter is devoted to questions of recurrence
and existence of limits. Such questions are usually grouped under the heading of
ergodic results. In rough terms, ergodicity refers to the equality of space averages and
time averages. It is exemplified best by the result of Corollary (2.22): in it, the left-
hand side is a time average, whereas the right-hand side is a weighted average over the
state space.*

*Of the ergodic results we listed, Theorem (2.26) is the most general. A similar
limit theorem holds, in fact, for chains with more than one irreducible class. We refer
to* CHUNG [1, p. 90 ff.] *for proofs and further results. In the case where the state space
is not discrete, results of this nature are usually referred to under headings such as*
DOEBLIN *ratio limit theorems or the* CHACON-ORNSTEIN *theorem. For such results
a good reference is the monograph by* OREY [1].

*Concerning the limiting behavior of transient Markov chains, all we have been
able to say is that such a chain moves out of every finite set and travels toward "infinity." Possible ways of achieving this and the behavior in the states near infinity are
subjects of investigation in the boundary theory of Markov chains. Interested readers
should read the paper by* FELLER [4], *or the book by* CHUNG [1, p. 112 ff.], *which has
a more probabilistic and clearer picture.*

The discovery of imbedded Markov chains in queues by KENDALL [1] *has had
a healthy influence on queueing theory and stimulated the interest in the theory of
Markov chains. Even then, the power of direct theoretical attacks is still underrated,
and there is too much reliance on old-fashioned mathematics. Interested readers
should compare our direct solution in Theorem (5.20) with the method of generating
functions which is generally employed. On the other hand, it is sometimes difficult
to avoid them. The theory of branching processes makes heavy use of such methods by
its very nature.*

*Our introduction to that topic is limited to the case where the time parameter is
discrete, the state space is countable, and there is only one type. Such a branching
process is called a* GALTON-WATSON *process after their pioneering work on the
problem of extinction of families amongst English peers. For the general theory we
refer the reader to the comprehensive and scholarly treatise by* HARRIS [1].

CHAPTER 7

Potentials, Excessive Functions, and Optimal Stopping of Markov Chains

This chapter is an introduction to the modern theory of Markov chains, which, happily, is also becoming increasingly important in applications. The two sections on optimal stopping illustrate an important class of applications.

This chapter can be read independently of Chapter 6, and the dependence on Chapter 5 is slight. As usual, Ω is a fixed sample space, P a probability measure on it; if X is a Markov chain on this probability space, then we write P_i for the probability measure which makes i the initial state, and denote by E_i the corresponding expectation.

1. Potentials

Let $X = \{X_n; n \in \mathbb{N}\}$ be a Markov chain with state space E and transition matrix P, and let g be a function defined on E and taking values in $\mathbb{R}$. Suppose that at each time n we are given a "reward" whose amount depends on the state X is in at that time: if X is in state j, then the reward is $g(j)$ units. For example, if the realization $\omega \in \Omega$ is such that $X_0(\omega) = 3$, $X_1(\omega) = 5$, $X_2(\omega) = 4$, $X_3(\omega) = 2, \ldots$, then the respective rewards obtained at times $0, 1, 2, 3, \ldots$ are $g(3), g(5), g(4), g(2), \ldots$ units. Furthermore, suppose that all future rewards are being discounted in such a way that one unit of reward at time n has the present worth of α^n units; here α is a number in $[0, 1]$. Then for the realization $\omega \in \Omega$, the successive states visited are $X_0(\omega)$, $X_1(\omega)$, $X_2(\omega), \ldots$, and the rewards received at times $0, 1, 2, \ldots$ have the respective present worths of $g(X_0(\omega)), \alpha g(X_1(\omega)), \alpha^2 g(X_2(\omega)), \ldots$. We are interested in the expected value of the total discounted return; starting at i, this quantity is

$$(1.1) \qquad R^\alpha g(i) = E_i\left[\sum_{n=0}^{\infty} \alpha^n g(X_n)\right], \quad i \in E.$$

For $\alpha \in [0, 1]$ and g non-negative, the function $R^\alpha g$ defined by (1.1) is called the α-*potential of the function* g. When $\alpha = 1$, instead of writing $R^1 g$, we simply write Rg and call Rg the *potential of* g; in other words,

$$(1.2) \qquad Rg(i) = E_i\left[\sum_{n=0}^{\infty} g(X_n)\right], \quad i \in E.$$

These definitions are extended to any g for which (1.1) and (or) (1.2) make sense. A function f is said to be an α-*potential* if $f = R^\alpha g$ for some function g. Similarly, f is said to be a *potential* if $f = Rg$ for some g.

(1.3) EXAMPLE. Let $g = 1_j$ for some fixed j, that is, let $g(k) = 1$ or 0 according as $k = j$ or $k \neq j$. Then $\sum_{n=0}^{\infty} g(X_n)$ is simply the total number of visits to j, and $Rg(i) = E_i[\sum g(X_n)]$ is the expected number of visits to j starting at i; in other words, for $g = 1_j$ we have $Rg(i) = R(i, j)$. Thus, for any j, the j-column of the matrix R is a potential. The same is true of a linear combination of any number of the columns of R. ☐

It is possible for a potential to take on infinite values; for example $R1_j(i)$ is infinite whenever j is recurrent and can be reached from i. On the other hand, if $\alpha < 1$ and g is bounded, then $R^\alpha g$ is also bounded. To see this, suppose g is bounded by some constant c; then the sum on the right-hand side of (1.1) is bounded by $c\sum \alpha^n = c/(1 - \alpha)$, and therefore the expectation of that sum is also bounded by the same number $c/(1 - \alpha)$.

(**1.4**) PROPOSITION. *For any* $\alpha \in [0, 1]$ *and* $g \geq 0$,

$$R^\alpha g(i) = \sum_{j \in E} R^\alpha(i, j)g(j), \quad i \in E,$$

where the matrix R^α is given by

$$R^\alpha = \sum_{n=0}^{\infty} \alpha^n P^n.$$

Proof. Noting that

$$E_i[g(X_n)] = \sum_{j \in E} g(j)P_i\{X_n = j\} = \sum_{j \in E} P^n(i, j)g(j),$$

we have

$$R^\alpha g(i) = \sum_{n=0}^{\infty} E_i[\alpha^n g(X_n)]$$

$$= \sum_{n=0}^{\infty} \sum_{j \in E} \alpha^n P^n(i, j) g(j)$$

$$= \sum_{j \in E} \left[\sum_{n=0}^{\infty} \alpha^n P^n(i, j) \right] g(j) = \sum_{j \in E} R^\alpha(i, j) g(j)$$

as claimed.   □

The matrix R^α is called the α-*potential matrix* of X. When $\alpha = 1$, instead of R^1 we simply write R and call it the *potential matrix* of X. The matrix R is precisely the matrix whose computation was discussed in Section 1 of Chapter 6. For $\alpha < 1$, we now supply the following computational result.

(1.5) PROPOSITION. Let $\alpha \in [0, 1)$, and let g be a bounded non-negative function on E. Then $f = R^\alpha g$ is the unique bounded solution of the system of linear equations

(1.6) $(I - \alpha P)f = g.$

Proof. Since $f = R^\alpha g = g + \alpha P g + \alpha^2 P^2 g + \cdots$, we have $\alpha P f = \alpha P g + \alpha^2 P^2 g + \alpha^3 P^3 g + \cdots$ and hence $f - \alpha P f = g$. Thus, $R^\alpha g$ is a solution of (1.6). To show that it is the only solution, let f be a bounded solution of (1.6), and put $h = f - R^\alpha g$. Since both f and $R^\alpha g$ are bounded, h is bounded; and since $f = g + \alpha P f$ and $R^\alpha g = g + \alpha P R^\alpha g$, h satisfies $h = \alpha P h$. Multiplying both sides by αP, we get $h = \alpha^2 P^2 h$; and repeating this, we obtain that

(1.7) $h = \alpha^n P^n h$

for all n. Since h is bounded, say by c, $P^n h$ is bounded by the same constant c. Therefore, since $\alpha < 1$, $\alpha^n P^n h \to 0$ as $n \to \infty$. This and (1.7) imply that $h = 0$, which completes the proof in view of the definition of h.   □

In particular, taking $g = 1_j$ in Proposition (1.5), we see that the j-column of the matrix R^α is the unique bounded solution of $(I - \alpha P)f = 1_j$. Thus, we have

(1.8) COROLLARY. For any $\alpha \in [0, 1)$, R^α is the unique solution of the system of linear equations

$$(I - \alpha P)R^\alpha = I.$$

In particular, if the state space E is finite,

$$R^\alpha = (I - \alpha P)^{-1}.$$

(1.9) EXAMPLE. Suppose X has the transition matrix

$$P = \begin{bmatrix} \frac{1}{3} & \frac{2}{3} \\ \frac{1}{2} & \frac{1}{2} \end{bmatrix},$$

and let $g = (1, 5)$. In this case, the α-potential matrix R^α is

$$R^\alpha = (I - \alpha P)^{-1} = \begin{bmatrix} 1 - \dfrac{\alpha}{3} & -\dfrac{2\alpha}{3} \\ -\dfrac{\alpha}{2} & 1 - \dfrac{\alpha}{2} \end{bmatrix}^{-1}$$

$$= \frac{1}{6 - 5\alpha - \alpha^2} \begin{bmatrix} 6 - 3\alpha & 4\alpha \\ 3\alpha & 6 - 2\alpha \end{bmatrix}.$$

Hence,

$$R^\alpha g = \frac{1}{6 - 5\alpha - \alpha^2} \begin{bmatrix} 6 - 3\alpha & 4\alpha \\ 3\alpha & 6 - 2\alpha \end{bmatrix} \begin{bmatrix} 1 \\ 5 \end{bmatrix}$$

$$= \frac{1}{6 - 5\alpha - \alpha^2} \begin{bmatrix} 6 + 17\alpha \\ 30 - 7\alpha \end{bmatrix}.$$

In particular, if $\alpha = 0.9$, this becomes $R^\alpha g = \left(\dfrac{710}{23}, \dfrac{790}{23}\right)$. $\square$

(1.10) EXAMPLE. Consider the remaining-lifetime problem of Example (5.1.17). Suppose $g(j)$ is the maintenance cost per year for a machine whose remaining useful life is j, $j \geq 1$; and interpret $g(0)$ as the difference between the price of the new machine and the salvage value of the old. Then $R^\alpha g(i)$ is the present value of all future costs of maintenance and replacement if the machine starting at time 0 has i years of remaining life.

If the firm's earnings average 20% of its investments, then 1 dollar today is worth 1.20 dollars next year, and the applicable discount factor α is $\alpha = 1/1.20 = 5/6$. To make a concrete example, let us suppose that the lifetime of a machine has the geometric distribution with average lifetime of 2.5 years; that is (since the mean of the geometric distribution pq^{n-1} is $1/p$), we have

$$p_i = (0.4)(0.6)^{i-1}, \quad i = 1, 2, \ldots.$$

Finally, let us suppose that $g = (1200, 600, 240, 100, 100, 100, \ldots)$ and that initially a new machine is installed at time zero.

Then the sum of all future costs of maintenance and replacement (discounted with factor $\frac{5}{6}$) has the expected value $f(0) = R^{5/6}g(0)$. To compute this we use Proposition (1.5). Now,

$$
I - \tfrac{5}{6}P =
\begin{bmatrix}
\frac{2}{3} & -\frac{1}{3}(0.6) & -\frac{1}{3}(0.6)^2 & -\frac{1}{3}(0.6)^3 & \cdots \\
-\frac{5}{6} & 1 & 0 & 0 & \cdots \\
 & -\frac{5}{6} & 1 & 0 & \cdots \\
 & & -\frac{5}{6} & 1 & \cdots \\
 & & & \ddots & \ddots
\end{bmatrix}
$$

so that the system of equations to be solved becomes, writing f_j for $f(j)$,

$$\frac{2}{3}f_0 - \frac{1}{3}(0.6)f_1 - \frac{1}{3}(0.6)^2 f_2 - \frac{1}{3}(0.6)^3 f_3 - \cdots = 1200,$$

$$-\frac{5}{6}f_0 + f_1 \qquad\qquad\qquad = 600,$$

$$-\frac{5}{6}f_1 + f_2 \qquad\qquad = 240,$$

$$-\frac{5}{6}f_2 + f_3 \qquad = 100,$$

$$-\frac{5}{6}f_3 + f_4 = 100, \ldots.$$

Solving the second, third, fourth, . . . equations recursively, we get

$$f_1 = 600 + \frac{5}{6}f_0,$$

$$f_2 = 240 + \frac{5}{6}600 + \left(\frac{5}{6}\right)^2 f_0,$$

$$f_3 = 100 + \frac{5}{6}\cdot 240 + \left(\frac{5}{6}\right)^2 600 + \left(\frac{5}{6}\right)^3 f_0,$$

$$f_4 = 100 + \frac{5}{6}\cdot 100 + \left(\frac{5}{6}\right)^2 240 + \left(\frac{5}{6}\right)^3 600 + \left(\frac{5}{6}\right)^4 f_0,$$

and so on. Now, the first equation becomes

$$2f_0 = 3600 + (0.5 + 0.5^2 + 0.5^3 + 0.5^4 + \cdots)f_0$$
$$+ (\ 1 + 0.5\ + 0.5^2 + 0.5^3 + \cdots)0.6\ \times 600$$
$$+ (\ 1 + 0.5\ + 0.5^2 + \cdots)0.6^2 \times 240$$
$$+ (\ 1 + 0.5\ + \cdots)0.6^3 \times 100$$
$$+ (\ 1 + \cdots)0.6^4 \times 100 + \cdots.$$

Carrying out the arithmetic involved, we find

$$f_0 = 3600 + 0.6 \times 600 \times 2 + 0.6^2 \times 240 \times 2$$
$$+ 0.6^3 \times 100 \times 2(1 + 0.6 + \cdots)$$
$$= 3600 + 720 + 172.80 + 108 = 4600.80$$

which was the quantity desired.   □

In certain applications one is interested in the reward obtained during a certain period. If this period is fixed, then the computation is simple: for example, the total discounted reward obtained during the first m transitions is $\sum_{n=0}^{m-1} \alpha^n g(X_n)$, whose expectation is, starting from i,

$$(1.11) \qquad E_i\left[\sum_{n=0}^{m-1} \alpha^n g(X_n)\right] = \sum_{n=0}^{m-1} \alpha^n P^n g(i).$$

Noting that

$$R^\alpha g = g + \cdots + \alpha^{m-1} P^{m-1} g + \alpha^m P^m g + \alpha^{m+1} P^{m+1} g + \cdots$$
$$= (g + \cdots + \alpha^{m-1} P^{m-1} g) + \alpha^m P^m (g + \alpha P g + \cdots)$$
$$= g + \cdots + \alpha^{m-1} P^{m-1} g + \alpha^m P^m R^\alpha g,$$

we see that the quantity in (1.11) is

$$(1.12) \qquad E_i\left[\sum_{n=0}^{m-1} \alpha^n g(X_n)\right] = R^\alpha g(i) - \alpha^m P^m R^\alpha g(i).$$

Since $P^m f(i) = E_i[f(X_m)]$ for any bounded function f, we can rewrite (1.12) as

$$(1.13) \qquad R^\alpha g(i) = E_i\left[\sum_{n=0}^{m-1} \alpha^n g(X_n)\right] + E_i[\alpha^m R^\alpha g(X_m)].$$

It is clear that the last term in (1.13) is the contribution of the rewards obtained at times m, $m + 1$, $m + 2$, ... This contribution is $\alpha^m[g(X_m) +$

$\alpha g(X_{m+1}) + \cdots]$, and to show that its expectation, given the initial state i, is $E_i[\alpha^m R^\alpha g(X_m)]$, we could have used Corollary (5.1.23). A third explanation of the result is the most intuitive: At time m, considering the future states of X, one is observing $\hat{X}_0 = X_m$, $\hat{X}_1 = X_{m+1}$, $\hat{X}_2 = X_{m+2}, \ldots$. The sequence $\hat{X}_0, \hat{X}_1, \ldots$ is again a Markov chain with the same transition probabilities as X; and the reward associated with it has the expectation $R^\alpha g(j)$ if $\hat{X}_0 = j$. In other words, for someone who looks into the future after m, that future looks the same as the future after time 0 would look to someone considering it at time 0. Hence, the expectation of all discounted rewards in the future after m has the present worth of $R^\alpha g(\hat{X}_0) = R^\alpha g(X_m)$ at time m. The expected value, starting at i, of the discounted value of this is $E_i[\alpha^m R^\alpha g(X_m)]$.

The considerations above also hold in the more general case where the fixed time m is replaced by a random time T. The results then follow from Proposition (5.1.25). Recall that a non-negative integer-valued random variable T is a stopping time if and only if, for each $n \in \mathbb{N}$, the occurrence or non-occurrence of the event $\{T \leq n\}$ can be determined once $X_0, \ldots, X_n$ are known (refer to the discussion preceding (5.1.24) for examples).

(1.14) THEOREM. For any stopping time T and non-negative function g,

$$R^\alpha g(i) = E_i\left[\sum_{n=0}^{T-1} \alpha^n g(X_n)\right] + E_i[\alpha^T R^\alpha g(X_T)].$$

Proof. Define $W_m = g(X_m) + \alpha g(X_{m+1}) + \cdots$ for $m \in \mathbb{N}$. We need to show that

$$E_i[\alpha^T W_T] = E_i[\alpha^T R^\alpha g(X_T)].$$

First, note that $E_j[W_0] = R^\alpha g(j)$, and therefore,

$$E_i[W_T \mid X_n; n \leq T] = R^\alpha g(X_T)$$

by the strong Markov property applied at T; see Proposition (5.1.25). Second, the values of the stopping time T are fixed by the history $\{X_n; n \leq T\}$; therefore,

$$E_i[\alpha^T W_T \mid X_n; n \leq T] = \alpha^T E_i[W_T \mid X_n; n \leq T] = \alpha^T R^\alpha g(X_T).$$

Now the desired result follows upon taking expectations on both sides and using Proposition (2.2.16). $\square$

(1.15) EXAMPLE. Consider the problem of Example (1.10) above. Suppose we are interested in the expected value of all discounted costs until the time of second failure, given that initially the machine has a lifetime of 3 years.

Let T be the time of second failure; then $T(\omega) = n$ if and only if $X_n(\omega) = 0$ and 0 appears in $X_0(\omega), \ldots, X_{n-1}(\omega)$ only once. Thus, $T(\omega) > n$ if $\{X_0(\omega), \ldots, X_n(\omega)\}$ contains at most one zero; and $T(\omega) \leq n$ otherwise. Therefore T is a stopping time; and what we are asked to compute is

$$a = E_3\left[\sum_{n=0}^{T-1} \alpha^n g(X_n)\right]$$

for α and g given in (1.10). By the preceding theorem,

$$a = R^\alpha g(3) - E_3[\alpha^T R^\alpha g(X_T)];$$

and $f = R^\alpha g$ was computed in Example (1.10) above. We had, first of all,

$$R^\alpha g(3) = f(3) = 100 + \frac{5}{6} \cdot 240 + \left(\frac{5}{6}\right)^2 600 + \left(\frac{5}{6}\right)^3 4600.8 = 3379.17.$$

By the definition of T, $X_T = 0$ provided that T is finite; and T is almost surely finite, since state 0 is recurrent (see Example (5.3.25) for this). Hence, $R^\alpha g(X_T) = R^\alpha g(0) = f(0) = 4600.80$, and

$$a = 3379.17 - 4600.80 E_3[\alpha^T].$$

Starting at 3, the first visit to 0 occurs at time 3, and the second at the time of failure of the new machine. Hence,

$$\begin{aligned} E_3(\alpha^T) &= \sum_{j=1}^{\infty} \alpha^{3+j}(0.4)(0.6)^{j-1} \\ &= (0.4)\left(\frac{5}{6}\right)^4 \sum_{j=1}^{\infty} \left(\frac{5}{6}(0.6)\right)^{j-1} = (0.8)\left(\frac{5}{6}\right)^4. \end{aligned}$$

Putting this result in the expression for a above, we have

$$a = 3379.17 - 1775.00 = 1604.17. \qquad \square$$

If g is bounded and $\alpha < 1$, then Theorem (1.14) is useful in computing

(1.16) $$h(i) = E_i[\sum_{n=0}^{T-1} \alpha^n g(X_n)], \quad i \in E,$$

especially when $R^\alpha g$ is already known as in the preceding example. If $\alpha = 1$, or if g is such that $R^\alpha g$ takes infinite values, or both, Theorem (1.14) might yield an expression such as $+\infty = h(i) + \infty$, which is worthless in computing $h(i)$. It is possible to compute h directly, without computing potentials, at least for a large class of stopping times, by solving a simple system of linear equations:

(1.17) THEOREM.   Let T be the time of first visit to a subset A of E, that is, let

$$T = \inf\{n \in \mathbb{N}: X_n \in A\};$$

and let h be defined by (1.16) for some bounded function g and some number $\alpha \in [0, 1]$. Then h satisfies

(1.18)
$$h(i) = \begin{cases} 0 & \text{if } i \in A, \\ g(i) + \alpha Ph(i) & \text{if } i \notin A. \end{cases}$$

Proof. Let $W = \sum_{n=0}^{T-1} \alpha^n g(X_n)$; W is a functional of the chain X, say, $W = f(X_0, X_1, \ldots)$ for some fixed function f. If $X_0(\omega) \in A$ then $T(\omega) = 0$ and hence $W(\omega) = 0$. So $h(i) = 0$ for $i \in A$.

Consider next an ω such that $X_0(\omega) \notin A$. Then $T(\omega) = 1 + \hat{T}(\omega)$ where $\hat{T} = \inf\{n \in \mathbb{N}: X_{1+n} \in A\}$. Therefore we can write $W(\omega) = g(X_0(\omega)) + \alpha \hat{W}(\omega)$ by letting

$$\hat{W} = \sum_{n=1}^{T-1} \alpha^{n-1} g(X_n) = \sum_{n=0}^{\hat{T}-1} \alpha^n g(X_{n+1}).$$

Note that $\hat{W}$ is the same functional of the sequence $(X_1, X_2, \ldots)$ as W is of $(X_0, X_1, \ldots)$; that is, $\hat{W} = f(X_1, X_2, \ldots)$ for the same f for which $W = f(X_0, X_1, \ldots)$. Therefore, by Corollary (5.1.23),

$$E_i[\hat{W} | X_0, X_1] = h(X_1).$$

Hence, for $i \notin A$, using Proposition (2.2.16), we obtain

$$h(i) = E_i[W] = E_i[g(X_0) + \alpha \hat{W}]$$
$$= g(i) + \alpha E_i[h(X_1)] = g(i) + \alpha \sum_j P(i, j)h(j),$$

which is as claimed by (1.18). $\square$

(1.19) EXAMPLE.   Let X be a Markov chain with state space $E = \{a, b, c, d\}$ and transition matrix

$$P = \begin{bmatrix} 0.3 & 0.7 & 0 & 0 \\ 0.4 & 0.3 & 0.3 & 0 \\ 0 & 0 & 0.5 & 0.5 \\ 0 & 0 & 1 & 0 \end{bmatrix}.$$

We are interested in computing

$$h(a) = E_a[\sum_{n=0}^{T-1} g(X_n)]$$

for $g = (5, 3, 1, 4)$ and $T = \inf\{n: X_n = d\}$.

It is easy to see that $Rg(i) = +\infty$ for each state i; therefore, Theorem (1.14) yields $+\infty = h(a) + \infty$, which is quite useless in obtaining $h(a)$. Using Theorem (1.17), we obtain

$$h(a) = 5 + 0.3h(a) + 0.7h(b),$$
$$h(b) = 3 + 0.4h(a) + 0.3h(b) + 0.3h(c),$$
$$h(c) = 1 + 0.5h(c) + 0.5h(d) = 1 + 0.5h(c).$$

Solving these, we get $h(c) = 2$, $0.7h(b) = 3.6 + 0.4h(a)$, and the desired number $h(a) = 8.6/(1 - 0.3 - 0.4) = \frac{86}{3}$.   □

2. Excessive Functions

Let X be a Markov chain with state space E and transition matrix P. Let f be a finite-valued function defined on E, and let α be a number in $[0, 1]$.

(2.1) DEFINITION. The function f is said to be α-*excessive* provided that $f \geq 0$ and $f \geq \alpha Pf$. If f is 1-excessive, then it is simply called *excessive*.   □

We may think of $f(j)$ as the reward promised for being in state j. Then $\alpha^n f(X_n)$ is the present worth of the reward promised for time n, and its expected value is $E_i[\alpha^n f(X_n)] = \alpha^n P^n f(i)$ if the initial state is i. Supposing that f is α-excessive, we have $f \geq \alpha Pf$, which implies $(\alpha P)f \geq (\alpha P)(\alpha Pf) = \alpha^2 P^2 f$; repeating this, we have $f \geq \alpha Pf \geq \alpha^2 P^2 f \geq \cdots \geq \alpha^n P^n f$. Hence, if f is α-excessive,

$$(2.2) \qquad f(i) \geq \alpha^n P^n f(i) = E_i[\alpha^n f(X_n)]$$

for any $i \in E$. In other words, no matter what the initial state i is, the reward we can get at time 0 is greater than or equal to the expected worth of any future reward.

The function 1 is excessive since $P1 = 1$. If a function f is α-excessive and $c > 0$ is a constant, then the function cf is also α-excessive. If f is α-excessive and if $\beta < \alpha$, then $f \geq \alpha Pf \geq \beta Pf$, which shows that f is also β-excessive. If f and g are two α-excessive functions, then $f + g \geq \alpha Pf + \alpha Pg = \alpha P(f + g)$, which implies that $f + g$ is also α-excessive.

Throughout the following, α is a number in $[0, 1]$ unless specified otherwise. In the next proposition, by $f \wedge g$ we mean the function whose i-entry is the smaller of $f(i)$ and $g(i)$.

(2.3) PROPOSITION. If f and g are α-excessive, then so is $f \wedge g$.

Proof. Since $f \wedge g \leq f$, $\alpha P(f \wedge g) \leq \alpha Pf$. Similarly, $\alpha P(f \wedge g) \leq \alpha Pg$.

Since f, g are α-excessive, $f \geq \alpha Pf \geq \alpha P(f \wedge g)$ and $g \geq \alpha Pg \geq \alpha P(f \wedge g)$. Thus, $f \wedge g \geq \alpha P(f \wedge g)$. Since $f, g \geq 0, f \wedge g \geq 0$ also. $\qquad\square$

(2.4) PROPOSITION. If f is the α-potential of a non-negative function and is finite, then f is α-excessive.

Proof. Suppose $g \geq 0$ and $f = R^{\alpha}g$. Then

$$f = g + \alpha Pg + \alpha^2 P^2 g + \cdots \geq \alpha Pg + \alpha^2 P^2 g + \cdots$$
$$= \alpha P(g + \alpha Pg + \cdots) = \alpha Pf;$$

and trivially, $f \geq 0$. $\qquad\square$

The preceding proposition provides a large number of α-excessive functions. In particular, for any $\alpha < 1$ and $j \in E$, the function f defined by $f(i) = R^{\alpha}(i, j)$ is an α-excessive function. Of course, any non-negative linear combination of such functions is also α-excessive.

The following theorem, called the *Riesz decomposition theorem*, shows that an excessive function can be written as the sum of a potential and a *harmonic* function (h is *harmonic* if $h \geq 0$, $h = Ph$).

(2.5) THEOREM. Any excessive function f has the form

$$(2.6) \qquad\qquad\qquad f = Rg + h,$$

where $g \geq 0$, $h \geq 0$, and h satisfies

$$(2.7) \qquad\qquad\qquad h = Ph.$$

Proof. Let f be excessive and put $g = f - Pf$. Since $f \geq Pf, g \geq 0$. We can write

$$f = g + Pf,$$

which gives, by replacing f on the right by $g + Pf$, $f = g + Pg + P^2 f$; iterating in this manner, we obtain

$$f = (g + Pg + \cdots + P^n g) + P^{n+1}f.$$

In the limit, as $n \to \infty$, the first term on the right converges to Rg. For the second term, noting that

$$f \geq Pf \geq P^2 f \geq \cdots,$$

we see that

$$h = \lim_{n} P^n f$$

exists, that $h \geq 0$, and that by the bounded convergence theorem

$$Ph = P(\lim_n P^n f) = \lim_n P^{n+1} f = h.$$

Hence, $f = Rg + h$, with g and h as claimed.   □

The next result is very useful in applications.

(2.8) PROPOSITION. Any excessive function f is an α-potential for every $\alpha < 1$.

Proof. Since f is excessive, $f \geq Pf \geq \alpha Pf$ for $\alpha < 1$. Then defining $g = f - \alpha Pf$, we see that $g \geq 0$ and that we can write

$$f = g + \alpha Pf.$$

Repeatedly replacing f on the right by $g + \alpha Pf$, we obtain

$$f = (g + \alpha Pg + \cdots + \alpha^n P^n g) + \alpha^{n+1} P^{n+1} f.$$

Since $f \geq Pf \geq \cdots \geq P^n f$, $0 \leq \alpha^{n+1} P^{n+1} f \leq \alpha^{n+1} f$. In the limit as $n \to \infty$, $\alpha^{n+1} f \to 0$, since $\alpha < 1$. Thus, as $n \to \infty$, $\alpha^{n+1} P^{n+1} f \to 0$, and we have

$$f = g + \alpha Pg + \cdots = R^\alpha g.$$ □

The same proof goes through for the following generalization:

(2.9) PROPOSITION. Suppose f is α-excessive. Then for any $\beta < \alpha$, f is the β-potential of some non-negative function g.

We had noted earlier that if a reward function f is excessive, then the expected worth of a reward at time n is less than the value of a reward at time 0 (see (2.2) for this). Here, the fixed time n can be replaced by certain random times. Again, recall that a random variable T is a stopping time if and only if knowledge of $X_0, \ldots, X_n$ completely determines whether $T \leq n$ or not ($n \in \mathbb{N}$). Also recall the conventions preceding (5.1.24).

(2.10) THEOREM. If f is α-excessive, then

$$f(i) \geq E_i[\alpha^T f(X_T)]$$

for all $i \in E$ and all stopping times T.

Proof. By Proposition (2.9), for any $\beta < \alpha$, we can write $f = R^\beta g$ for some $g \geq 0$. Thus, for any stopping time T, Theorem (1.14) yields

$$f(i) = E_i\left[\sum_{n=0}^{T-1} \beta^n g(X_n)\right] + E_i[\beta^T f(X_T)] \geq E_i[\beta^T f(X_T)].$$

Now taking limits as $\beta \uparrow \alpha$, we obtain the desired result. $\square$

Next suppose T and S are two stopping times such that $T(\omega) \leq S(\omega)$ for all $\omega \in \Omega$ (for example, T and S may be the respective times of the first and second visits to a set of states). Since the passage of time reduces the expected worth of a reward, when the reward function f is excessive, we would expect the expected value of a reward at time T to be more than that of a reward at time S. This is indeed the case, as the next theorem shows. (Note also that when $T = 0$ identically, this yields the preceding result (2.10).)

(2.11) THEOREM. If f is α-excessive, then for all i,

$$E_i[\alpha^T f(X_T)] \geq E_i[\alpha^S f(X_S)]$$

for any two stopping times T and S with $T \leq S$.

Proof. For $\beta < \alpha$ we can write $f = R^\beta g$ for some function $g \geq 0$ by using Proposition (2.9). Then by Theorem (1.14),

$$(2.12) \qquad E_i[\beta^T f(X_T)] = f(i) - E_i\left[\sum_{n=0}^{T-1} \beta^n g(X_n)\right]$$

$$\geq f(i) - E_i\left[\sum_{n=0}^{S-1} \beta^n g(X_n)\right] = E_i[\beta^S f(X_S)],$$

since

$$\sum_{n=0}^{T-1} \beta^n g(X_n) \leq \sum_{n=0}^{S-1} \beta^n g(X_n)$$

by the hypothesis that $T \leq S$. Now taking limits on both sides of (2.12), as $\beta \uparrow \alpha$, we obtain the desired result. $\square$

In the next section we will deal with computing excessive functions satisfying certain added conditions. Such computations are reduced to simple inspections in the case of irreducible recurrent Markov chains. The following is the cause of this simplification.

(2.13) PROPOSITION. Suppose X is irreducible recurrent. Then any excessive function f is constant.

Proof. By Theorem (2.10), $f(i) \geq E_i[f(X_T)]$ for any stopping time T. Let

T be the time of first visit to state j. Since X is irreducible recurrent, T is finite with probability one for any initial state i. By the definition of T, the process X is in state j at time T if T is finite. Hence, $X_T = j$ with probability one, and $E_i[f(X_T)] = f(j)$. We have thus shown that

$$f(i) \geq f(j).$$

Since this holds for all i and j, we must have $f(i) = c$ for some constant $c \geq 0$ for all $i \in E$. □

The preceding proposition applies to each irreducible recurrent class of a Markov chain separately.

(2.14) PROPOSITION.   For any excessive function f, $f(i) = f(j)$ for all i and j in the same irreducible recurrent class.

Proof is immediate from that of (2.13) once we note that if i and j are in the same irreducible recurrent class, then they can be reached from each other with probability one. □

3. Optimal Stopping

Let X be a Markov chain with state space E and transition matrix P, and let f be a bounded function defined on E.

The matrix P and the function f are known to us, and we are allowed to observe the process X as long as we desire. We may stop whenever we want; if at the time of stopping the process is in state j, we receive a payoff of $f(j)$ units, and the game is over. If we never stop, then the payoff is zero. We are interested in optimizing the payoff.

(3.1) EXAMPLE.   Consider the gambling machine pictured in Figure 7.3.1 When the specified number of coins are dropped, the machine starts and one of the five characters lights up; suppose it is $X_0(\omega) = J$. By pushing the STOP button, the player can get $f(J) = 3$ coins if he wants to. If he pushes the CONTINUE button, then either Q or K will light up with the probabilities indicated on the proper arrows. Noting that the payoff in either case is higher than at J, the player pushes the CONTINUE button, and character Q lights up (that is, $X_1(\omega) = Q$). Here he faces a harder decision: by stopping there he is sure to get $f(Q) = 4$ coins; whereas, if he continues, he has a chance to make 6 coins as well as running the risk of getting nothing. Since, after Q, A and 2 light up with probabilities $\frac{3}{4}$ and $\frac{1}{4}$ respectively, if he continues, the expected payoff is $(\frac{3}{4})f(A) + (\frac{1}{4})f(2) = 4.5$ coins. This amount being higher than $f(Q) = 4$, he chooses not to stop and pushes the CONTINUE button. It

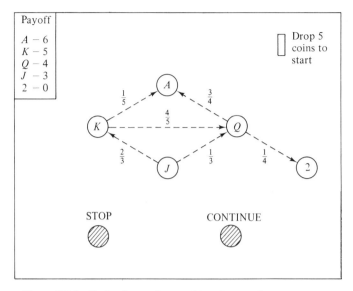

Figure 7.3.1 Optimal stopping machine. It pays the amount corresponding to the character which was lit at the time the STOP button was pushed.

turns out that $X_2(\omega) = A$, and he receives $f(A) = 6$ coins, and the machine turns off.

We will show shortly that the optimal strategy is to continue if the lit character is J or Q and to stop if the lit character is K, A, or 2 (in the case of the latter because the machine turns off by itself anyway). Thus, depending on chance, a player using the optimal strategy will end up with either 5, or 6, or 0 coins. For example, if the realization ω happens to be such that $X_0(\omega) = K$, then the optimal stopping time is $T(\omega) = 0$, and the payoff is $f(K) = 5$. If $X_0(\omega) = Q$, $X_1(\omega) = 2$, then the optimal stopping time is $T(\omega) = 1$, and the payoff is $f(2) = 0$. If $X_0(\omega) = Q$ and $X_1(\omega) = A$, the optimal stopping time is $T(\omega) = 1$ again, and the payoff is $f(A) = 6$. It is worth noting that the term "optimal," which we will make precise, does not necessarily yield the maximum payoff for every realization. □

It is clear that the time of stopping is a random variable T whose value is determined by the path $X_0, X_1, \ldots$ of the chain X. At each time n, the decision to stop or to continue must be made on the basis of the knowledge of the path available at that time. Hence, for any $\omega \in \Omega$, whether $T(\omega) = n$ or not must be determined completely by the path $X_0(\omega), X_1(\omega), \ldots, X_n(\omega)$ observed until that time n. Since this is true for every n, T is a stopping time.

At the time T of stopping the process is in state X_T, and therefore the payoff is $f(X_T)$. (Recall the convention introduced in the paragraph preceding (5.1.24): X_∞ is defined to be Δ, where Δ is a point not in E; any function f defined on E is automatically extended to $E \cup \{\Delta\}$ by setting $f(\Delta) = 0$. Hence, if $T(\omega) = \infty$, then the corresponding payoff is $f(X_{T(\omega)}(\omega)) = f(\Delta) = 0$.) The choice of a stopping time T corresponds to choosing a strategy. For different stopping times T, the expected payoffs $E_i[f(X_T)]$ will differ. We are interested in choosing a stopping time T for which this is maximized. We now state our problem:

(3.2)　PROBLEM.　(a) Compute the function

(3.3)
$$v(i) = \sup_T E_i[f(X_T)], \quad i \in E,$$

where the supremum is over all possible stopping times T.

(b) Find a stopping time T_0 such that

(3.4)
$$v(i) = E_i[f(X_{T_0})], \quad i \in E. \qquad \square$$

The function f is called the payoff function, and the function v defined by (3.3) is called the value of the game. A stopping time T_0 satisfying (3.4) is called an optimal stopping time.

(3.5)　EXAMPLE.　Returning to (3.1), we now formulate the problem. There, X is a Markov chain with state space $E = \{A, K, Q, J, 2\}$ and transition matrix

$$P = \begin{bmatrix} 1 & 0 & 0 & 0 & 0 \\ \frac{1}{5} & 0 & \frac{4}{5} & 0 & 0 \\ \frac{3}{4} & 0 & 0 & 0 & \frac{1}{4} \\ 0 & \frac{2}{3} & \frac{1}{3} & 0 & 0 \\ 0 & 0 & 0 & 0 & 1 \end{bmatrix}.$$

The payoff function is $f = (6, 5, 4, 3, 0)$. States A and 2 are absorbing; starting from either, the chain remains in the initial state. Therefore, $v(A) = 6$, $v(2) = 0$. Starting from the state Q, if we stop immediately the payoff is $f(Q) = 4$; if not the next step leads to either A or 2. Thus, not stopping gives an expected payoff of $(\frac{3}{4})6 = 4.5$. This being the best possible, $v(Q) = 4.5$. Starting from K, stopping immediately yields a payoff $f(K) = 5$; if not, the chain might move to A (with probability $\frac{1}{5}$ and a payoff $f(A) = 6$ there), or it might move to Q (with probability $\frac{4}{5}$, and starting from Q, our analysis shows an expected payoff $v(Q) = 4.5$). Thus, by not stopping at K, the optimal expected payoff becomes $\frac{1}{5}v(A) + \frac{4}{5}v(Q) = 4.8$; whereas stopping at K gave $f(K) = 5$. Hence, it is better to stop immediately at K; then the expected payoff is $v(K) = f(K) = 5$. Starting at J, it is silly to stop. By waiting one step more, either K lights up and we can make $v(K) = 5$, or Q lights up and the expected payoff starting there is $v(Q) =: 4.5$. Hence, starting at J, the optimal strategy is to wait; and then the expected payoff becomes $v(J) = \frac{2}{3}v(K) + \frac{1}{3}v(Q) = 4.833$.

Hence, for this example, the value of the game is given by $v = (6, 5, 4.5, 4.833, 0)$. The optimal strategy is to stop only when the lit character is either A, or K, or 2; in other words, T_0 is the time of first visit to the set of states $\{A, K, 2\}$. Note that this set is also the set of states j for which $f(j) = v(j)$.

□

(3.6) EXAMPLE. Let X be an irreducible recurrent Markov chain, and suppose $f \geq 0$. Let $c = \sup_i f(i)$, and suppose that there is at least one state j for which $f(j) = c$. (If the state space E is finite, c is the maximum value f attains, and there will always be at least one j for which $f(j)$ is equal to c.)

Let T_0 be the time of first visit to that state j such that $f(j) = c$. Since X is irreducible and j is recurrent, for any initial state i, $T_0 < \infty$ and $X_{T_0} = j$ with probability one. Hence, for any $i \in E$, $E_i[f(X_{T_0})] = c$, and since $f \leq c$, this makes the maximum possible for the expectation $E_i[f(X_T)]$ for any stopping time T. Hence,

$$v(i) = c$$

for all $i \in E$, and T_0 is an optimal stopping time.   □

(3.7) EXAMPLE. Let X be irreducible recurrent again, but with infinite state space $E = \{1, 2, 3, \ldots \}$. Suppose the payoff function f is given by $f(i) = 1 - 1/i$, $i \in E$.

For the stopping time T_j of the time of first visit to state j, we have $P_i\{T_j < \infty, X_{T_j} = j\} = 1$; hence

$$E_i[f(X_{T_j})] = f(j) = 1 - \frac{1}{j}.$$

Hence, since the supremum over all stopping times T must be greater than or equal to the supremum over the T_j alone,

$$v(i) = \sup_T E_i[f(X_T)] \geq \sup_{T_j} E_i[f(X_{T_j})]$$

$$= \sup_{j \in E} \left(1 - \frac{1}{j}\right) = 1,$$

for all i; and since $f \leq 1$, it follows trivially that $v \leq 1$. Hence, $v = 1$; but in this case there is no optimal stopping time. For any stopping time T, either $X_T = \Delta$ and $f(X_T) = 0$ or $X_T \in E$ and $f(X_T) = 1 - 1/X_T < 1$; thus $f(X_T) < 1$ and $E_i[f(X_T)] < 1 = v(i)$.

The difficulty here is caused by the fact that the state space E is infinite. We will later show that although there is no optimal stopping time T_0 in the sense of (3.4), for any $\varepsilon > 0$ we can find a stopping time T_ε satisfying

$$E_i[f(X_{T_\varepsilon})] \geq v(i) - \varepsilon.$$

Indeed, in this example, the time T_ε of first visit to the set of states $A = \{j: 1 - 1/j \geq 1 - \varepsilon\}$ is such a stopping time. $\square$

In the examples above we have been able to exploit the special features of the Markov chains involved to figure out the value v of the game and the optimal strategy if there is one. Examples (3.5) and (3.6) show that at least in the finite state space case, the optimal stopping time is the time of first visit to the set of states j for which $f(j) = v(j)$. This turns out to be true in general.

The following characterizes the value v of the game defined by (3.3).

(3.8) THEOREM. The value function v is the minimal excessive function greater than or equal to the payoff function f.

(3.9) COMPUTATIONAL NOTE. This theorem shows that in the case when the state space E is finite, the value of the game can be found by using the methods of linear programming. Since v is excessive, $v \geq 0$, $v \geq Pv$; since v majorizes f, $v \geq f$; and v is the minimal such function, so $\sum_i v(i)$ must be minimized. The resulting linear programming problem is:

$$\text{minimize} \sum_i v(i)$$

subject to

$$v(i) \geq \sum_j P(i, j)v(j),$$
$$v(i) \geq f(i),$$
$$v(i) \geq 0$$

for all $i \in E$.

Proof of Theorem (3.8). First we show that v is excessive. Since f is bounded, say by c, (3.3) shows that v is bounded by c as well; so v is finite-valued. Clearly, $v \geq 0$; to show that $v \geq Pv$, let $\varepsilon > 0$ be arbitrary. By the definition of $v(j)$, there exists a stopping time T_j (possibly depending on ε) such that

$$(3.10) \qquad E_j[f(X_{T_j})] > v(j) - \varepsilon.$$

Of course, T_j is a function of $j, X_1, X_2, \ldots$; say $T_j = t(j, X_1, \ldots)$. Now let

$T = 1 + t(X_1, X_2, \ldots)$; that is, T is the stopping time corresponding to the strategy which waits at $n = 0$, and if $X_1 = j$, then uses the strategy corresponding to T_j thereafter. Then for any $i \in E$, (3.10) gives

$$E_i[f(X_T)] = \sum_{j \in E} P(i, j) E_j[f(X_{T_j})]$$
$$\geq \sum_{j \in E} P(i, j)[v(j) - \varepsilon] = Pv(i) - \varepsilon.$$

Since v is the supremum of $E_i[f(X_T)]$ over all T, this shows that

$$v(i) \geq Pv(i) - \varepsilon, \quad i \in E.$$

Since ε was arbitrary, this means that $v \geq Pv$ and v is an excessive function.

Since $T = 0$ is a possible stopping time, $v(i) = \sup_T E_i[f(X_T)] \geq E_i[f(X_0)] = f(i)$ for all i.

Hence, v is excessive and $v \geq f$; to show that v is the minimal such function, let g be an excessive function and suppose $g \geq f$. Since g is excessive, by Theorem (2.10),

$$(3.11) \qquad\qquad g(i) \geq E_i[g(X_T)], \quad i \in E,$$

for any stopping time T. Since $g \geq f$, $g(X_T) \geq f(X_T)$, and (3.11) implies that

$$g(i) \geq E_i[f(X_T)], \quad i \in E,$$

for all stopping times T; hence, $g \geq v$ also. This completes the proof. $\quad\square$

Note that we have also proved the following not so obvious result.

(3.12) COROLLARY. For any bounded function f, there is a minimal excessive function majorizing it.

Combining Theorem (3.8) with the fact that for an irreducible recurrent Markov chain, all excessive functions are constants (cf. Proposition (2.13)), we obtain the following simple result.

(3.13) COROLLARY. If X is irreducible recurrent, then v is equal to the constant $c = \sup_j f(j)$.

This corollary applies separately to each irreducible recurrent set of states. In other words, using Theorem (3.8) in conjunction with Proposition (2.14), we obtain the following:

(3.14) COROLLARY. If C is an irreducible recurrent set of states, then for all $i \in C$, $v(i)$ is equal to the constant $c = \sup_{j \in C} f(j)$.

This corollary shows that in computing v, the only non-trivial part is the computation of $v(i)$ for transient states i. Before giving some numerical examples, we give the following main result characterizing the optimal stopping time.

(3.15) THEOREM. Suppose the state space E is finite. Then the time T_0 of the first visit to the set

$$(3.16) \qquad A = \{j \in E: v(j) = f(j)\}$$

is an optimal stopping time; that is, for any $\omega \in \Omega$,

$$(3.17) \qquad T_0(\omega) = \inf\{n \geq 0: f(X_n(\omega)) = v(X_n(\omega))\}.$$

To prove this we shall need the following

(3.18) LEMMA. Let T be the time of first visit to a fixed set A of states. If g is an excessive function, then the function h defined by

$$h(i) = E_i[g(X_T)], \quad i \in E,$$

is also excessive.

Proof. Let T be the time of first visit to A, and let S be the time of first visit to A at or after time 1; that is, for any $\omega \in \Omega$, let

$$T(\omega) = \inf\{n \geq 0: X_n(\omega) \in A\},$$
$$S(\omega) = \inf\{n \geq 1: X_n(\omega) \in A\}.$$

If $X_0(\omega) \in A$, then $T(\omega) = 0 < 1 \leq S(\omega)$; if $X_0(\omega) \notin A$, then $T(\omega) = \inf\{n \geq 1: X_n(\omega) \in A\} = S(\omega)$. Hence, for any $\omega \in \Omega$, $T(\omega) \leq S(\omega)$; and it is clear that both T and S are stopping times. Since g is excessive, Theorem (2.11) implies that

$$(3.19) \qquad h(i) = E_i[g(X_T)] \geq E_i[g(X_S)], \quad i \in E.$$

On the other hand, if $T = t(X_0, X_1, \ldots)$ for some function t, then $S = 1 + t(X_1, X_2, \ldots)$; hence by Corollary (5.1.23)

$$E_i[g(X_S) | X_1 = j] = E_j[g(X_T)] = h(j),$$

which implies that

$$E_i[g(X_S)] = \sum_{j \in E} P(i, j) h(j).$$

This together with (3.19) shows that $h \geq Ph$, and since $h \geq 0$ obviously, h is excessive. $\qquad\square$

Proof of Theorem (3.15). To show that T_0 defined by (3.17) is optimal, we need to show that

$$(3.20) \qquad h(i) = E_i[f(X_{T_0})]$$

is equal to $v(i)$ for each i. It is clear that $h \leq v$. To show the reverse inequality, we will first show that h is excessive and that $h \geq f$; then, since v is the minimal excessive function satisfying $v \geq f$ (cf. Theorem (3.8)), $v \leq h$ as well.

If the event $\{T_0 < \infty\}$ occurs, then $X_{T_0} \in A$ and $f(X_{T_0}) = v(X_{T_0})$ by the definition (3.16) of A; otherwise, if $\{T_0 = +\infty\}$ occurs, then $X_{T_0} = X_\infty = \Delta$ and $f(\Delta) = v(\Delta) = 0$. Hence $f(X_{T_0}) = v(X_{T_0})$, and (3.20) becomes

$$(3.21) \qquad h(i) = E_i[v(X_{T_0})], \quad i \in E.$$

The function v is excessive by Theorem (3.8), and by Lemma (3.18), (3.21) implies that h is also excessive.

Next we show that $h \geq f$. For $i \in A$, $P_i\{T_0 = 0\} = 1$, and therefore $h(i) = E_i[f(X_{T_0})] = f(i)$. Suppose for a moment that for some $i \notin A$, $h(i) < f(i)$. Since E is finite, there is a state $j \notin A$ at which the difference $f(i) - h(i)$ is maximized; let $c = f(j) - h(j)$ be this maximum value. Since h is excessive and the sum of two excessive functions is also excessive, the function $h + c$ is excessive. By the way c is picked, $h + c \geq f$. Since v is the minimal excessive function majorizing f (cf. Theorem (3.8)), we must have $v \leq h + c$, which implies that

$$v(j) \leq h(j) + c = h(j) + f(j) - h(j) = f(j).$$

Since $v \geq f$, this is possible only if $v(j) = f(j)$, in which case $j \in A$. This contradicts the assumption that $j \notin A$. Hence, $h(i) \geq f(i)$ for all i. This completes the proof. $\qquad\square$

If the payoff function f is excessive, then $v = f$, and the optimal stopping time is $T_0 = 0$. This is a further confirmation of the intuitive remarks made before about excessive functions.

If one were to draw the functions f and v on the same paper, the graphs of f and v would coincide at the points of A, and the graph of v would lie above that of f; in other words, at the points of A, the function f "supports" the graph of v. This picture suggests that the set $A = \{f = v\}$ be called the *support set*.

The following sometimes simplifies the task of computing v.

(3.22) CRITERION. A state j belongs to the support set A if and only if there exists an excessive function $h \geq f$ with $h(j) = f(j)$.

Proof. If $j \in A$, $v(j) = f(j)$, and v is excessive. In the other direction, let

h be an excessive function such that $h \geq f$ and $h(j) = f(j)$. Then by Theorem (3.8), $h \geq v$ and $v(j) \leq h(j) = f(j)$. Since $v \geq f$, $v(j) = h(j)$ and $j \in A$.

$\square$

(3.23) EXAMPLE.   Let X be the Markov chain of Example (6.1.25), and let the payoff function f be $f = (2, 0, 1, 3, 2, 4, 2.5)$. States 1 and 2 form an irreducible closed set, states 3, 4, and 5 form another, and states 6 and 7 are transient. By Corollary (3.14),

$$v(1) = v(2) = 2, \qquad v(3) = v(4) = v(5) = 3,$$

and by Criterion (3.22), since $h = (4, 4, \ldots, 4)$ is clearly excessive with $h \geq f$ and $h(6) = f(6)$, state 6 is in the support set, and

$$v(6) = f(6) = 4.$$

It remains to compute $v(7)$. From $v \geq Pv$ and $v \geq f$ we get

$$v(7) \geq (0.1 + 0.1)2 + (0.1 + 0 + 0.1)3 + 0.2 \times 4 + 0.4v(7),$$
$$v(7) \geq 2.5$$

respectively. The first inequality gives $v(7) \geq 3$; thus $v(7)$ is the smallest number greater than 3 and 2.5. Hence, the value of the game is given by

$$v = (2, 2, 3, 3, 3, 4, 3);$$

the support set is

$$A = \{f = v\} = \{1, 4, 6\},$$

and the optimal stopping time is

$$T_0 = \inf\{n : X_n \in A\},$$

namely, the time of first visit to the set $A = \{1, 4, 6\}$.

Let $X(\omega)$ denote the sequence of states realized by the Markov chain X corresponding to the outcome $\omega \in \Omega$. For ω_0 with $X(\omega_0) = (7, 5, 3, 4, 3, 5, \ldots)$, the *actual* time of stopping is $T_0(\omega_0) = 3$ and the *actual* payoff obtained is $f(X_3(\omega_0)) = f(4) = 3$. For ω_1 with $X(\omega_1) = (7, 7, 7, 7, 7, 7, 1, 1, 2, 1, \ldots)$ the stopping time is $T_0(\omega_1) = 6$ and the actual payoff is $f(X_6(\omega_1)) = f(1) = 2$ (imagine the poor gambler's disgust for turning down a payoff of 2.5 exactly 6 times to get the payoff of 2 at the end; but that is hindsight: he did have a good chance of getting 4 or 3). For ω_2 with $X(\omega_2) = (4, 3, 5, 3, \ldots)$, the stopping time is $T_0(\omega_2) = 0$, and the actual payoff is $f(4) = 3$.

(3.24) EXAMPLE.   Let X be a Markov chain with state space $E =$

$\{a, b, c, d, e\}$ and transition matrix

$$P = \begin{bmatrix} 1 & 0 & 0 & 0 & 0 \\ 0 & 1 & 0 & 0 & 0 \\ 0 & 0 & 0.2 & 0 & 0.8 \\ 0.3 & 0 & 0.3 & 0.2 & 0.2 \\ 0.2 & 0.4 & 0 & 0.3 & 0.1 \end{bmatrix},$$

and let the payoff function be $f = (5, 12, 7, 6, 8)$.

States a and b are absorbing; states c, d, and e are transient. By Corollary (3.14), $v(a) = 5$, $v(b) = 12$. Now, the equations to be satisfied by $v(c) = x$, $v(d) = y$, $v(e) = z$ become

$$x \geq 0.2x \qquad\qquad + 0.8z, \qquad\qquad x \geq 7;$$
$$y \geq 0.3x + 0.2y + 0.2z + 1.5, \qquad y \geq 6;$$
$$z \geq \qquad\quad 0.3y + 0.1z + 5.8, \qquad z \geq 8.$$

From the first equation we get $0.8x \geq 0.8z$, that is, $x \geq z$. Since $z \geq 8$, it follows that $x \geq z$ and $x \geq 7$ can both be satisfied by taking $x = z$. The remaining equations become

(3.25) $0.8y - 0.5z \geq 1.5;$

(3.26) $-0.3y + 0.9z \geq 5.8;$

(3.27) $y \geq 6; \qquad z \geq 8.$

The points (y, z) satisfying (3.25) are to the right of the line $G = \{(y, z): 0.8y - 0.5z = 1.5\}$, and the points (y, z) satisfying (3.26) are above the line $H = \{(y, z): -0.3y + 0.9z = 5.8\}$. Drawing these as in Figure 7.3.2, we see that $y + z$ is minimized at the intersection of lines G and H, namely, at the point $(y_0, z_0) = (\frac{850}{114}, \frac{1018}{114})$. Hence, the value of the game v is

$$v = \left(5, 12, 8\frac{53}{57}, 7\frac{26}{57}, 8\frac{53}{57}\right).$$

The support set is

$$A = \{a, b\};$$

and the optimal stopping time is the time of absorption to the set A. So the optimal strategy is to hope for absorption into b and bear the consequences. $\square$

(3.28) EXAMPLE. Let X be a Markov chain with state space $E =$

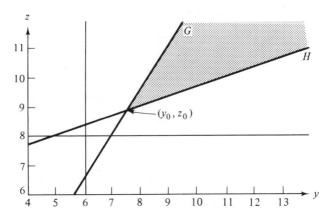

Figure 7.3.2 Points (x, y) satisfying (3.25) are to the right of line G, and those satisfying (3.26) are above line H. The sum $y + z$ is minimized at the point $(y_0, z_0) = (\frac{850}{114}, \frac{1018}{114})$.

$\{0, 1, 2, \ldots, 9\}$ and transition matrix

$$
P = \begin{bmatrix}
1 & 0 & & & & & \\
\frac{1}{2} & 0 & \frac{1}{2} & & & & 0 \\
 & \frac{1}{2} & 0 & \frac{1}{2} & & & \\
 & & \cdot & \cdot & \cdot & & \\
 & & & \cdot & \cdot & \cdot & \\
 & & & & \cdot & \cdot & \cdot \\
0 & & & & \frac{1}{2} & 0 & \frac{1}{2} \\
 & & & & & 0 & 1
\end{bmatrix}
$$

The payoff function is drawn in Figure 7.3.3 (with lines between the actual points for clarity).

States 0 and 9 are absorbing, the remaining states are transient. Clearly, $v(0) = v(9) = 0$. The equations for $v \geq Pv$ now become

$$
v(i) \geq \frac{1}{2}[v(i - 1) + v(i + 1)], \quad i = 1, \ldots, 8.
$$

In other words, the value $v(i)$ at the point i is above the line connecting $v(i - 1)$ and $v(i + 1)$. Thus, the function v is concave. So v is the minimum concave function lying above f. This function is drawn by dashed lines in Figure 7.3.3. One can see that the support A literally supports v. ☐

In the case of infinite state spaces, we have seen in Example (3.7) that it

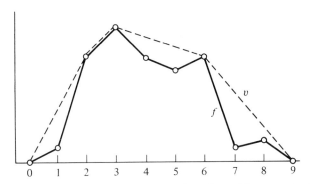

Figure 7.3.3 Function f supports v at the points of the support set $\{0, 2, 3, 6, 9\}$.

is possible for the game not to have an optimal stopping time. However, for any $\varepsilon > 0$, there is a strategy which yields an expected payoff of at least $v(i) - \varepsilon$ for all i. Such a strategy is called ε-*optimal*.

(3.29) THEOREM. Let T_ε be the time of first visit to the set

$$(3.30) \qquad A_\varepsilon = \{j: f(j) \geq v(j) - \varepsilon\}.$$

Then

$$(3.31) \qquad E_i[f(X_{T_\varepsilon})] \geq v(i) - \varepsilon, \quad i \in E.$$

Proof. The function h defined by

$$h(i) = E_i[v(X_{T_\varepsilon})], \quad i \in E,$$

is excessive by Lemma (3.18), since v is excessive and T_ε is the time of first visit to a set.

Let $c = \sup_i[f(i) - h(i)]$, and suppose $c > 0$. Then the function $h + c \geq f$, and is excessive, and therefore by Theorem (3.8) $h + c \geq v$. As $c > 0$, there is a state j for which

$$(3.32) \qquad f(j) - h(j) > 0, \quad f(j) - h(j) > c - \varepsilon.$$

Now $f(j) > h(j) + c - \varepsilon \geq v(j) - \varepsilon$ implies that $j \in A_\varepsilon$, which in turn implies that, starting at j, $T_\varepsilon = 0$ and $X_{T_\varepsilon} = j$ almost surely. Therefore, $h(j) = v(j)$; and since $v \geq f$, $h(j) \geq f(j)$, which is a contradiction to (3.32). So c must not be positive; that is, $h \geq f$. Therefore, by Theorem (3.8), $h \geq v$.

On the event $\{T_\varepsilon < \infty\}$, $X_{T_\varepsilon} \in A_\varepsilon$, and therefore $f(X_{T_\varepsilon}) \geq v(X_{T_\varepsilon}) - \varepsilon$. The same also holds on $\{T_\varepsilon = +\infty\}$, since there, $X_{T_\varepsilon} = \Delta$ and $f(\Delta) = v(\Delta)$

$= 0$. Thus,

$$E_i[f(X_{T_\varepsilon})] \geq E_i[v(X_{T_\varepsilon}) - \varepsilon]$$
$$= h(i) - \varepsilon \geq v(i) - \varepsilon$$

as desired. $\qquad\square$

4.   Games with Discounting and Fees

In applications of the optimal stopping problem, in addition to the payoff at the end, there are costs and penalties for waiting.

(4.1)   EXAMPLE.   A man is selling his house, for which he is asking \$47,000, but he will also consider lower offers. Arrivals of prospective buyers form a Poisson process with rate $\lambda = 2$ per month. As far as the nth customer is concerned, the worth of the house is Y_n, so unless the house is sold before his arrival, he offers \$47,000 or Y_n, whichever is smaller. If the offer is acceptable to the owner, the sale is completed. If not, the owner rejects the offer and the customer leaves never to return. The owner is able to make 15% per year on his investments, and the rate of inflation is about 6% per year.

Supposing that $Y_0, Y_1, \ldots$ are independent and identically distributed random variables with values in $\{35,000, \ldots, 50,000\}$, the successive offers made form a Markov chain X, where $X_n = \min(Y_n, 47,000)$. The problem is an optimal stopping problem with two added features:

(a) Because of inflation, an offer of y by the *next* customer to arrive is actually worth

$$\alpha y = y \int_0^\infty e^{-0.06t/12} \lambda e^{-\lambda t} dt = \frac{2}{2 + 0.005} y.$$

(b) By rejecting an offer of x today, the owner is losing the opportunity to earn (noting that the net earning rate is $15 - 6 = 9\%$)

$$g(x) = x \int_0^\infty e^{0.09t/12} \lambda e^{-\lambda t} dt - x = \frac{2x}{2 - 0.0075} - x = \frac{0.0075x}{2 - 0.0075}$$

between now and the time of arrival of the next customer.

Taking these two factors into account, a present offer x must be compared with the expected discounted value of the next offer less the lost earnings due to rejection. $\qquad\square$

First we take up the effect of discounting. Let X be a Markov chain with state space E and transition matrix P, and let f be a bounded function defined on E. Suppose the discount factor is $\alpha \in [0, 1]$. Then if T is the time of stopping, the payoff is $f(X_T)$, and the present worth of this payoff is $\alpha^T f(X_T)$. We are interested in maximizing the expected value of this by choosing the stopping time T properly. Let

(4.2) $v(i) = \sup_T E_i[\alpha^T f(X_T)], \quad i \in E,$

where the supremum is taken over all possible stopping times of the Markov chain X. As before, v is called the value of the game. The next theorem shows that v is the minimal α-excessive function majorizing f. The proof is exactly the same as that of Theorem (3.8) and can be obtained from the latter by replacing "excessive" by "α-excessive" and "P" by "αP" throughout.

(4.3) THEOREM.   The value v of the game is the minimal α-excessive function which is greater than or equal to f.

Computationally, if E is finite, this means that v is the solution of the linear programming problem

$$\text{minimize } \sum_i v(i)$$

subject to

$$v(i) \geq \alpha \sum_j P(i, j)v(j), \quad i \in E,$$

$$v(i) \geq f(i), \qquad\qquad i \in E,$$

$$v(i) \geq 0, \qquad\qquad i \in E.$$

A note of warning may be in order here. When $\alpha < 1$, we no longer have Corollary (3.14) to aid in the computation of v. In general, an α-excessive function of an irreducible recurrent chain is not constant.

Just as in the undiscounted case, the optimal stopping time is the time of first entrance to the support set $A = \{f = v\}$. We will omit the proofs of the next two theorems, as they can be obtained easily from the proofs of their respective analogs.

(4.4) THEOREM.   If E is finite, the optimal stopping time is the time of first entrance to the set $A = \{j: f(j) = v(j)\}$; in other words, if $T = \inf\{n \geq 0: X_n \in A\}$, then

$$E_i[\alpha^T f(X_T)] = v(i), \quad i \in E.$$

(4.5) THEOREM.   For any $\varepsilon > 0$ let $A_\varepsilon = \{j: f(j) \geq v(j) - \varepsilon\}$. Then the time of first entrance to the set A_ε is ε-optimal. In other words, if $T = \inf\{n: X_n \in A_\varepsilon\}$, then

$$E_i[\alpha^T f(X_T)] \geq v(i) - \varepsilon, \quad i \in E.$$

(4.6) EXAMPLE. Let X be a Markov chain with state space $E = \{a, b, c\}$ and transition matrix

$$P = \begin{bmatrix} 0.2 & 0.8 & 0 \\ 0.6 & 0.4 & 0 \\ 0.4 & 0.2 & 0.4 \end{bmatrix}.$$

Suppose $f = (6, 5, 3)$ and $\alpha = 0.75$. In open form, $v = (x, y, z)$ must satisfy

$$x \geq 0.15x + 0.60y, \qquad x \geq 6,$$

(4.7) $$y \geq 0.45x + 0.30y, \qquad y \geq 5,$$

(4.8) $$z \geq 0.30x + 0.15y + 0.30z, \quad z \geq 3.$$

Criterion (3.22) applies to give $x = 6$, since $(6, 6, 6)$ is α-excessive. Then (4.7) implies that $y = 5$, and (4.8) implies that $z = 3.643$. Hence,

$$v = (6, 5, 3.643),$$

and the optimal stopping time is the time of first entrance to the set

$$A = \{a, b\}.$$

Note the effect of discounting in this problem. If there were no discounting, the support set would have been simply $\{a\}$. Because of discounting, α is small enough to make it optimal to stop at b as well. ☐

Finally, we consider the case where in addition to discounting, there is a continuation cost to be paid before the next state is revealed. At each time n, if the game has not stopped at or before n, then a fee of $g(X_n)$ is to be paid. Here g is a bounded non-negative function on E. Hence, the cost of playing the game is, when discounted,

(4.9) $$\sum_{n=0}^{T-1} \alpha^n g(X_n),$$

and the quantity to be maximized is the expected value of the discounted payoff diminished further by this amount. We define, for $\alpha \in [0, 1)$, $f \geq 0$, and $g \geq 0$,

(4.10) $$v(i) = \sup_T E_i \left[\alpha^T f(X_T) - \sum_{n=0}^{T-1} \alpha^n g(X_n) \right]$$

to be the maximum value of the discounted net gain. Now our problem is to compute this function v and find an optimal stopping time.

We will first reduce this seemingly more complex problem to the case with no fees. Then, using Theorems (4.3), (4.4), and (4.5) above, we will obtain both computational formulas for v and a characterization of the optimal time to stop.

By Theorem (1.14), for any stopping time T,

$$E_i\left[\sum_{n=0}^{T-1} \alpha^n g(X_n)\right] = R^\alpha g(i) - E_i[\alpha^T R^\alpha g(X_T)],$$

where $R^\alpha g$ is the α-potential of g. Hence, if we put

(4.11) $\hat{f} = f + R^\alpha g, \quad \hat{v} = v + R^\alpha g,$

then we can write from (4.10) that

(4.12) $\hat{v}(i) = \sup_T E_i[\alpha^T \hat{f}(X_T)].$

It is clear from this that $\hat{v}$ is the value of a game with no fees and with payoff function $\hat{f}$, and that the optimal stopping time for the game with payoff function $\hat{f}$ and no fees is equal to the optimal stopping time with payoff function f and fee schedule g.

By Theorem (4.3), $\hat{v}$ is the minimal α-excessive function majorizing $\hat{f}$; that is, $\hat{v}$ is the minimal solution of

(4.13) $\hat{v} \geq \alpha P\hat{v}, \quad \hat{v} \geq \hat{f}, \quad \hat{v} \geq 0.$

In view of the definition of $\hat{v}$,

$$\alpha P\hat{v} = \alpha Pv + \alpha PR^\alpha g = \alpha Pv + R^\alpha g - g$$

which shows that $\hat{v} \geq \alpha P\hat{v}$ if and only if

(4.14) $v \geq -g + \alpha Pv.$

It is clear from (4.11) that $\hat{v} \geq \hat{f}$ if and only if $v \geq f$, and the inequality $\hat{v} \geq 0$ implies that $v \geq -R^\alpha g$. (This last relation $v \geq -R^\alpha g$ always holds if f and g are non-negative.) We have thus proved the following theorems:

(4.15) THEOREM. The function v defined by (4.10) is the minimal solution of

$$v(i) \geq -g(i) + \alpha \sum_j P(i,j)v(j), \quad i \in E,$$
$$v(i) \geq f(i), \qquad\qquad\qquad i \in E.$$

(4.16) THEOREM. If E is finite, then the optimal stopping time is the time of first entrance to the set

$$A = \{j \in E: f(j) = v(j)\}.$$

(4.17) THEOREM. If T is the time of first entrance to the set

$$A_\varepsilon = \{j \in E : f(j) \geq v(j) - \varepsilon\},$$

then

$$E_i\left[\alpha^T f(X_T) - \sum_{n=0}^{T-1} \alpha^n g(X_n)\right] \geq v(i) - \varepsilon, \quad i \in E.$$

(4.18) EXAMPLE. Let X be a Markov chain with state space $E = \{1, 2, 3, 4, 5\}$ and transition matrix

$$P = \begin{bmatrix} 0 & 0 & 0 & 1 & 0 \\ 0 & 0 & 0 & \frac{1}{2} & \frac{1}{2} \\ 0 & \frac{2}{3} & 0 & 0 & \frac{1}{3} \\ \frac{1}{3} & 0 & 0 & \frac{2}{3} & 0 \\ 0 & 0 & 0 & 1 & 0 \end{bmatrix}.$$

Suppose the fee schedule is $g = (0, 1, 1, 2, 0)$, the payoff function $f = (17, 5, 2, 4, 3)$, and the discounting factor $\alpha = \frac{3}{4}$.

The inequalities to be considered are

$$v(1) \geq \quad (\tfrac{3}{4})v(4), \qquad\qquad v(1) \geq 17;$$
$$v(4) \geq -2 + (\tfrac{3}{4})[v(1)/3 + 2v(4)/3], \quad v(4) \geq 4;$$
$$v(5) \geq \quad (\tfrac{3}{4})v(4), \qquad\qquad v(5) \geq 3;$$
$$v(2) \geq -1 + (\tfrac{3}{4})[v(4)/2 + v(5)/2], \quad v(2) \geq 5;$$
$$v(3) \geq -1 + (\tfrac{3}{4})[2v(2)/3 + v(5)/3], \quad v(3) \geq 2.$$

It is obvious that $v(1) = 17$; putting this in the second line, we see that $v(4) = \frac{9}{2}$; this put in the third line gives $v(5) = \frac{27}{8}$; the fourth line now gives $v(2) = 5$; and finally, the last line implies that $v(3) = \frac{75}{32}$. So

$$v = (17, 5, \tfrac{75}{32}, \tfrac{9}{2}, \tfrac{27}{8}),$$

and the optimal strategy is to stop at the time of first entrance to the set $A = \{1, 2\}$.

For example, if the outcome ω turns out to be such that $X_0(\omega) = 5$, $X_1(\omega) = 4$, $X_2(\omega) = 4$, $X_3(\omega) = 1, \ldots$, then this strategy would stop the game at time $T(\omega) = 3$. The fees paid are $g(5) = 0$ at $n = 0$, and $g(4) = 2$ at times $n = 1, 2$. The net gain is, therefore,

$$\alpha^3 f(1) - g(5) - \alpha g(4) - \alpha^2 g(4) = 291/64 \simeq 4.2.$$

If the outcome ω turned out to make $X_0(\omega) = 5$, $X_1(\omega) = 4$, $X_2(\omega) = 4$,

$\ldots$, $X_8(\omega) = 4$, $X_9(\omega) = 1, \ldots$, then the strategy would yield an actual discounted net gain of

$$\alpha^9 f(1) - g(5) - \alpha g(4) - \cdots - \alpha^8 g(4)$$

$$= \alpha^9 f(1) - \left(\frac{1 - \alpha^9}{1 - \alpha} - 1\right) g(4)$$

$$= 17\left(\frac{3}{4}\right)^9 - 6\left[1 - \left(\frac{3}{4}\right)^8\right] \cong -4.2.$$

Knowing that the player had the opportunity to receive $f(5) = 3$, or $\alpha f(4) - g(5) = 3, \ldots$, we may think him foolish, but $\ldots$. $\square$

There is a slightly more complex version of the last problem that is worth mentioning. In some applications, cost of continuation is likely to depend not only on the present state but also on the next. Suppose $\hat{g}$ is a bounded function on $E \times E$, and suppose that a fee $\hat{g}(i, j)$ is collected after every transition from i to j. Hence, if the stopping time $T(\omega) = 5$, the fees collected are $\hat{g}(X_0(\omega), X_1(\omega))$ at time $n = 1$, $\hat{g}(X_1(\omega), X_2(\omega))$ at $n = 2, \ldots$, $\hat{g}(X_4(\omega), X_5(\omega))$ at $n = 5$. Then the problem is to optimize

(4.19) $$E_i\left[\alpha^T f(X_T) - \sum_{n=0}^{T-1} \alpha^{n+1} \hat{g}(X_n, X_{n+1})\right].$$

Noting that

$$E_i[\alpha \hat{g}(X_n, X_{n+1})] = \alpha \sum_j \sum_k P^n(i, j) P(j, k) \hat{g}(j, k)$$

$$= \sum_j P^n(i, j) g(j) = P^n g(i)$$

with

(4.20) $$g(j) = \alpha \sum_{k \in E} P(j, k) \hat{g}(j, k),$$

we have

$$E_i\left[\sum_{n=0}^{\infty} \alpha^{n+1} \hat{g}(X_n, X_{n+1})\right] = R^\alpha g(i),$$

and using Proposition (5.1.25),

(4.21) $$E_i\left[\sum_{n=0}^{T-1} \alpha^{n+1} \hat{g}(X_n, X_{n+1})\right] = R^\alpha g(i) - E_i[\alpha^T R^\alpha g(X_T)]$$

$$= E_i\left[\sum_{n=0}^{T-1} \alpha^n g(X_n)\right].$$

Putting (4.21) into (4.19), we see that our problem becomes the optimization of

$$(4.22) \qquad E_i\left[\alpha^T f(X_T) - \sum_{n=0}^{T-1} \alpha^n g(X_n)\right],$$

where g is given by (4.20). But this is precisely the problem whose solution is described by Theorems (4.15), (4.16), and (4.17).

5. Exercises

(5.1) Compute the potential of g for the Markov chain of
  (a) Exercise (6.8.1a) for $g = (1, 3, 0)$.
  (b) Exercise (6.8.1b) for $g = (0, 0, 1, -2)$.
  (c) Exercise (6.8.1c) for $g = (1, 3, 1, 1, 2)$.
  (d) Exercise (6.8.1d) for $g = (0, 0, 1, -1, 3)$.

(5.2) Compute the α-potential of g for the Markov chain of
  (a) Exercise (6.8.1a) for $\alpha = 0.8$ and $g = (1, 3, 0)$.
  (b) Exercise (6.8.1b) for $\alpha = 0.75$ and $g = (1, -1, 1, -2)$.
  (c) Exercise (6.8.1c) for $\alpha = 0.6$ and $g = (1, 2, 0, 0, 0)$.

(5.3) Consider the Markov chain of Example (5.2.5), and let $g = (0, 5, 15)$.
  (a) Compute the α-potential of g for arbitrary $\alpha \leq 1$.
  (b) Let T be the time of first visit to state 3. Compute

$$E_2\left[\sum_{n=0}^{T-1} (0.8)^n g(X_n)\right].$$

(5.4) Consider a Markov chain X with state space $E = \{a, b, c\}$ and transition matrix P given in Exercise (6.8.1a), and let $g = (1, 3, 4)$.
  (a) Compute $E_a[\sum_{n=0}^{T-1} g(X_n)]$ for the stopping time T of the time of first visit to state c.
  (b) Note that the result is the same as $R\hat{g}(a)$ as computed in (5.1) above for $\hat{g} = (1, 3, 0)$. Explain the reason for this, and use it to compute $E_a[\sum_{n=0}^{T-1} g(X_n)]$ for the stopping time T of the time of first visit to the set $\{b, c\}$.

(5.5) *Generalization.* Let X be a Markov chain with state space E and transition matrix P, and let T be the time of first visit to a subset A of E. For each $n \in \mathbb{N}$ and $\omega \in \Omega$ define

$$Y_n(\omega) = \begin{cases} X_n(\omega) & \text{if } n < T(\omega), \\ \Delta & \text{if } n \geq T(\omega), \end{cases}$$

where Δ is a point not in E.

  (a) Show that $Y = \{Y_n; n \in \mathbb{N}\}$ is a Markov chain with state space $A^c \cup \{\Delta\}$.
  (b) Compute the transition matrix $\hat{P}$ of Y.
  (c) Show that for any function $g \geq 0$ defined on E, we have (note that

$g(\Delta) = 0$, as always)

$$E_i\left[\sum_{n=0}^{T-1} \alpha^n g(X_n)\right] = E_i\left[\sum_{n=0}^{\infty} \alpha^n g(Y_n)\right]$$

for every $i \in A^c$ and $\alpha \in [0, 1]$.

(5.6) Consider the Markov chain Y of Exercise (5.4.10); in other words, Y_n is the age of the machine used at time n where every failure is immediately replaced by a new one. Suppose the probabilities p_j are as given in Example (1.10) so that, in the notation of (5.4.10) we have $q_j = 0.6$ for all j.

    (a)   Show that for any $\alpha \in [0, 1]$ and $i = 0, 1, \ldots,$

$$R^\alpha(i, j) = \begin{cases} \dfrac{0.4\alpha}{1-\alpha}(0.6\alpha)^j & \text{if } j < i, \\[2mm] \dfrac{0.4\alpha}{1-\alpha}(0.6\alpha)^j + (0.6\alpha)^{j-i} & \text{if } j \geq i. \end{cases}$$

    (b)   Suppose the cost of (buying and) maintaining a machine of age j is given by $g(j)$, where $g = (3000, 200, 200, 500, 600, 400, 200, 100, 100, 100, \ldots)$, and suppose the discount factor is $\alpha = 0.8$. If the age of the machine initially in use is $X_0 = 3$, compute the expected value of the total discounted cost of maintenance and replacement.

(5.7) *Continuation.* In the problem of Exercise (5.6) let T be the time of the fifth failure.

    (a)   Find the expected value of the total discounted cost before T; in other words, compute

$$E_3\left[\sum_{n=0}^{T-1} \alpha^n g(Y_n)\right]$$

for $\alpha = 0.8$ and g as given in (5.6).

    (b)   Find the expected value of the undiscounted cost for the same period, namely

$$E_3\left[\sum_{n=0}^{T-1} g(Y_n)\right],$$

by first computing the discounted value for arbitrary α and then letting $\alpha \longrightarrow 1$.

(5.8) Show that $f = (2, 5, 2, 4, 2)$ is an excessive function for the Markov chain X of Exercise (5.3.10).

(5.9) Show that $f = (7, 6, 7, 5)$ is not excessive for the Markov chain X of Exercise (6.8.1b), but that it is α-excessive for $\alpha = 0.8$.

(5.10) Consider the Markov chain X with transition function as in Exercise (6.8.1a). Show that any excessive function is of form $f = a(1, 1, 0) + b(3, 2, 0) + c(1, 1, 1)$, where a, b, and c are arbitrary non-negative numbers.

(5.11) Use Theorem (2.5) to show that an excessive function f is a potential if and only if $P^n f \longrightarrow 0$ as $n \longrightarrow \infty$.

(5.12) Prove that if f is excessive and g is a non-negative potential, then $h = f \wedge g$ is also a potential.

(5.13) Show that any excessive function of a transient Markov chain is the limit of a non-decreasing sequence of potentials. (Hint: Supposing the state space is $E = \{0, 1, \ldots\}$, let $g_n(i) = n \sum_{j=0}^n R(i, j)$. Then $f_n = f \wedge g_n$ is the required sequence of potentials.)

(5.14) Consider the optimal stopping problem of Examples (3.1) and (3.5). What are the actual payoffs obtained by using the optimal strategy for the realizations ω for which

(a) $X(\omega) = (J, Q, A, A, \ldots)$,
(b) $X(\omega) = (J, Q, 2, 2, \ldots)$,
(c) $X(\omega) = (K, Q, A, A, \ldots)$,
(d) $X(\omega) = (K, Q, 2, 2, \ldots)$?

(5.15) Consider the Markov chain of Example (5.2.5), and let the payoff function be $f = (0, 2, 6)$.

(a) Compute the value of the game of optimal stopping of this chain. What is the optimal stopping time?

(b) What is the actual payoff obtained by using the optimal strategy if the outcome ω were such that $X(\omega) = (2, 2, 3, 2, 1, 1, 1, \ldots)$? What is the actual payoff if $X(\omega) = (2, 2, 2, 1, 1, \ldots)$?

(5.16) Consider the problem of optimal stopping of the Markov chain of Example (5.3.10) given the payoff function $f = (1, 5, 2, 4, 3)$.

(a) Compute the value function v. Find the optimal stopping time T_0.

(b) Compute the probability distribution

$$P_d\{f(X_{T_0}) = k\}, \qquad k = 1, 2, 3, 4, 5,$$

of the actual payoff received given that the initial state is d.

(c) Compute $v(d) = E_d[f(X_{T_0})]$ by using the distribution obtained in (b) (and check the result with that obtained in (a)).

(5.17) Consider the problem of optimal stopping of the Markov chain of Example (5.3.23) with the payoff function $f = (0, 2, 1, 2, 4, 0, 0, 3, 1, 1)$. Find the optimal stopping time and the expected value of the payoff using the optimal strategy.

(5.18) Let X be a Markov chain with two states, and let the rows of its transition matrix be $(\frac{1}{2}, \frac{1}{2})$ and $(\frac{2}{3}, \frac{1}{3})$. Suppose the payoff function is $f = (5, 3)$, and suppose a discount factor of $\alpha = 0.6$ is applicable. Find the optimal stopping time, and compute the expected value of the discounted payoff under the optimal strategy.

(5.19) Let X be a Markov chain whose transition matrix has the rows $(\frac{1}{2}, \frac{1}{4}, \frac{1}{4})$, $(\frac{1}{2}, \frac{1}{2}, 0)$, $(\frac{1}{4}, \frac{1}{4}, \frac{1}{2})$. Consider the problem of optimal stopping of this chain when the discount factor is $\alpha = 0.8$ and the payoff function $f = (4, 2, 1)$. Find the value of this game and the optimal stopping time.

(5.20) Consider the optimal stopping problem of the Markov chain X with state space $E = \{a, b\}$, transition matrix P, payoff function f, and fee function g given as

$$P = \begin{bmatrix} \frac{1}{2} & \frac{1}{2} \\ 1 & 0 \end{bmatrix}, \qquad f = \begin{bmatrix} 5 \\ 3 \end{bmatrix}, \qquad g = \begin{bmatrix} 2 \\ 1 \end{bmatrix}.$$

Find the optimal stopping time and the value of the game.

(5.21) Solve the problem of Exercise (5.20) above in the presence of discounting with $\alpha = 0.7$.

(5.22) Consider the problem of optimal stopping of the Markov chain X with state space $E = \{a, b, c\}$, transition matrix P, payoff function f, and fee function g, where

$$P = \begin{bmatrix} \frac{1}{2} & 0 & \frac{1}{2} \\ 1 & 0 & 0 \\ 0 & \frac{2}{3} & \frac{1}{3} \end{bmatrix}, \quad f = \begin{bmatrix} 7 \\ 2 \\ 3 \end{bmatrix}, \quad g = \begin{bmatrix} 0 \\ 3 \\ 1 \end{bmatrix}.$$

Find the optimal stopping time.

(5.23) Consider the problem of Exercise (5.22) again, but now with discounting. What is the smallest value of the discount factor α under which the optimal stopping time remains the same as in the case with no discounting?

(5.24) Consider the same problem as in Exercise (5.22) above but now with data

$$P = \begin{bmatrix} \frac{1}{2} & \frac{1}{4} & \frac{1}{4} \\ \frac{1}{2} & \frac{1}{2} & 0 \\ \frac{1}{4} & \frac{1}{4} & \frac{1}{2} \end{bmatrix}, \quad f = \begin{bmatrix} 4 \\ 2 \\ 1 \end{bmatrix}, \quad g = \begin{bmatrix} 0 \\ 1 \\ \frac{1}{2} \end{bmatrix}.$$

    (a)   Find the optimal stopping time in the case of no discounting.
    (b)   Do the same with discounting when $\alpha = 0.8$.

(5.25) Consider the same problem as in Exercise (5.22) above, but now the state space is $E = \{1, 2, 3, 4, 5\}$ and the data are

$$P = \begin{bmatrix} 0 & 0 & 0 & 1 & 0 \\ 0 & 0 & 0 & \frac{1}{2} & \frac{1}{2} \\ 0 & \frac{2}{3} & 0 & 0 & \frac{1}{3} \\ \frac{1}{3} & 0 & 0 & \frac{2}{3} & 0 \\ 0 & 0 & 0 & 1 & 0 \end{bmatrix}, \quad f = \begin{bmatrix} 5 \\ 4 \\ 3 \\ 7 \\ 2 \end{bmatrix}, \quad g = \begin{bmatrix} 3 \\ 1 \\ 2 \\ 4 \\ 3 \end{bmatrix}.$$

    (a)   Find the optimal stopping time with no discounting.
    (b)   Find the optimal stopping time with discounting with $\alpha = 0.8$.

(5.26) Consider a two-state Markov chain with transition probabilities $P(a, a) = \frac{1}{2}$, $P(b, a) = \frac{2}{3}$ (the other two numbers are $\frac{1}{2}$ and $\frac{2}{3}$). Consider the problem of optimal stopping of this chain when the payoff function is $f = (5, 3)$, no discounting, a fee of 1 dollar for every step, *and* the added feature that the player has only 2 dollars (so that if he has not stopped before, he must stop at $n = 2$).

    (a)   Show that the optimal stopping time T_0 is given by

$$T_0(\omega) = \begin{cases} 0 & \text{if } X_0(\omega) = a; \\ 1 & \text{if } X_0(\omega) = b, X_1(\omega) = a; \\ 2 & \text{otherwise} \end{cases}$$

for every ω.

(b)   Compute the value of the game

$$v(i) = \sup_{T \leq 2} E_i[f(X_T) - T], \qquad i = a, b.$$

(c)   Compare the result of part (a) with the case where the player has an unlimited amount of capital, so that he does not have to stop at any predetermined time.

(5.27)   *Selling an asset.* Find the optimal strategy for selling the house in Example (4.1).

(5.28)   *Time-Dependent Payoffs.* Let X be a Markov chain with state space E and transition matrix P. Suppose we have a non-negative function f_n defined on E for each $n \in \mathbb{N}$. Consider the problem of optimal stopping of X if the payoff is $f_n(j)$ for stopping at time n in state j. ($f_n = \alpha^n f$ gives the special case of discounting; $f_n = f$ for $n < v$ and $f_n = 0$ for $n \geq v$ gives the "finite horizon" case of Exercise (5.26).)

(a)   Let $\hat{X}$ be the space–time Markov chain associated with X described in Exercise (5.4.12); $\hat{X}$ has the state space $\hat{E} = E \times \mathbb{N}$. Show that the problem under consideration is equivalent to that of optimal stopping of the chain $\hat{X}$ with payoff function $\hat{f}$ on $\hat{E}$ defined by $\hat{f}(i, m) = f_m(i)$.

(b)   Let $\hat{v}$ be the value of the game of stopping $\hat{X}$, and define $v_m(i) = \hat{v}(i, m)$. Show that the v_m are the maximal solutions of

$$v_m \geq 0, \qquad v_m \geq f_m, \qquad v_m \geq Pv_{m+1}, \qquad m = 0, 1, \ldots.$$

(c)   Suppose that E is finite and that $f_v = f_{v+1} = \cdots = 0$ for some integer v. Then show that

$$T_0 = \inf\{n : f_n(X_n) = v_n(X_n)\}$$

is an optimal stopping time.

(5.29)   Consider the optimal stopping problem with time-dependent payoff functions f_n for a Markov chain X with transition matrix P, where

$$P = \begin{bmatrix} \frac{1}{2} & \frac{1}{2} & 0 \\ 0 & \frac{1}{4} & \frac{3}{4} \\ \frac{1}{3} & \frac{1}{3} & \frac{1}{3} \end{bmatrix}, \quad f_0 = \begin{bmatrix} 3 \\ 4 \\ 3 \end{bmatrix}, \quad f_1 = \begin{bmatrix} 3 \\ 4 \\ 2 \end{bmatrix}, \quad f_2 = \begin{bmatrix} 1 \\ 3 \\ 3 \end{bmatrix}, \quad f_3 = \begin{bmatrix} 1 \\ 3 \\ 2 \end{bmatrix},$$

and $f_4 = f_5 = \cdots = 0$ (state space is $E = \{a, b, c\}$).

(a)   Find the optimal stopping time T_0.

(b)   Find the value of $T_0(\omega)$ for each one of the realizations $X(\omega) = (b, b, c, a, b, \ldots)$, $X(\omega) = (c, c, b, b, c, \ldots)$, $X(\omega) = (c, a, a, b, \ldots)$, $X(\omega) = (c, c, a, a, \ldots)$, $X(\omega) = (a, b, b, c, \ldots)$.

(5.30)   Consider the problem of the preceding exercise, and further suppose that a fee of $g_n(i)$ must be paid at time n if the state X_n is i in order to continue the game. That is, we want to optimize

$$E_i\left[f_T(X_T) - \sum_{n=0}^{T-1} g_n(X_n) \right]$$

by choosing the stopping time T properly.

Find the optimal stopping time T_0 with the data P, f_n as given before in (5.29), and

$$g_0 = \begin{bmatrix} 0 \\ 0 \\ 1 \end{bmatrix}, \qquad g_1 = \begin{bmatrix} 1 \\ 0 \\ 0 \end{bmatrix}, \qquad g_2 = \begin{bmatrix} 1 \\ 1 \\ -2 \end{bmatrix}, \qquad g_3 = g_4 = \cdots = 0.$$

(5.31)   Consider the problem of the preceding exercise with discounting with factor $\alpha = 0.9$. Find the optimal stopping time and compute the value of the game.

*   *   *

Until very recently, topics such as potentials and excessive functions were considered to be of only theoretical interest. Fortunately, this is no longer the case, as the applications to optimal stopping problems alone illustrate. We have been able to give only a sketchy introduction to these very important topics. The interested reader will find an extensive treatment in the book by KEMENY, SNELL, *and* KNAPP [1].

It is usual with many authors to introduce potentials by referring to the classical potential theory and by drawing analogies with the amount of work which must be done in order to bring a particle of unit charge from infinity to a point x in the presence of other charged particles here and there. Considering the background of the modern student of stochastic processes, this attempt reminds us of the following.

A man was delegated the responsibility of explaining the facts of life to his nine-year-old nephew, presumably because the father was too shy to do it himself.

"You know how it is with the birds and the bees, don't you?" starts the uncle hopefully.

"What about the birds and the bees?" asks the young boy, quite puzzled.

"Well, you know, the birds and the bees . . ." starts the uncle again but has to stop seeing the confusion in junior's face. After several such unsuccessful attempts, he exclaims:

"You know how human beings reproduce, don't you?!"

"Of course," answers the boy, "everybody knows that!"

"Well," says the uncle, quite relieved, "it is the same with the birds and the bees."

Well, it is the same with classical potential theory.

Our formulation and development of the optimal stopping problem follows that given in DYNKIN *and* YUSHKEVICH [1]. *That book also contains a number of other very interesting topics all explained in a clear, readable style.*

CHAPTER 8

Markov Processes

In considering Markov chains, we thought in terms of a system which moved from one state to the next in such a manner that the future of the process was conditionally independent of its past provided that the present state be known. There we had concentrated on such things as the chances of moving from one state to another in a fixed number of steps, or the number of steps it takes for the process to move from a state into another, or the number of times the process visited a certain state during the first so many transitions. In applications, when considering processes which are running on "real time," this meant concentrating on the transitions but ignoring the actual times spent in between the transitions.

The processes we will introduce in this chapter will remedy this by allowing us to take into account not only the changes of state, but also the actual times spent in between.

In Section 1 we give an introduction to Markov processes and discuss the Markov and strong Markov properties. Section 2 is on the sample path behavior of Markov processes. Times spent in a state, jumps between states, and times between jumps are examined in Section 3. These structural properties are used in Section 4 to obtain computational results on potentials and infinitesimal generators. Limiting properties of the transition functions and ratios of additive functionals are discussed in Section 5 in terms of the similar results for (discrete time) chains. Finally, in Section 6, birth and death processes are discussed briefly and a number of examples from queueing theory are given.

1. Markov Processes

Let E be a countable set, and for each $\omega \in \Omega$ and $t \in \mathbb{R}_+ = [0, \infty)$, let $Y_t(\omega)$ be an element of E. For fixed t, this defines a random variable Y_t taking values in E, and the collection of all these random variables is a stochastic process $Y = \{Y_t; t \in \mathbb{R}_+\}$ with a continuous time parameter. The set E is called the state space of Y, and the process is said to be in state j at time t if $Y_t = j$. If an outcome $\omega \in \Omega$ is fixed, then the values $Y_t(\omega)$ define a function $t \longrightarrow Y_t(\omega)$ from $\mathbb{R}_+ = [0, \infty)$ into E. A typical realization of the process Y, namely the function $t \longrightarrow Y_t(\omega)$ for a typical fixed ω, will be as in Figure 8.1.1. Our aim will be the clarification of the stochastic structure of Y and of its functionals.

The following definition is the continuous time version of the definition of a Markov chain.

(1.1)   DEFINITION.   The stochastic process $Y = \{Y_t; t \in \mathbb{R}_+\}$ is said to be a *Markov process* with state space E provided that for any $t, s \geq 0$ and $j \in E$,

$$P\{Y_{t+s} = j \,|\, Y_u; u \leq t\} = P\{Y_{t+s} = j \,|\, Y_t\}. \qquad \square$$

The conditional probability appearing above may, in general, depend on both t and s (in addition to j and the value of Y_t). When

$$(1.2) \qquad\qquad P\{Y_{t+s} = j \,|\, Y_t = i\} = P_s(i, j)$$

is independent of $t \geq 0$ for all $i, j \in E$ and $s \geq 0$, the process Y is said to be a *time-homogeneous* Markov process. For fixed $i, j \in E$, the function $t \longrightarrow P_t(i, j)$ is called a *transition function*; and the family of matrices P_t, $t \geq 0$,

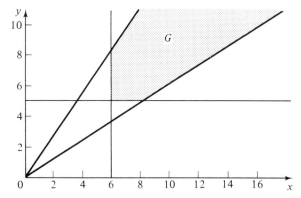

Figure 8.1.1   A possible realization of a continuous time process.

of the transition functions $P_t(i, j)$ is simply called the *transition function* of the Markov process Y. Throughout, we will limit ourselves to time-homogeneous Markov processes and will omit mentioning the qualifier "time-homogeneous."

Concerning the transition functions, we observe that

(1.3) $$P_t(i, j) \geq 0,$$

(1.4) $$\sum_{k \in E} P_t(i, k) = 1,$$

(1.5) $$\sum_{k \in E} P_t(i, k) P_s(k, j) = P_{t+s}(i, j).$$

for every $i, j \in E$ and $t, s \geq 0$. Of these, (1.3) and (1.4) are self-evident, and (1.5) is the *Chapman–Kolmogorov equation* for the continuous time case (cf. (5.1.8) for the discrete-parameter case). To show (1.5), we use (1.2) to write the left-hand side of (1.5) as

$$\sum_{k \in E} P\{Y_t = k \mid Y_0 = i\} P\{Y_{t+s} = j \mid Y_t = k\}.$$

By the Markov property,

$$P\{Y_{t+s} = j \mid Y_t = k\} = P\{Y_{t+s} = j \mid Y_t = k, Y_0 = i\}.$$

Hence,

$$\sum_{k \in E} P_t(i, k) P_s(k, j) = \sum_{k \in E} P\{Y_{t+s} = j \mid Y_t = k, Y_0 = i\} P\{Y_t = k \mid Y_0 = i\}$$

$$= \sum_{k \in E} P\{Y_{t+s} = j, Y_t = k \mid Y_0 = i\}$$

$$= P\{Y_{t+s} = j \mid Y_0 = i\} = P_{t+s}(i, j).$$

Corresponding to Theorem (5.1.5) we have the following.

(1.6) PROPOSITION. For any integer $n \in \mathbb{N}$, numbers $0 \leq t_0 < t_1 < \cdots < t_n$ in $[0, \infty)$, and states $i_0, i_1, \ldots, i_n$ in E, we have

$$P\{Y_{t_1} = i_1, \ldots, Y_{t_n} = i_n \mid Y_{t_0} = i_0\} = P_{t_1 - t_0}(i_0, i_1) \cdots P_{t_n - t_{n-1}}(i_{n-1}, i_n).$$

Hence, the joint distribution of $Y_{t_0}, \ldots, Y_{t_n}$ is specified by the initial distribution π of Y_0 and the transition function (P_t) of the process. This result also enables one to go in the converse direction: Suppose we are given a probability distribution π and a transition function (P_t) over E satisfying (1.3), (1.4), and (1.5). Then it is always possible to construct a sample space Ω, random variables Y_t from Ω into E, and a probability measure P on Ω in such a manner that over (Ω, P), the process Y is a Markov process with state space E, initial distribution π, and transition function (P_t). In such constructions it is usual to take Ω to be the set of all functions ω from $[0, \infty)$ into E, to define $Y_t(\omega) = \omega(t)$ for every $t \geq 0$ and $\omega \in \Omega$, and first specify the probabilities of events of the form

$$A = \{Y_{t_0} = i_0, Y_{t_1} = i_1, \ldots, Y_{t_n} = i_n\}$$

by

$$P(A) = \sum_{i \in E} \pi(i) P_{t_0}(i, i_0) P_{t_1 - t_0}(i_0, i_1) \cdots P_{t_n - t_{n-1}}(i_{n-1}, i_n),$$

and then extend the definition of $P(A)$ to other events A by using the axioms (1.1.6) and the various available results on the possibility of such extensions. This way every outcome $\omega \in \Omega$ is also the realization of Y corresponding to that ω. In thinking of such matters we will think of ω as the realization even when that is not strictly true.

(1.7) EXAMPLE. *Poisson Process.* Let $N = \{N_t; t \geq 0\}$ be a Poisson process. Then by axiom (4.1.1b),

$$P\{N_{t+s} = j \,|\, N_u; u \leq t\} = P\{N_{t+s} = j \,|\, N_t\},$$

which implies that N is a Markov process. By Corollary (4.1.15), for some λ,

$$P\{N_{t+s} = j \,|\, N_t = i\} = \begin{cases} 0 & \text{if } j < i, \\ \dfrac{e^{-\lambda s}(\lambda s)^{j-i}}{(j - i)!} & \text{if } j \geq i. \end{cases}$$

Hence, the transition function (P_t) of this Markov process has the form

$$P_t = \begin{bmatrix} p_t(0) & p_t(1) & p_t(2) & \cdots \\ & p_t(0) & p_t(1) & \cdots \\ & & p_t(0) & \cdots \\ 0 & & & \ddots \end{bmatrix}$$

where p_t is the Poisson distribution with parameter λt, i.e.,

$$p_t(k) = \frac{e^{-\lambda t}(\lambda t)^k}{k!}, \quad k = 0, 1, \ldots. \qquad \Box$$

(1.8) EXAMPLE. *Compound Poisson Processes.* Let Y be a compound Poisson process (cf. Definition (4.6.1)) taking integer values. Then (4.6.1b) implies that Y is a Markov process with state space $E = \{\ldots, -1, 0, 1, \ldots\}$. Further, by (4.6.1b) and (4.6.1c),

$$P\{Y_{t+s} = j \,|\, Y_t = i\} = p_s(j - i)$$

where

$$p_t(k) = P\{Y_t - Y_0 = k\}, \quad k = 0, \pm 1, \ldots.$$

Hence, the transition function (P_t) for this Markov process is

$$P_t = \begin{bmatrix} \cdot & \cdot & \cdot & \cdot & \cdot & \cdot & \cdot & \cdot & \cdot & \cdot \\ \cdot & \cdot & \cdot & \cdot & \cdot & \cdot & \cdot & \cdot & \cdot \\ \cdot & \cdot & \cdot & \cdot & \cdot & \cdot & \cdot & \cdot \\ \cdot & \cdot & \cdot & p_t(0) & p_t(1) & p_t(2) & \cdot & \cdot & \cdot \\ \cdot & \cdot & \cdot & p_t(-1) & p_t(0) & p_t(1) & \cdot & \cdot & \cdot \\ \cdot & \cdot & \cdot & p_t(-2) & p_t(-1) & p_t(0) & \cdot & \cdot & \cdot \\ \cdot & \cdot & \cdot & \cdot & \cdot & \cdot & \cdot & \cdot \\ \cdot & \cdot & \cdot & \cdot & \cdot & \cdot & \cdot & \cdot \\ \cdot & \cdot & \cdot & \cdot & \cdot & \cdot & \cdot & \cdot & \cdot \end{bmatrix}.$$

$\square$

(1.9) EXAMPLE. *Markov chains subordinated to Poisson processes.* Let $\hat{X} = \{\hat{X}_n : n \in \mathbb{N}\}$ be a Markov chain with state space E and transition matrix K, and let $N = \{N_t; t \geq 0\}$ be a Poisson process with rate λ which is independent of $\hat{X}$. Consider a system which moves from one state of E to another in such a way that the successive states visited form the Markov chain $\hat{X}$ and the times of jumps (times at which the system changes its state) form the Poisson process N. Then, N_t being the number of jumps in $(0, t]$, the system is in state $Y_t = \hat{X}_{N_t}$ at time t. In other words, if $T_1, T_2, \ldots$ are the arrival times of N and $T_0 = 0$, then $Y_t = \hat{X}_n$ for all t in the time interval $[T_n, T_{n+1})$. Then $Y = \{Y_t; t \geq 0\}$ is a Markov process. The intuitive argument for this is as follows. Suppose we are given the past history of Y before t and are asked to predict Y_{t+s}. Knowing $Y_t = i$, if we know further that there will be $N_{t+s} - N_t = n$ transitions during $(t, t + s]$, then the probability that $Y_{t+s} = j$ becomes the probability that the Markov chain $\hat{X}$ moves from the state $Y_t = i$ to the state j in exactly n steps. This conditional probability is $K^n(i, j)$ where K^n is the nth power of K. Furthermore, the number of transitions to take place during $(t, t + s]$ is independent of the past history (by the fundamental property of Poisson processes, it is independent of the knowledge about N_u, $u \leq t$, which $\{Y_u, u \leq t\}$ may provide; and by the independence of N and $\hat{X}$, it is independent of the knowledge about $\hat{X}$ which $\{Y_u, u \leq t\}$ might provide). Hence, given all the past history before t, the probability of n transitions during $(t, t + s]$ is simply $e^{-\lambda s}(\lambda s)^n/n!$. Thus, if the event $\{Y_t = i\}$ has occurred, then

$$P\{Y_{t+s} = j \mid Y_u; u \leq t\} = \sum_{n=0}^{\infty} \frac{e^{-\lambda s}(\lambda s)^n}{n!} K^n(i, j).$$

Somewhat more formally, the arguments above can be written as

$$
\begin{aligned}
P\{Y_{t+s} &= j \,|\, Y_u; u \le t\} \\
&= E[P\{Y_{t+s} = j \,|\, Y_u; u \le t; N_{t+s} - N_t\} \,|\, Y_u; u \le t] \\
&= E[K^{N(t+s)-N(t)}(Y_t, j) \,|\, Y_u; u \le t] \\
&= E\left[\sum_{n=0}^{\infty} \frac{e^{-\lambda s}(\lambda s)^n}{n!} K^n(Y_t, j) \,\Big|\, Y_u; u \le t \right] \\
&= \sum_{n=0}^{\infty} \frac{e^{-\lambda s}(\lambda s)^n}{n!} K^n(Y_t, j).
\end{aligned}
$$

Here, the first equality sign is justified by using Theorem (2.2.19); the second equality follows from the construction of Y and the fact that $\hat{X}$ is a Markov chain independent of N; the third follows from the independence of $N_{t+s} - N_t$ from $\{Y_u; u \le t\}$ and from the fact that its distribution is Poisson; and the final equality sign uses Corollary (2.2.22).

In fact, in addition to showing that Y is a Markov process, we have also computed its transition function: it is given by

$$
P_t = \sum_{n=0}^{\infty} \frac{e^{-\lambda t}(\lambda t)^n}{n!} K^n, \quad t \ge 0.
$$

We will return to this example again. In a sense, this example has all the characteristics of all the Markov processes with finite state spaces. We will make this "sense" more precise later in Section 4 (see Theorem (4.31) there).

For a concrete example, suppose $\hat{X}$ is the Markov chain of Example (5.1.10). Then the state space is $E = \{1, 2\}$, and

$$
K^n = \begin{bmatrix} \frac{3}{8} + \frac{5}{8}(0.2)^n & \frac{5}{8} - \frac{5}{8}(0.2)^n \\ \frac{3}{8} - \frac{3}{8}(0.2)^n & \frac{5}{8} + \frac{3}{8}(0.2)^n \end{bmatrix}.
$$

Suppose further that the Poisson process has rate $\lambda = 15$. Then $P_t(1, 1)$ becomes

$$
\begin{aligned}
P_t(1, 1) &= \sum_{n=0}^{\infty} \frac{e^{-15t}(15t)^n}{n!} \left[\frac{3}{8} + \frac{5}{8}(0.2)^n \right] \\
&= \frac{3}{8} + e^{-15t} \frac{5}{8} \sum_{n=0}^{\infty} \frac{[(0.2)15t]^n}{n!} = \frac{3}{8} + \frac{5}{8} e^{-12t}.
\end{aligned}
$$

Similar computations for the other $P_t(i, j)$ yield

$$
P_t = \begin{bmatrix} \frac{3}{8} + \frac{5}{8}e^{-12t} & \frac{5}{8} - \frac{5}{8}e^{-12t} \\ \frac{3}{8} - \frac{3}{8}e^{-12t} & \frac{5}{8} + \frac{3}{8}e^{-12t} \end{bmatrix}, \quad t \ge 0. \qquad \square
$$

(1.10) EXAMPLE. *M/M/1 queueing system.* Suppose the arrivals of customers at a service facility form a Poisson process. If a customer arrives to
find the server idle, his service starts immediately; otherwise, if he finds the
server busy, he waits until it is his turn to be served. The time spent serving
a customer is independent of all the other service times and of the arrival
process. Service times are exponentially distributed. We are interested in the
process $\{Y_t; t \geq 0\}$, where Y_t is the number of customers in the system
(waiting or being served) at time t.

Suppose we have been observing the process until time t and know, in
particular, the number of customers Y_t that are in the system. To predict
the number Y_{t+s} of customers that are there at time $t + s$, we think as
follows: Y_{t+s} is equal to the sum of Y_t and the number of arrivals during
$(t, t + s]$ less the number of services completed during $(t, t + s]$. The number
of arrivals during $(t, t + s]$ is independent of everything else that went on
before t. As to the services, because of the exponential distribution's lack of
memory, the remaining service time of the customer being served at time
t (if any) is independent of how long that service has been going on; and it
is clear that the other service times are completely independent of the past.
Hence, the number of services completed during $(t, t + s]$ can depend only
on Y_t and the arrivals during $(t, t + s]$, and this dependence is only because
one cannot serve nonexistent customers. This analysis shows that Y is a
Markov process. Unlike the preceding examples, its transition function cannot be displayed explicitly with easy computations.

We will return to this example a number of times and will give various
extensions of it later in Section 6.   □

Just as with Markov chains, the definition of Markov processes implies
several stronger statements. The next theorem states that the future and the
past are conditionally independent given the present state.

(1.11) THEOREM. Let Y be a Markov process, and let $t \geq 0$ be fixed. Then
$\{Y_s; s \geq t\}$ and $\{Y_u; u \leq t\}$ are conditionally independent given Y_t.

Proof. From the definition of conditional independence (cf. Definition
(2.2.25)), what this means is that for any $n, m \in \mathbb{N}, 0 \leq t_1 < \cdots < t_n = t \leq t_{n+1} < \cdots < t_{n+m}$, and any bounded function g on E^m we have

$$E[g(Y_{t_{n+1}}, \ldots, Y_{t_{n+m}}) \mid Y_{t_1}, \ldots, Y_{t_n}] = E[g(Y_{t_{n+1}}, \ldots, Y_{t_{n+m}}) \mid Y_{t_n}].$$

Now let $X_k = Y_{t_k}$ for $k = 1, \ldots, n + m$. The resulting statement to be
proved is Theorem (5.1.19), whose proof goes through word for word.   □

For the computation of expectations, the following is useful. It is equivalent to (5.1.22) and (5.1.23), whose proofs are easy to adapt to the present case.

(1.12) PROPOSITION.   For any $m \in \mathbb{N}$, numbers $t \geq 0$ and $0 \leq s_1 < \cdots < s_m$, and any bounded function f on E^m,

$$E[f(Y_{t+s_1}, \ldots, Y_{t+s_m}) | Y_u; u \leq t] = g(Y_t),$$

where

(1.13) $g(i) - E[f(Y_{s_1}, \ldots, Y_{s_m}) | Y_0 = i]$

$$= \sum_{i_1 \in E} \cdots \sum_{i_m \in E} P_{s_1}(i, i_1) \cdots P_{s_m - s_{m-1}}(i_{m-1}, i_m) f(i_1, \ldots, i_m).$$

Proof is omitted. The next question to raise is whether "the independence of the past and the future given the present " holds when random times play the role of "present." It turns out that it holds for a certain class of random times called stopping times. A random variable T taking values in $[0, +\infty]$ is said to be a *stopping time* for Y provided that for every $t \geq 0$ finite, the occurrence or non-occurrence of the event $\{T \leq t\}$ can be determined from the history $\{Y_u; u \leq t\}$ of the process until that time t. The following is, then, the analog of Theorems (5.1.24) and (4.2.18). In interpreting the theorem, we again make the conventions that an extra state Δ is added to E, $Y_\infty(\omega) = \Delta$ for all $\omega \in \Omega$, and if f is a function on E it is automatically extended to $E \cup \{\Delta\}$ by setting $f(\Delta) = 0$. Proof will be omitted.

(1.14) THEOREM.   For any $m \in \mathbb{N}$, numbers $0 \leq s_1 < \cdots < s_m$, and any positive function f on E^m, we have

$$E[f(Y_{T+s_1}, \ldots, Y_{T+s_m}) | Y_u; u \leq T] = g(Y_T)$$

for all stopping times T; here g is as defined by (1.13).

In particular, by taking $m = 1$ and $f = 1_j$ for some fixed $j \in E$, we obtain

(1.15) $P\{Y_{T+s} = j | Y_u; u \leq T\} = P_s(Y_T, j).$

This is called the *strong Markov property* at T.

2.   Sample Path Behavior

Let Y be a Markov process with a countable state space E and transition function (P_t). In this section we are interested in the properties of the sample paths $t \rightarrow Y_t(\omega)$ for typical ω.

To discuss such things as smoothness for a sample path, we need to specify a topology on the state space first. Therefore, throughout the remainder of this chapter we assume that E is given the *discrete topology*, that is, we assume that every subset of E is open. Then for any $\omega \in \Omega$ and sequence t_n decreasing to t,

$$(2.1) \qquad \lim_{n \to \infty} Y_{t_n}(\omega) = i$$

for some $i \in E$ if and only if $Y_{t_n}(\omega) = i$ for all $n \geq n_0$ for some integer n_0 possibly depending on ω. In particular, if E is a subset of the integers, then the discrete topology on E coincides with the ordinary one.

Arbitrary Markov processes can have highly irregular sample paths. To insure the existence of a version with some regularity properties, we impose a continuity condition on the transition function: We assume that

$$(2.2) \qquad \lim_{t \downarrow 0} P_t(i, j) = I(i, j), \quad i, j \in E;$$

then (P_t) is said to be a *standard transition function*.

When (2.2) holds, $P_t(i, i) \to 1$ as $t \downarrow 0$ for every $i \in E$; conversely, this property implies (2.2), since $\sum_j P_t(i, j) = 1$ for all t. It follows from (1.5) that

$$P_{t+s}(i, i) \geq P_t(i, i) P_s(i, i)$$

for all t and s, which by iteration implies that $P_t(i, i) \geq [P_u(i, i)]^n$ for $u = t/n$. Since $P_t(i, i) \to 1$ as $t \downarrow 0$, $P_u(i, i) > 0$ for u small enough, and therefore, $P_t(i, i) \geq [P_u(i, i)]^n > 0$. Hence, as a particular consequence of (2.2) we have that $P_t(i, i) > 0$ for all $t \geq 0$. The following is a more complete account. We omit the proof.

(2.3) PROPOSITION. Fix $i, j \in E$; the function $t \longrightarrow P_t(i, j)$ vanishes either everywhere or nowhere in $(0, \infty)$; in either case, it is continuous everywhere.

For a Poisson process N, the probability that a fixed point t_0 is a time of arrival was zero; in other words, for any fixed t, $P\{N_t = N_s\}$ goes to one in the limit as s approaches t. This property is called *stochastic continuity* and is also true for more general Markov processes.

(2.4) PROPOSITION. Y is stochastically continuous; that is, for any $t > 0$,

$$\lim_{u \downarrow 0} P\{Y_{t-u} \neq Y_t\} = \lim_{u \downarrow 0} P\{Y_t \neq Y_{t+u}\} = 0.$$

Proof. For any initial state i,

$$P\{Y_t \neq Y_{t+u} \mid Y_0 = i\} = \sum_{j \in E} P_t(i, j)[1 - P_u(j, j)].$$

As $u \downarrow 0$, (2.2) implies that $1 - P_u(j, j) \rightarrow 0$, which in turn implies that $P\{Y_t \neq Y_{t+u} \mid Y_0 = i\} \rightarrow 0$ by the bounded convergence theorem. Since this is true for any i, a second application of the bounded convergence theorem yields the result that $P\{Y_t \neq Y_{t+u}\} \rightarrow 0$ as $u \rightarrow 0$.

Next we consider the limit from left. For $i \in E$,

$$P\{Y_{t-u} = Y_t \mid Y_0 = i\} = \sum_{j \in E} P_{t-u}(i, j) P_u(j, j)$$

$$\geq \sum_{j \in A} P_{t-u}(i, j) P_u(j, j)$$

for any finite subset A of E. As $u \downarrow 0$, $P_u(j, j) \rightarrow 1$ by (2.2), and $P_{t-u}(i, j) \rightarrow P_t(i, j)$ by Proposition (2.3). Hence, as $u \downarrow 0$, the right-hand side approaches $\sum_{j \in A} P_t(i, j)$, which in turn increases to one as A increases to E. Thus, $P\{Y_{t-u} = Y_t \mid Y_0 = i\} \rightarrow 1$ as $u \downarrow 0$, and the proof is complete since this holds for all i. □

Stochastic continuity does not imply the continuity of the sample paths. For example, a Poisson process N is stochastically continuous, but $t \rightarrow N_t(\omega)$ has infinitely many jumps for any typical ω. Usually, for almost all ω, the path $t \rightarrow Y_t(\omega)$ of a Markov process Y will have many discontinuities (see Figure 8.1.1 for a typical picture). In general, stochastic continuity does not even imply almost sure continuity at a fixed point: for fixed t_0, the set Ω_0 of all ω for which $t \rightarrow Y_t(\omega)$ is continuous at t_0 may be such that $P(\Omega_0) < 1$. However, $P(\Omega_0) = 1$ if Y has no instantaneous states (see (2.12) for the definition). Nevertheless, stochastic continuity enables us to modify the paths $t \rightarrow Y_t(\omega)$ to obtain certain desirable properties. $\hat{Y}$ is said to be a *modification* of Y if $P\{\hat{Y}_t = Y_t\} = 1$ for every t.

The next theorem summarizes what can be done. We need to add an extra point Δ to the state space so that the modified version has the state space $E \cup \{\Delta\}$. If $E = \{0, 1, \ldots\}$, then Δ should be $+\infty$; otherwise, in any ordering of the elements of E, the extra point Δ should correspond to $+\infty$.

(2.5) THEOREM. It is possible to modify Y, without altering its probability law, in such a way that the resulting version has the following property: For any $\omega \in \Omega$ and $t \in \mathbb{R}_+$, either

 (a) $Y_t(\omega) = \Delta$ and $Y_{t_n}(\omega) \rightarrow \Delta$ for any sequence $t_n \rightarrow t$;

or

 (b) $Y_t(\omega) = i$ for some $i \in E$, there are sequences $t_n \downarrow t$ such that $Y_{t_n}(\omega) \rightarrow i$, and for any sequence $t_n \downarrow t$ such that $Y_{t_n}(\omega)$ has a limit in E, we have $\lim Y_{t_n}(\omega) = i$.

Proof is omitted. From here on we let Y denote the modified version. Since Y is a modification of the original version Y^0, we have $Y_t = Y_t^0$ almost surely for any t. Therefore, in particular,

$$(2.6) \qquad P\{Y_t = \Delta \,|\, Y_0 = i\} = 1 - P\{Y_t^0 \in E \,|\, Y_0^0 = i\}$$
$$= 1 - \sum_{j \in E} P_t(i, j) = 0.$$

So the addition of Δ to the state space is for purposes of smoothness: for any ω, if $Y_{t_0}(\omega) = \Delta$, then $t \to Y_t(\omega)$ is continuous at t_0. In fact, even for other states we almost have right continuity:

(2.7) COROLLARY. For any $\omega \in \Omega$ and $t \in \mathbb{R}_+$, if there is a sequence $t_n \downarrow t$ such that $\lim Y_{t_n}(\omega) = i \in E$, then $Y_t(\omega) = i$.

Proof. Since $\lim Y_{t_n}(\omega)$ is in E, case (a) does not hold, and $Y_t(\omega)$ is in E. Then (b) implies that $\lim Y_{t_n}(\omega) = Y_t(\omega)$. $\qquad\qquad\square$

Note that in case (2.5b), we have not excluded the possibilities that for some sequences $t_n \downarrow t$ the sequence $Y_{t_n}(\omega)$ does not have a limit or has the limit Δ. These possibilities cannot arise if $Y_t(\omega) = i$ implies $Y_{t+s}(\omega) = i$ for all $s \in (0, \varepsilon)$ for some $\varepsilon > 0$; because, then, $\lim Y_{t_n}(\omega) = i$ for all sequences $t_n \downarrow t$. We are thus led to examine the length of time Y spends in a state after entering it.

We define W_t as the length of time the process Y remains in the state being occupied at the instant t; that is, for every $\omega \in \Omega$ and $t \geq 0$,

$$(2.8) \qquad W_t(\omega) = \inf\{s > 0 : Y_{t+s}(\omega) \neq Y_t(\omega)\}.$$

We then have the following important

(2.9) THEOREM. For any $i \in E$ and $t \geq 0$,

$$P\{W_t > u \,|\, Y_t = i\} = e^{-\lambda(i)u}, \quad u \geq 0,$$

for some number $\lambda(i) \in [0, +\infty]$ (if $\lambda(i) = +\infty$, then $e^{-\lambda(i)u} = 0$ for all $u \geq 0$).

Proof. By the time homogeneity of Y, the conditional probability in question is independent of t. For fixed i, then, let it be denoted by $f(u)$. Since the event $\{W_t > u + v\}$ is equal to the event $\{W_t > u, W_{t+u} > v\}$, we have

$$(2.10) \qquad f(u + v) = P\{W_t > u + v \,|\, Y_t = i\}$$
$$= P\{W_t > u, W_{t+u} > v \,|\, Y_t = i\}$$
$$= P\{W_t > u \,|\, Y_t = i\} P\{W_{t+u} > v \,|\, Y_t = i, W_t > u\}.$$

If $Y_t(\omega) = i$ and $W_t(\omega) > u$, then $Y_{t+u}(\omega) = i$, and that knowledge makes all the past history useless as far as predicting the future is concerned. Hence, the second factor on the right is equal to

$$(2.11) \qquad P\{W_{t+u} > v \,|\, Y_{t+u} = i\} = f(v).$$

The function f is bounded by 0 and 1, and (2.10) and (2.11) show that it also satisfies $f(u + v) = f(u)f(v)$ for all $u, v \geq 0$. Therefore, it must be of the form $f(u) = e^{-cu}$ for some constant $c \geq 0$, possibly $c = +\infty$; and this constant c might depend on the state i being fixed. ☐

(2.12) DEFINITION. A state i is called *absorbing* if $\lambda(i) = 0$, *stable* if $0 < \lambda(i) < \infty$, and *instantaneous* if $\lambda(i) = +\infty$. ☐

If i is absorbing and if $\{Y_t = i\}$ occurs, then $Y_{t+s} = i$ for all $s \geq 0$; in other words, if Y enters i, then it stays there forever.

If i is a stable state, then

$$(2.13) \qquad P\{0 < W_t < \infty \,|\, Y_t = i\} = 1;$$

in other words, if Y is in i at time t, it stays there a positive but finite time further.

Finally, if i is instantaneous,

$$(2.14) \qquad P\{W_t = 0 \,|\, Y_t = i\} = 1,$$

which means that the process jumps out of an instantaneous state as soon as it enters it.

The following theorem clarifies (2.5) further.

(2.15) THEOREM. For almost all $\omega \in \Omega$ and any $t \in \mathbb{R}_+$, one of the following three possibilities holds:
(a) $Y_t(\omega) = \Delta$; then $\lim Y_{t_n}(\omega) = \Delta$ for any sequence $t_n \to t$.
(b) $Y_t(\omega) = i$ where $i \in E$ is either stable or absorbing; then $\lim Y_{t_n}(\omega) = i$ for any sequence $t_n \downarrow t$.
(c) $Y_t(\omega) = i$ where $i \in E$ is instantaneous; then $\lim Y_{t_n}(\omega) = i$ for some sequences $t_n \downarrow t$, $\lim Y_{t_n}(\omega) = \Delta$ for some other sequences $t_n \downarrow t$, and $\lim Y_{t_n}(\omega)$ does not exist for some still other sequences $t_n \downarrow t$.

Proof. Case (a) is the same as in (2.5). In case (b), if $Y_t(\omega) = i$ for some stable or absorbing state i, then by (2.13) there is an interval $[t, t + \varepsilon)$ of positive length ε throughout which $Y_u(\omega) = i$. It follows that $\lim Y_u(\omega) = i$ as $u \downarrow t$.

In case (c), it follows from Theorem (2.5) that for some sequence $t_n \downarrow t$, we have $\lim Y_{t_n}(\omega) = i$. Since i is instantaneous, (2.14) implies that there are instants s arbitrarily close to t at which $Y_s(\omega) \neq i$. To show the second possibility, let $s_n \downarrow t$ be selected so that $Y_{s_n}(\omega) \neq i$. Since $E \cup \{\Delta\}$ is compact, there is some subsequence (t_n) of (s_n) such that $\lim Y_{t_n}(\omega)$ exists. If the limit is in E, then it must be i by part (b) of Theorem (2.5), which implies that $Y_{t_n}(\omega) = i$ for infinitely many t_n by (2.1), which in turn contradicts the fact that $Y_{t_n}(\omega) \neq i$ for any t_n. Hence, $\lim Y_{t_n}(\omega)$ is not in E, that is, $\lim Y_{t_n}(\omega) = \Delta$. The third possibility in case (c) follows from the other two. $\qquad\square$

The sample path behavior described by part (c) above is not easy to visualize. The process jumps out of an instantaneous state i as soon as it enters i, but also returns to i infinitely often within arbitrarily short times. In fact, returns are often enough to make the amount of time spent in i positive: this follows upon observing that for any $\varepsilon > 0$

$$(2.16) \qquad E\left[\int_t^{t+\varepsilon} 1_i(Y_u)\, du \mid Y_t = i\right] = \int_t^{t+\varepsilon} P\{Y_u = i \mid Y_t = i\}\, du$$

$$= \int_t^{t+\varepsilon} P_{u-t}(i, i)\, du = \int_0^\varepsilon P_s(i, i)\, ds,$$

which is positive since $P_s(i, i) > 0$ for all $s \geq 0$.

We will soon restrict ourselves to Markov processes with no instantaneous states. It is worth mentioning, however, that this is not as harmless as it might look at first. The presence of an instantaneous state will show itself in the observational data. Moreover, there are examples of Markov processes with one, two, or any number of instantaneous states; and there are examples where all states are instantaneous.

On the other hand, in at least two important cases there can be no instantaneous states: when there are only finitely many states, and when $t \to Y_t$ is right continuous.

(2.17) PROPOSITION. If E is finite, then no state is instantaneous.

Proof. Let t_n be a sequence decreasing to t. By (2.6), $Y_{t_n} \in E$ almost surely for each n and therefore for all n; that is, for almost all ω, $Y_{t_n}(\omega) \in E$ for all n. This implies that $Y_{t_n}(\omega) = i$ for some $i \in E$ for infinitely many n, since there are only finitely many states. Hence, there is some subsequence (s_n) such that $s_n \downarrow t$ and $Y_{s_n}(\omega) = i$ for all s_n. Therefore, $\lim Y_{t_n}(\omega) \neq \Delta$ no matter how the sequence t_n is chosen. This shows that case (c) of Theorem (2.15) cannot arise. $\qquad\square$

(2.18) PROPOSITION. The mapping $t \to Y_t(\omega)$ is right continuous for almost all ω if and only if there are no instantaneous states.

Proof. If there are no instantaneous states, for almost all ω, either case

(a) or case (b) of (2.15) holds at every t_0; and in either case $t \rightarrow Y_t(\omega)$ is right continuous at t_0. If there is an instantaneous state i, then for $Y_0 = i$, case (c) of (2.15) applies to imply that $Y_s(\omega)$ has two possible limits as $s \downarrow 0$; and thus $t \rightarrow Y_t(\omega)$ is not right continuous. □

In fact, if there are no instantaneous states, then $t \rightarrow Y_t$ is right continuous and has left-hand limits everywhere.

Next we consider the set of times spent in a fixed stable state. For a stable state i, let

$$(2.19) \qquad G_i(\omega) = \{t : Y_t(\omega) = i\}.$$

By (2.13) the set $G_i(\omega)$ is composed of intervals each one of which is of positive length. By Theorem (2.9), the lengths of these intervals are independent and identically distributed with an exponential distribution; in other words, if the component intervals of G_i were laid next to each other, then we would obtain a Poisson process with parameter $\lambda(i)$. Therefore, in a finite interval $[0, t]$ there are only finitely many intervals belonging to G_i. Moreover, by (2.15b), these intervals are all of the form [ ), that is, closed on the left and open on the right. We have thus proved the following (see Figure 8.2.1 for the typical picture):

$$0 \qquad\qquad\qquad\qquad\qquad\qquad\qquad\qquad\qquad\qquad t$$

Figure 8.2.1 The set $G_i(\omega)$ of times spent in a stable state i is a union of disjoint intervals of positive length.

(2.20) THEOREM. Let i be a stable state. Then for almost all $\omega \in \Omega$, the time set $G_i(\omega)$ is a countable union of disjoint intervals each of which has positive length; each component interval has the form [ ); and there are only finitely many components in $G_i(\omega) \cap [0, t]$.

Finally, suppose $Y_t(\omega) = i$ where i is stable. Then $t \in G_i(\omega)$, and by the preceding theorem, there is a maximal interval $[S(\omega), T(\omega))$ which includes t. Therefore, for any sequence $t_n \uparrow T(\omega)$, we have $\lim_n Y_{t_n}(\omega) = i$. Thus, case (a) of Theorem (2.15) cannot hold at time $T(\omega)$, and $Y_{T(\omega)}(\omega) = j$ for some $j \in E$. Note that $T(\omega) = t + W_t(\omega)$ with W_t defined by (2.8), and what we have observed can be reworded as follows.

(2.21) PROPOSITION. If $Y_t = i$ where i is stable, then the next state Y_{t+W_t} to be entered is j for some $j \in E$.

Consequently, if the process Y enters Δ at a finite time ζ, it must do so by going over a sequence of stable states X_n with limit Δ and not by a jump from a stable state into Δ. If the state space $E = \{0, 1, 2, \ldots\}$, then Proposition (2.21) states that all jumps are of finite magnitude.

3. Structure of a Markov Process

Throughout this section Y is a Markov process with state space E and standard transition function (P_t). We assume that E has the discrete topology, and that the extra point Δ is the "point at infinity" if E is not finite. Finally, we assume that $t \longrightarrow Y_t(\omega)$ is right continuous for almost all $\omega \in \Omega$. By Proposition (2.18) this is equivalent to assuming that there are no instantaneous states. Until page 250 we further suppose that all states are stable.

For almost all $\omega \in \Omega$, the path $t \longrightarrow Y_t(\omega)$ can be described as follows (we write $Y(t, \omega) = Y_t(\omega)$ for typographical ease): Initially, at time $T_0(\omega) = 0$, the process is in some state $X_0(\omega) = Y(0, \omega)$. It remains in that state for some positive time; and at the end, at some time $T_1(\omega)$, it jumps to a new stable state $X_1(\omega) = Y(T_1(\omega), \omega)$. The process remains in the new state for some positive time; and at some time $T_2(\omega)$ jumps to another state $X_2(\omega) = Y(T_2(\omega), \omega)$. Then it stays there until some later time $T_3(\omega)$, at which it jumps again, to some state $X_3(\omega) = Y(T_3(\omega), \omega)$; and so on (see Figure 8.3.1 for the picture). In this section we will examine the underlying structure of Y as defined by these X_n and T_n.

In terms of the waiting time W_t from t until the instant of the next change of state (see (2.8) for the definition), we define

$$(3.1) \qquad\qquad T_0 = 0; \quad T_{n+1} = T_n + W_{T_n}, \quad n \in \mathbb{N},$$

$$(3.2) \qquad\qquad X_n = Y(T_n), \qquad\qquad n \in \mathbb{N}.$$

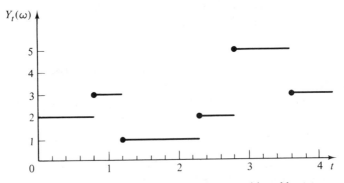

Figure 8.3.1 A realization of a Markov process with stable states. For the realization ω drawn, the successive states visited are $X_0(\omega) = 2$, $X_1(\omega) = 3$, $X_2(\omega) = 1$, $X_3(\omega) = 2, \ldots$, and the successive jump times are $T_1(\omega) = 0.8$, $T_2(\omega) = 1.2$, $T_3(\omega) = 2.3, \ldots$.

The times $T_0, T_1, T_2, \ldots$ are the instants of transitions for the process Y, and $X_0, X_1, X_2, \ldots$ are the successive states visited by Y. If $X_n = i$, the interval $[T_n, T_{n+1})$ is said to be a *sojourn* interval in i. Since all states are stable, almost surely, $0 < W_{T_n} < \infty$ for each n by (2.13), and each X_n is in E by Proposition (2.21).

The following theorem shows that $\{X_n; n \in \mathbb{N}\}$ is a Markov chain, and $T_{n+1} - T_n$ has an exponential distribution with the parameter depending on X_n. This is an important result.

(3.3) THEOREM. For any $n \in \mathbb{N}, j \in E$, and $u \in \mathbb{R}_+$ we have

$$(3.4) \quad P\{X_{n+1} = j, T_{n+1} - T_n > u \,|\, X_0, \ldots, X_n; T_0, \ldots, T_n\} = Q(i, j)e^{-\lambda(i)u}$$

if $\{X_n = i\}$ occurs. Here, $Q(i, j) \geq 0$, $Q(i, i) = 0$, and $\sum_j Q(i, j) = 1$.

Proof. Fix j, n, and u. First we note that T_n is a stopping time: this follows since $T_n(\omega) \leq t$ if and only if there are $0 < s_1 < s_2 < \cdots < s_n \leq t$ such that $Y_0(\omega) \neq Y_{s_1}(\omega) \neq \cdots \neq Y_{s_n}(\omega)$, and the existence of such $s_1, \ldots, s_n$ can be determined once $Y_s(\omega)$ is known for all $s \leq t$.

We put $T = T_n$ for simplicity. Then $T_{n+1} - T_n = W_T$ and $X_{n+1} = Y(T + W_T)$. Clearly, knowing $X_0, \ldots, X_n, T_0, \ldots, T_n$ is equivalent to knowing Y_t for all $t \leq T_n = T$. Hence, the left-hand side of (3.4) is

$$P\{Y(T + W_T) = j, W_T > u \,|\, Y_t; t \leq T\}.$$

By the strong Markov property (cf. Theorem (1.14)), this in turn is equal to $g(Y_T) = g(X_n)$, where

$$(3.5) \qquad g(i) = P\{Y(W_0) = j, W_0 > u \,|\, Y_0 = i\}.$$

To complete the proof we need to show that $g(i)$ has the form

$$(3.6) \qquad g(i) = Q(i, j)e^{-\lambda(i)u}.$$

We can write (3.5) as

$$(3.7) \qquad g(i) = P\{W_0 > u \,|\, Y_0 = i\}P\{Y(W_0) = j \,|\, Y_0 = i, W_0 > u\}.$$

First, by Theorem (2.9),

$$(3.8) \qquad P\{W_0 > u \,|\, Y_0 = i\} = e^{-\lambda(i)u}.$$

On the other hand, the event $\{Y_0 = i, W_0 > u\}$ is the same as $\{Y_s = i; s \leq u\}$, and given that $W_0 > u$ we have $W_0 = u + W_u$. Thus,

(3.9)
$$P\{Y(W_0) = j \mid Y_0 = i, W_0 > u\}$$
$$= P\{Y(u + W_u) = j \mid Y_s = i; s \leq u\}$$
$$= P\{Y(u + W_u) = j \mid Y_u = i\}$$
$$= P\{Y(W_0) = j \mid Y_0 = i\} = Q(i, j)$$

independent of u by Proposition (1.12). Putting (3.8) and (3.9) into (3.7) yields (3.6). $\qquad\qquad\square$

Putting $u = 0$ in (3.4) yields the following

(3.10) COROLLARY. *The sequence $X_0, X_1, X_2, \ldots$ of successive states visited form a Markov chain with the transition matrix Q.*

It follows from Theorem (3.3) that

$$P\{T_{n+1} - T_n > u \mid X_n = i, X_{n+1} = j\} = e^{-\lambda(i)u}$$

independent of j; in other words, the distribution of a sojourn time depends only on the state in which that sojourn passes and not on the next state. By iteration, we can obtain the following

(3.11) COROLLARY. *For any $n \in \mathbb{N}$, $i_0, \ldots, i_n \in E$, and $u_1, \ldots, u_n \in \mathbb{R}_+$ we have*

$$P\{T_1 - T_0 > u_1, \ldots, T_n - T_{n-1} > u_n \mid X_0 = i_0, \ldots, X_{n-1} = i_{n-1}, X_n = i_n\}$$
$$= e^{-\lambda(i_0)u_1} \cdots e^{-\lambda(i_{n-1})u_n}.$$

In other words, the times between transitions are conditionally independent of each other given the successive states being visited, and each such sojourn time has an exponential distribution with the parameter dependent on the state being visited. This and the fact that the successive states visited form a Markov chain clarify the structure of a Markov process.

(3.12) EXAMPLE. Consider the queueing process M/M/1 introduced in Example (1.10). There, the times of transitions are the instants at which the queue size changes. If the change is due to an arrival, then the queue size increases by one; and if the change is due to a departure, then it decreases by one.

If $\{X_n = 0\}$ occurs, then the next change must be due to an arrival, which means that $X_{n+1} = 1$ with probability one, and furthermore, $T_{n+1} - T_n$ has the same distribution as the time until the next arrival, which is exponential with parameter a. Thus,

$$\lambda(0) = a; \quad Q(0, 1) = 1, \quad Q(0, 0) = Q(0, 2) = \cdots = 0.$$

Suppose next that $\{X_n = i\}$, $i \geq 1$, occurs. Let A be the time from T_n until the instant of next arrival, and let S be the time from T_n until the instant of completion of service being carried on. Then A and S are exponentially distributed random variables with respective parameters a and b and are independent of each other. Furthermore, $T_{n+1} - T_n = \min(A, S)$, which means that

$$P\{T_{n+1} - T_n > t \mid X_n = i\} = P\{A > t, S > t\} = e^{-at}e^{-bt};$$

in other words,

$$(3.13) \qquad\qquad \lambda(i) = a + b, \quad i = 1, 2, \ldots .$$

Concerning the queue size after the next change of state, we note that $X_{n+1} = i + 1$ if $S > A$ and $X_{n+1} = i - 1$ if $S < A$ (the third possibility, $A = S$, has probability zero). Thus,

$$P\{X_{n+1} = i + 1 \mid X_n = i\} = P\{S > A\}.$$

Using Proposition (2.2.16), we get

$$P\{S > A\} = E[P\{S > A \mid A\}]$$
$$= E[e^{-bA}] = \int_0^\infty ae^{-as}e^{-bs}\,ds = \frac{a}{a + b}.$$

Thus, for $i \geq 1$,

$$(3.14) \qquad\qquad Q(i, j) = \begin{cases} \dfrac{a}{a + b} & \text{if } j = i + 1, \\[2mm] \dfrac{b}{a + b} & \text{if } j = i - 1, \\[2mm] 0 & \text{otherwise.} \end{cases}$$

To summarize, then, $(\lambda(0), \lambda(1), \ldots) = (a, a + b, a + b, \ldots)$, and (writing $p = a/(a + b)$, $q = b/(a + b)$ for simplicity)

$$Q = \begin{bmatrix} 0 & 1 & & & 0 \\ q & 0 & p & & \\ & q & 0 & p & \\ & & \cdot & \cdot & \cdot \\ 0 & & & \cdot & \cdot \end{bmatrix}.$$

$\square$

In applications, as the preceding example illustrates, the data for a Markov process are usually given in terms of the parameters $\lambda(i)$ and the transition probabilities $Q(i, j)$. We will shortly see, in the next section, that the structure outlined above is also the starting point in many computations. There are, however, two further points to be attended to: namely, the problem of absorbing states and whether the Markov chain X and the transition times T_n specify Y completely.

We take up absorbing states first. Suppose $X_n(\omega)$ and $T_n(\omega)$ are defined by (3.1) and (3.2) for $n = 0, 1, \ldots, m$, and suppose $X_m(\omega) = i$ where i is absorbing. Then, we know that $Y_t(\omega) = i$ for all $t \geq T_m(\omega)$. Using the definition (2.8) of W_t, we get $T_{m+1}(\omega) = +\infty$, but (3.1) cannot be used further to get $T_{m+2}(\omega)$, etc. (since W_∞ is not defined), and more seriously, (3.2) gives $X_{m+1}(\omega) = Y_\infty(\omega) = \Delta$. To limit $X_n(\omega)$ to the state space E, and to define $X_n(\omega)$ and $T_n(\omega)$ beyond m, we now introduce the following definitions.

(3.15) $$T_0 = 0; \qquad X_0 = Y_0; \qquad W_\infty = +\infty;$$

and

(3.16) $$T_{n+1} = T_n + W_{T_n};$$

(3.17) $$X_{n+1} = \begin{cases} Y(T_{n+1}) & \text{if } T_{n+1} < \infty, \\ X_n & \text{if } T_{n+1} = \infty, \end{cases}$$

for each $n \in \mathbb{N}$. If there are no absorbing states, these definitions reduce to (3.1) and (3.2). It is easy to see that the following holds.

(3.18) REMARK. If Y is right continuous, and if (X_n, T_n) are defined by (3.15)–(3.17), then Theorem (3.3) and its corollaries (3.10) and (3.11) remain true. If i is an absorbing state, then it is an absorbing state for the Markov chain X as well. Thus, the transition matrix Q of the Markov chain can be used to classify the states as absorbing or stable: $Q(i, i) = 1$ for an absorbing state i, and $Q(i, i) = 0$ for a stable state i. $\square$

Next we take up the question of whether Y can be defined in terms of the X_n and T_n. It follows from the definitions (3.15)–(3.17) that

(3.19) $$Y_t(\omega) = X_n(\omega) \quad \text{for } t \in [T_n(\omega), T_{n+1}(\omega))$$

for any $\omega \in \Omega$. Hence, Y can be defined by (3.19) in terms of X and the T_n provided that any $t \geq 0$ belong to the interval $[T_n(\omega), T_{n+1}(\omega))$ for some n; in other words, provided that

(3.20) $$\zeta(\omega) = \sup_n T_n(\omega) = +\infty$$

for almost all ω. Note that, in general,

$$(3.21) \qquad \zeta(\omega) = \sup_n T_n(\omega) = \inf \{t : Y_t(\omega) = \Delta\}.$$

If $\zeta(\omega) < \infty$, case (a) of Theorem (2.15) applies, and there are infinitely many transitions within any ε neighborhood of $\zeta(\omega)$. Therefore, if $\zeta < \infty$ with a positive probability, it is impossible to talk of a next jump after ζ, and we cannot define Y constructively beyond ζ. Such questions as the behavior of Y at Δ are the subject matter of the *boundary theory* of Markov processes. We will not go into it here but instead will give criteria for recognizing whether ζ is finite or not.

(3.22) Definition. The process Y is said to be a *regular* Markov process provided that for almost all $\omega \in \Omega$,
  (a) $t \longrightarrow Y_t(\omega)$ is right continuous, and
  (b) $\zeta(\omega) = \sup_n T_n(\omega) = +\infty$. $\qquad\qquad\square$

The following are some easily applied criteria for determining regularity.

(3.23). Theorem. If $\lambda(i) \leq c$ for all $i \in E$ for some constant $c < \infty$, then Y is regular.

Proof. From Corollary (3.11) we have, since $e^{-\lambda(i)t} \geq e^{-ct}$ under the hypothesis,

$$P\{T_1 \leq t_1, \ldots, T_n - T_{n-1} \leq t_n\} \leq (1 - e^{-ct_1}) \cdots (1 - e^{-ct_n}).$$

The right side is the joint distribution of the first n interarrival times in a Poisson process with rate c. Thus,

$$P\{T_n \leq t\} \leq \sum_{k=n}^{\infty} \frac{e^{-ct}(ct)^k}{k!}.$$

Since $T_0 \leq T_1 \leq \ldots$, we have $\{T_0 \leq t\} \supset \{T_1 \leq t\} \supset \cdots$ and the intersection is $\cap_n \{T_n \leq t\} = \{\sup T_n \leq t\}$. Hence, by Corollary (1.1.12),

$$P\{\zeta \leq t\} \leq \lim_{n \to \infty} \sum_{k=n}^{\infty} \frac{e^{-ct}(ct)^k}{k!} = 0.$$

So, $\zeta < \infty$ with probability zero. $\qquad\qquad\square$

If there are only finitely many states, then by Proposition (2.17) there are no instantaneous states and by (2.18) Y is right continuous. Then picking c to be the largest of the $\lambda(i)$, we see that the hypothesis of the preceding theorem is satisfied. We thus have the following pleasant

(3.24) COROLLARY.   If E is finite, then Y is regular.

If the initial state i is recurrent in the Markov chain X, then it will be visited infinitely often, and each visit will last an exponential time with parameter $\lambda(i)$. The sum of an infinite number of random variables with the same exponential distribution is infinite (this follows by taking the random variables in question as the interarrival times in a Poisson process), and therefore, the total time spent in that recurrent state i alone is infinite; and lim T_n must be larger than that. Hence we have proved the following.

(3.25) THEOREM.   Suppose there are no instantaneous states, and suppose every state is recurrent in the Markov chain X. Then Y is a regular Markov process.

The arguments leading up to the preceding theorem remain good if the Markov chain X does not stay forever in a set of transient states:

(3.26) COROLLARY.   Suppose the probability is zero that the Markov chain X stays forever in a set of transient states. If there are no instantaneous states, then Y is regular.

Finally, to show that not all processes are regular, we list the following theorem without proof.

(3.27) THEOREM.   Suppose there exists a set A of transient states such that starting in A, the chain X has a positive probability of staying forever in A. Let $R = \sum_n Q^n$, and suppose

$$\sum_{j \in A} \frac{R(i, j)}{\lambda(j)} < \infty$$

for some $i \in A$. Then Y is *not* regular.

The reason for the conclusion can be seen as follows. Chain X, and therefore Y, can stay in A forever. The expected number of visits to j is $R(i, j)$, and each visit lasts an exponential time with mean $1/\lambda(j)$. Hence, the sum appearing in (3.27) is the expected amount of time spent in *all* the states of A; and since this is finite, the time spent itself must not be infinite.

(3.28) EXAMPLE.   *Pure Birth Process.* This is used to depict the size of a bacterial colony; Y_t is the "size" of the population at time t. There are no "deaths," and the size increases without end. The question is whether the population size can become infinite in a finite time.

Stochastically, this is a slight generalization of the Poisson process where the time between the nth and the $(n + 1)$th arrivals is exponentially distributed as before but now with a mean $1/\lambda(n)$ depending on n. (In the population model, it is argued that the more individuals there are in the population, the shorter the time until the next birth.)

Hence the Markov chain $\{X_n\}$ is very simple; if the initial size is $X_0(\omega) = i$, then $X_1(\omega) = i + 1$, $X_2(\omega) = i + 2, \ldots$. Thus, if $X_0 = i$, the time T_n of the nth transition is the sum of n exponentially distributed independent random variables with means $1/\lambda(i)$, $1/\lambda(i + 1), \ldots, 1/\lambda(i + n - 1)$. Hence,

$$E[T_n \,|\, Y_0 = i] = \sum_{m=0}^{n-1} \frac{1}{\lambda(i + m)},$$

so that, by the monotone convergence theorem (2.1.34),

(3.29) $$E[\lim_n T_n \,|\, Y_0 = i] = \sum_{m=0}^{\infty} \frac{1}{\lambda(i + m)}.$$

If the latter sum is finite, that is, if

(3.30) $$\sum_{n=0}^{\infty} \frac{1}{\lambda(n)} < \infty,$$

then (3.29) implies that

$$\zeta = \lim_{n \to \infty} T_n < \infty$$

almost surely no matter what the initial state i is. If, however, the sum in (3.30) is infinite, then it can be shown that $\lim_n T_n = +\infty$ almost surely. $\square$

4. Potentials and Generators

This section is devoted to the computational aspects of Markov processes and to the relationships between the transition function and the other, more elementary, data such as the $\lambda(i)$ and $Q(i, j)$ of Theorems (2.9) and (3.3).

Throughout this section we take $Y = \{Y_t; t \geq 0\}$ to be a regular Markov process with state space E and transition functions $P_t(i, j)$. We will write $P_i\{A\}$ for the conditional probability $P\{A \,|\, Y_0 = i\}$, and we will denote the corresponding expectations by E_i.

We have shown in Theorem (3.3) and its corollaries (3.10) and (3.11) that the successive states visited by Y form a Markov chain $X = \{X_n; n \in \mathbb{N}\}$ and that the sojourn time $T_{n+1} - T_n$ in state X_n has an exponential distribution with parameter $\lambda(i)$ if $X_n = i$ (see also Remark (3.18)). Because Y is regular, for any t there is a sojourn interval $[T_n, T_{n+1})$ which contains t, and therefore (3.19) defines Y completely in terms of X and the T_n. Let Q be the transition matrix of X, and let $\lambda(i)$ be the parameter of the exponential distribution of a sojourn time in i.

The following results relate the transition functions $P_t(i, j)$ to the data $Q(i, j)$ and $\lambda(i)$ and vice versa.

(4.1) LEMMA. For any $i, j \in E$ and $t \in \mathbb{R}_+$,

$$P_t(i, j) = e^{-\lambda(i)t}I(i, j) + \int_0^t \lambda(i)e^{-\lambda(i)s} \sum_k Q(i, k)P_{t-s}(k, j)\, ds.$$

Proof. If i is absorbing, then $\lambda(i) = 0$ and $P_t(i, j) = I(i, j)$, so the claim holds. Suppose i is stable and first write

(4.2) $$P_t(i, j) = P_i\{Y_t = j, T_1 > t\} + P_i\{Y_t = j, T_1 \leq t\}.$$

If $T_1 > t$, then $Y_t = Y_0$, so

(4.3) $$P_i\{Y_t = j, T_1 > t\} = I(i, j)P_i\{T_1 > t\} = e^{-\lambda(i)t}I(i, j).$$

If $\{T_1 \leq t\}$ occurs, then the strong Markov property (1.15) at T gives

$$P_i\{Y_t = j \mid X_1, T_1\} = P_{t-T_1}(X_1, j),$$

which yields

(4.4) $$P_i\{Y_t = j, T_1 \leq t\} = \int_0^t \lambda(i)e^{-\lambda(i)s} \sum_k Q(i, k)P_{t-s}(k, j)\, ds.$$

The proof is completed by putting (4.3) and (4.4) into (4.2). $\qquad\square$

The following is the main result relating the $P_t(i, j)$ to $Q(i, j)$ and $\lambda(i)$.

(4.5) THEOREM. For any $i, j \in E$ the function $t \longrightarrow P_t(i, j)$ is differentiable and the derivative is continuous. At $t = 0$ the derivative is

(4.6) $$A(i, j) = \begin{cases} -\lambda(i) & \text{if } i = j, \\ \lambda(i)Q(i, j) & \text{if } i \neq j. \end{cases}$$

For arbitrary $t \geq 0$,

(4.7) $$\frac{d}{dt}P_t(i, j) = \sum_{k \in E} A(i, k)\, P_t(k, j) = \sum_{k \in E} P_t(i, k)A(k, j).$$

Proof. In the expression of Lemma (4.1) we first make a change of variable $t - s = u$ to write

(4.8) $$P_t(i, j) = e^{-\lambda(i)t}[I(i, j) + \int_0^t \lambda(i)e^{\lambda(i)u} \sum_k Q(i, k)P_u(k, j)\, du].$$

By proposition (2.3), $u \rightarrow P_u(k, j)$ is continuous. Now the bounded convergence theorem applies to show that $u \rightarrow \sum_k Q(i, k) P_u(k, j)$ is continuous. Thus, the integral on the right defines a function which has a continuous derivative.

Taking derivatives, we obtain

(4.9)
$$\frac{d}{dt} P_t(i, j) = -\lambda(i) P_t(i, j) + e^{-\lambda(i)t} \cdot \lambda(i) e^{+\lambda(i)t} \sum_k Q(i, k) P_t(k, j)$$
$$= \sum_k [-\lambda(i) I(i, k) + \lambda(i) Q(i, k)] P_t(k, j).$$

As $t \downarrow 0$, this becomes

$$A(i, j) = -\lambda(i) I(i, j) + \lambda(i) Q(i, j),$$

which is the same as (4.6). Putting this back in (4.9), we obtain the first equality in (4.7). To obtain the second equality in (4.7) we use $P_{t+s} = P_t P_s$, differentiate with respect to s, and evaluate it at $s = 0$. ☐

If the transition function (P_t) is known, then its derivative at $t = 0$ is A, and this determines Q and λ: namely, $\lambda(i) = -A(i, i)$; if $\lambda(i) > 0$ then $Q(i, j) = A(i, j)/\lambda(i)$; and if $\lambda(i) = 0$ then $Q(i, j) = 0$ for all $j \neq i$. Of course, $Q(i, i) = 0$ if i is stable and 1 if i is absorbing.

Consider next the converse problem where Q and λ, and therefore A, are known and P_t is to be computed. One method is to solve the infinite system of differential equations (given by (4.7))

(4.10)
$$\frac{d}{dt} P_t = A P_t$$

or

(4.11)
$$\frac{d}{dt} P_t = P_t A.$$

If (P_t) were a numerical-valued function, the solution of either would have been e^{At}. A similar result holds in general with the interpretation of e^{At} as

(4.12)
$$e^{tA} = \sum_{n=0}^{\infty} \frac{t^n}{n!} A^n.$$

This we put next without proof.

(4.13) THEOREM. For any $t \geq 0$ we have

$$P_t = e^{tA}.$$

The differential equations (4.10) and (4.11) are called, respectively, *Kolmogorov's backward and forward equations*. The fact that (P_t) is defined by its derivative at $t = 0$ makes these results interesting. For this reason A is called the *generator* of the process Y.

If the state space E is finite, there are certain matrix-theoretic methods of attack for computing P_t from A by using (4.13) (cf. Appendix). However, unless A has some special structure, such computations are very difficult in the case of infinite E. In this case, an alternate line of attack is provided by considering the potentials, defined for $\alpha \geq 0$,

$$(4.14) \qquad U^\alpha(i, j) = \int_0^\infty e^{-\alpha t} P_t(i, j)\, dt, \quad i, j \in E.$$

Since $P_t(i, j)$ is continuous in t, its Laplace transform $U^\alpha(i, j)$ determines it uniquely. Thus, if $U^\alpha(i, j)$ is known for all α in some interval, then $P_t(i, j)$ can be obtained by inverting the Laplace transform.

However, this is not the real reason for our interest in U^α. Let g be a non-negative function defined on the state space E, and let $\alpha \geq 0$ be a fixed number. We now think of $g(j)$ as the rate of reward being received when the process Y is in state j, and of α as the discount rate such that one unit received at time t has the present worth of $e^{-\alpha t}$. Then the rate of reward at time t is $g(Y_t)$, and the total discounted reward is

$$\int_0^\infty e^{-\alpha t} g(Y_t)\, dt.$$

We are interested in the expected value of this random variable given the initial state i. Motivated by this, we define, for any non-negative function g,

$$(4.15) \qquad U^\alpha g(i) = E_i\left[\int_0^\infty e^{-\alpha t} g(Y_t)\, dt\right], \quad i \in E.$$

The function $U^\alpha g$ defined by (4.15) is called the α-*potential* of g for the process Y. If $\alpha = 0$, instead of $U^0 g$ we simply write Ug and call it the *potential* of g for Y. The operator U^α for a fixed $\alpha \geq 0$ is called the α-*potential operator* of Y. The family $\{U^\alpha; \alpha \geq 0\}$ is called the *resolvent* of (P_t).

Passing the expectation in (4.15) inside the integral, and noting that

$$E_i[g(Y_t)] = \sum_{j \in E} P_t(i, j) g(j) = P_t g(i),$$

we see that

$$U^\alpha g(i) = \sum_{j \in E}\left[\int_0^\infty e^{-\alpha t} P_t(i, j)\, dt\right] g(j) = \sum_{j \in E} U^\alpha(i, j) g(j),$$

where U^α is the matrix defined by (4.14). Thus, to obtain the α-potential of g, we need to multiply the matrix U^α by the column vector g.

In various applications the following analog of (7.1.14) for the continuous time turns out to be useful.

(4.16) THEOREM. For any stopping time T,

$$U^\alpha g(i) = E_i\left[\int_0^T e^{-\alpha t} g(Y_t)\, dt\right] + E_i[e^{-\alpha T} U^\alpha g(Y_T)],$$

for all $i \in E$, $\alpha \geq 0$, and $g \geq 0$.

Proof. We need to show that if $f = U^\alpha g$,

$$(4.17) \qquad E_i\left[\int_T^\infty e^{-\alpha t} g(Y_t)\, dt\right] = E_i[e^{-\alpha T} f(Y_T)].$$

Making a change of variable $s = t - T$, we write

$$Z = \int_T^\infty e^{-\alpha t} g(Y_t)\, dt = e^{-\alpha T}\int_0^\infty e^{-\alpha s} g(Y_{T+s})\, ds.$$

Now using (2.2.16) and (1.14) in succession, we get

$$E_i[Z] = E_i\left[E_i\left[e^{-\alpha T}\int_0^\infty e^{-\alpha s} g(Y_{T+s})\, ds\,|\, Y_u;\, u \leq T\right]\right]$$

$$= E_i\left[e^{-\alpha T} E_i\left[\int_0^\infty e^{-\alpha s} g(Y_{T+s})\, ds\,|\, Y_u;\, u \leq T\right]\right] = E_i[e^{-\alpha T} f(Y_T)].$$

This is the desired result (4.17).   ☐

A particular application of the preceding theorem reduces the computation of $U^\alpha g$ to that of solving a system of linear equations.

(4.18) THEOREM. Let g be a bounded function and $\alpha > 0$. Then $f = U^\alpha g$ is the unique solution of the system of linear equations

$$(\alpha I - A)f = g.$$

Proof. In Theorem (4.16) let the stopping time T be the time of the first jump T_1. Then $Y_t = Y_0$ for all $t < T$, so

$$(4.19) \quad E_i\left[\int_0^T e^{-\alpha t} g(Y_t)\, dt\right] = g(i) E_i\left[\int_0^T e^{-\alpha t}\, dt\right]$$

$$= g(i) E_i\left[\frac{(1 - e^{-\alpha T})}{\alpha}\right]$$

$$= \frac{1}{\alpha} g(i)\left[1 - \int_0^\infty \lambda(i) e^{-\lambda(i) t} e^{-\alpha t}\, dt\right] = \frac{g(i)}{\alpha + \lambda(i)}.$$

On the other hand, $Y_T = X_1$, and we have

$$(4.20) \quad E_i[e^{-\alpha T} f(Y_T)] = \sum_j Q(i, j) f(j)\int_0^\infty \lambda(i) e^{-\lambda(i) t} e^{-\alpha t}\, dt = \frac{\lambda(i)}{\alpha + \lambda(i)} Q f(i).$$

Adding (4.19) and (4.20) side by side, noting that by (4.16) the left side is $f(i)$, and rearranging the terms, we get

(4.21) $$[\alpha + \lambda(i)]f(i) - \lambda(i)Qf(i) = g(i).$$

Using (4.6) to put $\lambda(i)I(i, j) - \lambda(i)Q(i, j) = -A(i, j)$ yields the equation f is claimed to satisfy.

To show that $f = U^\alpha g$ is the only solution, let $\hat{f}$ be another solution and note that, then, $h = f - \hat{f}$ satisfies $(\alpha I - A)h = 0$; that is,

$$Ah = \alpha h.$$

Using (4.11), we now have

$$\frac{d}{dt}(e^{-\alpha t}P_t h) = -\alpha e^{-\alpha t}P_t h + e^{-\alpha t}P_t Ah = 0.$$

Integrating this from 0 to s, we have

$$e^{-\alpha s}P_s h - P_0 h = e^{-\alpha s}P_s h - h = 0.$$

Thus, $h = e^{-\alpha s}P_s h$. Since h is bounded and P_s is a Markov matrix, $P_s h$ is bounded by some constant. Thus, letting $s \to \infty$, we see that $h = 0$ since $e^{-\alpha s} \to 0$. $\qquad\square$

(4.22) EXAMPLE. Suppose Y is a Markov process with state space $E = \{a, b, c\}$ and generator

$$A = \begin{bmatrix} -0.5 & 0.5 & 0 \\ 1.2 & -1.2 & 0 \\ 0.6 & 0.2 & -0.8 \end{bmatrix}.$$

If $\alpha = 0.1$ and $g = (2, 5, 4)$, then the α-potential of g satisfies $(\alpha I - A)f = g$, which in this instance becomes (we write $f = (x, y, z)$)

$$\begin{bmatrix} 0.6 & -0.5 & 0 \\ -1.2 & 1.3 & 0 \\ -0.6 & -0.2 & 0.9 \end{bmatrix}\begin{bmatrix} x \\ y \\ z \end{bmatrix} = \begin{bmatrix} 2 \\ 5 \\ 4 \end{bmatrix}.$$

The solution is

$$f = (\tfrac{170}{6}, 30, 30).$$ $\qquad\square$

In particular, if $g = 1_j$, then $f = U^\alpha g$ is the j-column of the matrix U^α. Thus, the matrix U^α can be obtained by solving

(4.23) $$(\alpha I - A)U^\alpha = I.$$

If there are only finitely many states, (4.23) implies that

$$(4.24) \qquad\qquad U^\alpha = (\alpha I - A)^{-1}.$$

Most of these computations can be simplified by a device which is applicable to a large class of Markov processes; namely, those for which $\lambda(i) \leq c$ for all i for some constant c. In particular, this class includes all Markov processes with finitely many states. The idea involved is that of reducing the problems considered to the case of Markov processes of the type discussed in Example (1.9). We take this up once more.

Suppose $\hat{X}$ is a Markov chain with a transition matrix K, and suppose N is a Poisson process with rate c and is independent of $\hat{X}$. Define $\hat{Y}_t = \hat{X}_{N_t}$ for all t. Then we know from Example (1.9) that $\hat{Y}$ is a Markov process with transition function

$$(4.25) \qquad\qquad P_t = \sum_{n=0}^{\infty} \frac{e^{-ct}(ct)^n}{n!} K^n.$$

Then U^α is easy to compute: since

$$\int_0^\infty e^{-\alpha t} \frac{c(ct)^{n-1} e^{-ct}}{(n-1)!} dt = \left(\frac{c}{\alpha + c}\right)^n,$$

$$c \int_0^\infty e^{-\alpha t} P_t \, dt = \frac{c}{\alpha + c} I + \left(\frac{c}{\alpha + c}\right)^2 K + \left(\frac{c}{\alpha + c}\right)^3 K^2 + \cdots,$$

which gives

$$(4.26) \qquad\qquad (\alpha + c) U^\alpha = \sum_{n=0}^{\infty} \left(\frac{c}{\alpha + c}\right)^n K^n.$$

The right side of this is $R^\beta = \sum \beta^n K^n$ with $\beta = c/(\alpha + c)$, and the computation of R^β was discussed earlier in Chapter 7.

Now consider the generator A corresponding to the process $\hat{Y}$. If $\hat{Y}_0 = \hat{X}_0 = i$, then the chain stays in i for exactly n steps with probability $K(i, i)^{n-1}$ $(1 - K(i, i))$, and given that it is moving out of i, the probability that it moves into j is $K(i, j)/(1 - K(i, i))$. Therefore, a sojourn of $\hat{Y}$ in state i is the sum of n interarrival times of N with probability pq^{n-1}, where $p = 1 - K(i, i)$. From our results of Section 5 of Chapter 4, then, a sojourn in state i has the exponential distribution with parameter

$$\lambda(i) = cp = c[1 - K(i, i)].$$

The successive states visited by $\hat{Y}$ form a Markov chain X which moves through the same states as $\hat{X}$ but without repetitions of any state. The transition probabilities for X are, if i is not absorbing,

$$(4.27) \qquad Q(i, j) = \begin{cases} 0 & \text{if } i = j, \\ \dfrac{K(i, j)}{1 - K(i, i)} & \text{if } i \neq j. \end{cases}$$

Hence, the generator A of $\hat{Y}$ is

$$(4.28) \qquad A(i, j) = \begin{cases} -c[1 - K(i, i)] & \text{if } i = j, \\ cK(i, j) & \text{if } i \neq j, \end{cases}$$

which in matrix notation becomes

$$(4.29) \qquad A = cK - cI.$$

Starting with a given generator A, we can reverse the above computations provided that (4.29) holds for some constant $c > 0$ and Markov matrix K. It is clear from (4.28) that this is always possible if $\lambda(i) \leq c$ for all i, for then the matrix

$$(4.30) \qquad K = I + \frac{1}{c} A$$

is a Markov matrix. We thus have proved the following

(4.31) THEOREM. Suppose $\lambda(i) \leq c$ for all i for some finite constant c. Then (4.25) holds with K defined by (4.30).

In other words, if Y is a Markov process with $\lambda(i) \leq c$ for all i, then we may think of Y as if it were the process $\hat{Y}$. Since both Y and $\hat{Y}$ have the same transition function P_t,

$$P_i\{Y_{t_1} = i_1, \ldots, Y_{t_n} = i_n\} = P_i\{\hat{Y}_{t_1} = i_1, \ldots, \hat{Y}_{t_n} = i_n\},$$

which in turn implies that the expectation of any random variable defined in terms of Y_t, $t \geq 0$, is equal to that obtained by replacing the Y_t's by $\hat{Y}_t$'s.

(4.32) EXAMPLE. Let Y be a Markov process with state space $E = \{r, w, b\}$ and generator

$$A = \begin{bmatrix} -2 & 1 & 1 \\ 1 & -3 & 2 \\ 1 & 0 & -1 \end{bmatrix}.$$

Now $\lambda(r) = 2$, $\lambda(w) = 3$, $\lambda(b) = 1$. For c we can choose any number greater than or equal to 3. Let us pick $c = 3$. Then

$$K = I + \frac{1}{3}A = \begin{bmatrix} \frac{1}{3} & \frac{1}{3} & \frac{1}{3} \\ \frac{1}{3} & 0 & \frac{2}{3} \\ \frac{1}{3} & 0 & \frac{2}{3} \end{bmatrix},$$

and

$$P_t = \sum_{n=0}^{\infty} \frac{e^{-3t}(3t)^n}{n!} K^n.$$

Consider now the distribution of the first passage time from state w into state r. For the Markov chain $\hat{X}$ with transition matrix K, this takes place at exactly the nth step with probability $F_n(w, r)$, where

$$F_n(w, r) = \left(\frac{2}{3}\right)^{n-1}\left(\frac{1}{3}\right), \quad n = 1, 2, 3, \ldots.$$

From the way P_t is expressed, then, the first passage time from w to r is the time of the nth arrival with probability $F_n(w, r)$, where the arrivals form a Poisson process N with parameter $c = 3$. Hence, letting T be the time of first entrance to r, we have

$$P_w\{T \leq t\} = \sum F_n(w, r) \cdot P\{N_t \geq n\}$$
$$= \sum_{n=1}^{\infty} \left(\frac{1}{3}\right)\left(\frac{2}{3}\right)^{n-1} \sum_{k=n}^{\infty} \frac{e^{-3t}(3t)^k}{k!} = 1 - e^{-t}. \qquad \square$$

5. Limit Theorems

Let $Y = \{Y_t; t \in \mathbb{R}_+\}$ be a regular Markov process with state space E and transition function (P_t). Let X be the underlying Markov chain, and let Q denote its transition matrix (see (3.3) ff. for the precise definitions).

For a fixed state $j \in E$, let the lengths of successive sojourns in j be denoted by $W_1, W_2, \ldots$. These are independent and identically distributed random variables with the common distribution being exponential with parameter $\lambda(j)$. Thus, were these laid end to end, one would obtain a Poisson process with parameter $\lambda(j)$; that is, $W_1, W_1 + W_2, \ldots$ form a Poisson process with rate $\lambda(j)$.

If state j is transient for the Markov chain X, then the total number N of visits to j is finite with probability one, and starting in state i, its distribution is (cf. Proposition (5.2.8))

$$(5.1) \qquad P_i\{N = k\} = \begin{cases} 1 - F(i, j) & \text{if } k = 0, \\ F(i, j)F(j, j)^{k-1}[1 - F(j, j)] & \text{if } k = 1, 2, \ldots, \end{cases}$$

where $F(i, j)$ is the probability of ever reaching j from i. Then the total time spent in state j is the sum of N exponential random variables:

$$(5.2) \qquad \int_0^\infty 1_j(Y_s(\omega)) \, ds = \begin{cases} 0 & \text{if } N(\omega) = 0, \\ W_1(\omega) + \cdots + W_{N(\omega)}(\omega) & \text{if } N(\omega) \geq 1; \end{cases}$$

and this is finite with probability one. In fact we have the following

(5.3) PROPOSITION. If state j is transient, then

$$P_i \left\{ \int_0^\infty 1_j(Y_s) \, ds \leq t \right\} = 1 - F(i, j) \exp[-t\lambda(j)(1 - F(j, j))].$$

Proof. The number of visits N to j is determined by the Markov chain X, whereas the sojourn times $W_1, W_2, \ldots$ are independent of X. Hence, from (5.2),

$$P_i \left\{ \int_0^\infty 1_j(Y_s) \, ds \leq t \right\} = \sum_{k=0}^\infty P_i\{N = k\} P\{W_1 + \cdots + W_k \leq t\}.$$

Using (5.1) and the fact that $W_1, W_1 + W_2, \ldots$ form a Poisson process (we write $p(t, n)$ for the probability of n arrivals during $[0, t]$), we see that the right side is equal to

$$\sum_{k=0}^\infty P_i\{N = k\} \sum_{n=k}^\infty p(t, n) = \sum_{n=0}^\infty p(t, n) \sum_{k=0}^n P_i\{N = k\}$$

$$= \sum_{n=0}^\infty \frac{e^{-t\lambda(j)}[t\lambda(j)]^n}{n!} [1 - F(i, j)F(j, j)^n]$$

$$= 1 - F(i, j) \exp[-t\lambda(j)(1 - F(j, j))],$$

which is as claimed. $\qquad\qquad\square$

In other words, the total amount of time spent in a transient state j is finite with probability one; and it has the exponential distribution with parameter $\lambda(j)(1 - F(j, j))$ provided that the initial state is j. From this it is easy to compute the expected value of the total time spent in j:

(5.4) COROLLARY. If j is transient, then

$$U(i, j) = E_i \left[\int_0^\infty 1_j(Y_s) \, ds \right] = \frac{R(i, j)}{\lambda(j)} < \infty$$

where $R(i, j)$ is the expected number of visits to j, starting from i.

Computation of R was discussed in Chapter 6; once this is done, the expected total time spent in j is easy to obtain. It is clear from the foregoing that we have the following

(5.5) PROPOSITION. If state j is transient, then

$$\lim_{t \to \infty} P_i\{Y_t = j\} = \lim_{t \to \infty} P_t(i, j) = 0.$$

In the case of a recurrent state j, (5.2) still holds for the total time spent in j. By Corollary (5.2.11), the number of visits to j is infinite provided that it ever be reached. If j is reached, then the total time spent there is infinite with probability one. That is,

$$(5.6) \qquad P_i\left\{\int_0^\infty 1_j(Y_s)\, ds = +\infty\right\} = F(i, j) \geq 0,$$

and

$$(5.7) \qquad P_i\left\{\int_0^\infty 1_j(Y_s)\, ds = 0\right\} = 1 \quad \text{if } F(i, j) = 0.$$

Throughout the remainder of this section we will limit ourselves to Markov processes Y which are *irreducible recurrent*, that is, to the processes whose underlying Markov chains are irreducible recurrent. In view of the preceding remarks, this is without any loss of generality.

The limiting behavior of the transition functions $P_t(i, j)$ as $t \to \infty$ is just as in the case of Markov chains except that it is made simpler by the disappearance of periodicity.

Since the underlying Markov chain X is irreducible recurrent, by Theorem (6.2.25), there is a unique (up to multiplication by a constant) solution to

$$(5.8) \qquad v = vQ$$

which is strictly positive. The ratio $v(i)/v(j)$ had the interpretation of being the expected number of visits to i in between two visits to j. Now consider a time interval starting with an entrance to the state j and ending with the next entrance to j. This cycle starts with an interval of expected length $1/\lambda(j)$ spent visiting j, and then there are other intervals spent in states $i \neq j$; the expected number of visits to i is $v(i)/v(j)$, and each visit to i lasts an expected time of $1/\lambda(i)$ units. Hence, the expected length of the cycle considered is

$$(5.9) \qquad \frac{1}{\lambda(j)} + \frac{1}{v(j)} \sum_{i \neq j} \frac{v(i)}{\lambda(i)} = \frac{1}{v(j)} \sum_{i \in E} \frac{v(i)}{\lambda(i)};$$

therefore the *ratio* of the expected value of the time spent in j to that of the total length of the cycle is

$$(5.10) \qquad \pi(j) = \frac{v(j)/\lambda(j)}{\sum_{i \in E} v(i)/\lambda(i)}.$$

Hence, the proportion of time the process Y spends in j is equal to this value $\pi(j)$, and as should be expected, this is also the probability that the process is in state j at time t for very large t. These arguments prove (see (10.5.22) for the missing link) the following basic limit

(5.11) THEOREM. Let Y be an irreducible recurrent Markov process. Then for any $j \in E$,

$$(5.12) \qquad\qquad \pi(j) = \lim_{t \to \infty} P\{Y_t = j\}$$

exists and is independent of the distribution of Y_0.

The limiting probabilities $\pi(j)$ are as given by (5.10). Therefore, either all $\pi(j)$ are zero, or else every $\pi(j)$ is positive.

In particular, if E is finite, the limiting probabilities are all positive. In the case where E is not finite, if the denominator in (5.10) is infinite, then $\pi(j) = 0$ for all j, and the process Y is said to be null. Otherwise, if the denominator in (5.10) is finite, then $\pi(j) > 0$ for every j, and $\sum \pi(j) = 1$. In this case, Y is said to be non-null. Note that the nullness of Y is independent of the corresponding notion for the underlying Markov chain X. Also note that by the same token, a positive limiting distribution for Y might exist even though X has no such positive distribution, and vice versa.

If the generator A of the process Y is readily available, then we can obtain π either by (5.10) and (5.8) or, more directly, from the following

(5.13) COROLLARY. Let Y be irreducible recurrent. Then the system of linear equations

$$(5.14) \qquad\qquad \mu A = 0$$

has a unique (up to multiplication by a constant) solution μ which is strictly positive. The limiting probabilities $\pi(j)$ of (5.12) are then given by

$$(5.15) \qquad\qquad \pi(j) = \frac{\mu(j)}{\sum_i \mu(i)}, \quad j \in E$$

(where we interpret $1/+\infty$ as 0).

Proof. It follows from the definition (4.6) of A that (5.14) holds if and only if the vector v defined by

$$(5.16) \qquad\qquad v(j) = \lambda(j)\mu(j), \quad j \in E,$$

satisfies (5.8). The existence, uniqueness, and positivity of μ now follow from the corresponding properties for v. The formula (5.15) follows from (5.10), (5.12), and (5.16).   □

By Theorem (4.5), the derivative of (P_t) at t is equal to $AP_t = P_tA$. Since the limits (5.12) exist, in particular

$$(5.17) \qquad \pi(j) = \lim_{t \to \infty} P_t(i, j) = P_\infty(i, j)$$

exists and is as given by (5.10). In fact, this could be obtained from noting that the derivative at t must become zero in the limit so that $\lim_t P_tA = \lim_t AP_t = 0$. The preceding corollary further shows that the limits can be passed inside the summations implied, and

$$P_\infty A = AP_\infty = 0.$$

It is also possible to reverse these arguments to prove (5.13), but our method is better.

It follows from (4.13) relating P_t to A that a row vector μ satisfies

$$(5.18) \qquad \mu P_t = \mu$$

for all $t \geq 0$ if and only if $\mu A = 0$. A vector μ satisfying (5.18) is called an *invariant* measure, and Corollary (5.13) above implies the following

(5.19) PROPOSITION. Let Y be irreducible recurrent. Then there exists a strictly positive invariant measure μ. Any other invariant measure is a constant multiple of μ.   □

If the sum $\sum_j \mu(j) < \infty$, then by dividing every entry by this sum if necessary, we may assume that μ is also a probability distribution. Then $\mu = \pi$, where π is the vector of limiting probabilities defined in (5.12). Indeed, if Y_0 has the distribution π, then (5.18) implies that

$$P\{Y_t = j\} = \pi(j), \quad j \in E,$$

for all t. This is to explain the term "invariant distribution."

(5.20) EXAMPLE. Let Y be a Markov process with generator

$$A = \begin{bmatrix} -5 & 2 & 3 \\ 2 & -3 & 1 \\ 2 & 4 & -6 \end{bmatrix}.$$

Then the equations for $\mu A = 0$ become, with $\mu = (x, y, z)$,

$$-5x + 2y + 2z = 0$$
$$2x - 3y + 4z = 0$$
$$3x + y - 6z = 0.$$

One of these equations is a linear combination of the other two; and we may throw one out, say the third. Choosing $x = 42$, the remaining two equations become

$$2y + 2z = 210$$
$$-3y + 4z = -84,$$

which can be solved easily to yield $z = 33$, $y = 72$. Hence, one particular solution of $\mu A = 0$ and $\mu P_t = \mu$ is $\mu = (42, 72, 33)$, and any other solution is a constant multiple of this. In particular, $c\mu$ with $c = \frac{1}{147}$ is

$$\pi = (\tfrac{14}{49}, \tfrac{24}{49}, \tfrac{11}{49}),$$

and this is the vector of limiting probabilities.   □

For other examples of these computations we refer the reader to Section 6 on applications.

Throughout the preceding discussion we had been working with the generator A or the transition function P_t. In some applications, the data concerning A are hard to collect because such data are obtained by observing the process continuously in time. The next theorem shows that, as far as computing the invariant vector μ and the limiting probabilities $\pi(j)$ is concerned, it is sufficient to observe the process Y at the instants of arrivals of a Poisson process independent of Y. We recall that U^α denotes the α-potential operator of Y defined by (4.14).

(5.21)   THEOREM. $\mu A = 0$ if and only if

$$(5.22) \qquad\qquad\qquad \mu U^1 = \mu.$$

Proof. By (4.7) we have $P_t A = A P_t$, which implies that $U^1 A = A U^1$ upon taking integrals on both sides. This fact along with (4.23) implies

$$A U^1 = U^1 - I = U^1 A.$$

If $\mu A = 0$, then $\mu U^1 - \mu = 0$. If $\mu U^1 = \mu$, then $\mu A = \mu U^1 A = \mu(U^1 - I)$ $= \mu - \mu = 0$.   □

To see the significance of the preceding theorem, first note that by the definition of U^α,

$$U^1(i, j) = \int_0^\infty e^{-t} P_t(i, j)\, dt = P_i\{Y_S = j\}$$

where S is a random variable independent of Y having the exponential distribution

$$P\{S \leq t\} = 1 - e^{-t}$$

with parameter 1. It is clear that $U^1(i, j) \geq 0$ and $\sum_j U^1(i, j) = 1$, that is, U^1 is a Markov matrix.

Now let $S_1, S_2, \ldots$ be the times of arrivals in a Poisson process with rate $\lambda = 1$, and suppose this Poisson process is independent of the Markov process. If we define $S_0 = 0$ and put

(5.23) $$\hat{X}_n = Y_{S_n}, \quad n = 0, 1, \ldots,$$

then it follows that the stochastic process $\hat{X} = \{\hat{X}_n; n \in \mathbb{N}\}$ is a Markov *chain* with transition probabilities

(5.24) $$P\{\hat{X}_{n+1} = j \mid \hat{X}_n = i\} = U^1(i, j).$$

Since any state j is recurrent in Y, the amount of time spent in state j is infinite with probability one; that is, the set $G_j = \{t: Y_t = j\}$ has infinite "length." Therefore, the number of the S_n that fall in this set G_j must also be equal to infinity. But this number is exactly the number of times $\hat{X}_n$ becomes equal to j. This argument shows that every state j is a recurrent state for the Markov chain $\hat{X}$ as well. Similarly, $\hat{X}$ is also irreducible. Therefore, by Theorem (6.2.25), there exists an invariant measure μ for this Markov chain $\hat{X}$; that is, $\mu = \mu U^1$. What the preceding theorem states is that this μ is also invariant for the Markov process Y.

This result becomes more striking when μ is an invariant *probability distribution*. Then $\mu(j)$ is equal to the proportion of time spent in state j by the Markov process Y. According to the theorem above, $\mu(j)$ is also the frequency with which $\hat{X}$ visits j.

Finally, we remark that the theorem remains unaltered when U^1 is replaced there by αU^α for any fixed $\alpha > 0$.

(5.25) EXAMPLE. *Work Sampling.* Successive tasks performed by an office worker form a Markov chain with finitely many states, and each task takes an exponential time. It is desired to find the percentages of time spent by this office worker in performing his various duties.

By Theorem (5.21), it is sufficient to record the tasks he was working on at the instants $S_0, S_1, S_2, \ldots$ where these "sampling times" are picked in advance, without knowledge of the worker's schedule, in such a way that $S_1 - S_0, S_2 - S_1, \ldots$ are independent and exponentially distributed random variables with parameter α. If the number of times the worker was seen performing the task j during the first n observations at the times $S_1, \ldots, S_n$ is equal to $N_j(n)$, then $N_j(n)/n$ is approximately equal to the percentage $\pi(j)$ of time he spends in performing the task j.   □

Let f and g be two functions defined on E. As usual, we interpret $f(j)$ and $g(j)$ as the rates of rewards obtained at times t where $Y_t = j$. The question we now address ourselves to is, if we are given a choice, which of the two reward rate schedules f and g should we pick in order to optimize our return? The total rewards obtained during $[0, t]$ using f and g are, respectively,

$$(5.26) \qquad A_t = \int_0^t f(Y_s)\, ds, \quad B_t = \int_0^t g(Y_s)\, ds.$$

ıl. if Y is irreducible recurrent and f and g are non-negative, both A_t and B_t will increase to infinity as $t \to \infty$. But the rates at which they do so can be compared by considering the ratios A_t/B_t or $E_i[A_t]/E_i[B_t]$ for large t. Below we will show that both of these ratios go, in the limit, to the same number.

Suppose Y is irreducible recurrent non-null. Then it has a limiting probability distribution μ, and we may interpret $\mu(j)$ as the expected value of the time spent in j during one unit of time in the long run. In other words,

$$(5.27) \qquad \mu(j) = \lim_{t \to \infty} E_i\left[\frac{1}{t}\int_0^t 1_j(Y_s)\, ds\right].$$

Since we can write

$$A_t = \int_0^t f(Y_s)\, ds = \sum_j f(j) \int_0^t 1_j(Y_s)\, ds,$$

(5.27) implies that

$$\lim_{t \to \infty} \frac{1}{t} E_i[A_t] = \sum_j \mu(j) f(j) = \mu f$$

(note that μ is a row vector, f a column vector, and μf the scalar product of the two). A similar result holds for B_t, and therefore we have

$$(5.28) \qquad \lim_{t \to \infty} \frac{E_i[A_t]}{E_i[B_t]} = \frac{\mu f}{\mu g}.$$

The following theorem states that this result is true even when μ is merely an invariant measure (and not necessarily a probability distribution).

(5.29) THEOREM. Suppose $g \geq 0$ and μg is positive and finite. Then for any $f \geq 0$ and any $i, j \in E$,

$$\lim_{t \to \infty} \frac{E_i \left[\int_0^t f(Y_s) \, ds \right]}{E_j \left[\int_0^t g(Y_s) \, ds \right]} = \frac{\mu f}{\mu g}$$

independent of i and j. Moreover,

$$\lim_{t \to \infty} \frac{\int_0^t f(Y_s) \, ds}{\int_0^t g(Y_s) \, ds} = \frac{\mu f}{\mu g}$$

with probability one, independent of the initial state.

The processes $A = \{A_t; t \geq 0\}$ and $B = \{B_t; t \geq 0\}$ defined in (5.26) are examples of a larger class of *functionals* of the Markov process known as *additive functionals*. In general, $A = \{A_t; t \geq 0\}$ is said to be an *additive functional* of the Markov process Y provided that

(a) $A_0 = 0$ and $t \longrightarrow A_t$ is right continuous and non-decreasing almost surely;

(b) for any $t \in \mathbb{R}_+$, the value A_t is completely determined by the past history $\{Y_s; s \leq t\}$ of Y before t;

(c) for any $t \geq 0$ and $s > 0$, the value of the increment $A_{t+s} - A_t$ depends only on the values $\{Y_u; t \leq u \leq t + s\}$ and this only through the shape of the function $u \longrightarrow Y_u$ in that interval $[t, t + s]$.

(5.30) EXAMPLE. For any non-negative function f defined on E,

$$(5.31) \qquad\qquad A_t = \int_0^t f(Y_s) \, ds, \quad t \geq 0,$$

defines an additive functional A of Y. In this case $t \longrightarrow A_t$ is continuous. Conversely, if Y is a regular Markov process, then every *continuous* additive functional A is of this form.   □

(5.32) EXAMPLE. Let f be a non-negative function defined on $E \times E$, and suppose $f(i, i) = 0$ for every i. Then, if we denote by Y_{s-} the value of Y just before s, $f(Y_{s-}, Y_s)$ is zero for all s except the instants s of jumps from one state to another. Let

$$(5.33) \qquad\qquad B_t = \sum_{s \leq t} f(Y_{s-}, Y_s)$$

be the sum of all the jumps of the function $s \longrightarrow f(Y_{s-}, Y_s)$ recorded during $[0, t]$. In other words,

$$B_t = \sum_{n=1}^{\infty} f(X_{n-1}, X_n)I_{(0,t]}(T_n),$$

where X is the underlying Markov chain and the T_n the times of jumps for Y. Then B is an additive functional. In this case, $t \longrightarrow B_t$ stays constant except at the jump points T_n of Y, at which times it jumps by the respective amounts $f(X_{n-1}, X_n)$.

In particular, if we take $f(i, j) = 1$ for all $i \neq j$, then B_t becomes the number of jumps of Y during $(0, t]$. If we take $f(i, j) = 0$ for all i, j except for $f(i_0, j_0) = 1$, then B_t becomes the number of times Y jumps from i_0 into j_0 during $(0, t]$. By taking $f(i, j) = 0$ for all i, j except for $f(i, j) = 1$ when $i \neq j$ and $j \in B$ for some set $B \subset E$, we obtain the number of visits to states j in B during the time interval $(0, t]$.   ☐

Though we are not able to prove this here, it is known that any additive functional of a regular Markov process is the sum of two additive functionals A and B, where A is of the form (5.31) and B is of the form (5.33).

The following extends Theorem (5.29) to all additive functionals. We omit the proof.

(5.34)   THEOREM.   If A and B are any two additive functionals of the process Y, then

$$\lim_{t \to \infty} \frac{E_i[A_t]}{E_i[B_t]} = \frac{\mu_A}{\mu_B},$$

where

$$\mu_A = \sum_j \mu(j)E_j[A_1],$$

and similarly for μ_B. Moreover,

$$\lim_{t \to \infty} \frac{A_t}{B_t} = \frac{\mu_A}{\mu_B}$$

with probability one for any initial state i.

We end this section by giving a computational formula for μ_A when Y has an invariant distribution. As we had remarked earlier, any additive functional of a regular Markov process Y is of the form

$$(5.35) \qquad\qquad A_t = \int_0^t f(Y_s) \, ds + \sum_{s \leq t} g(Y_{s-}, Y_s),$$

where $f \geq 0$ is a function defined on E, and where $g \geq 0$ is a function defined on $E \times E$ satisfying $g(i, i) = 0$ for all i. Then, if μ is the invariant distribution of Y and if A is its generator defined by (4.6), then

$$(5.36) \qquad \mu_A = \sum_i \mu(i)[f(i) + \sum_j A(i, j)g(i, j)]$$

(the subscript A in μ_A refers to the additive functional $A = \{A_t; t \geq 0\}$ and not to the generator!).

6. Birth and Death Processes

This is an important class of Markov processes with applications in biology, demography, and queueing theory. Here Y_t depicts the size of a population at time t, and the state space is $E = \{0, 1, \ldots\}$. A "birth" increases the size by one; a "death" decreases it by one. Let V_t and W_t be respectively the lengths of times from t until the instants of the next birth and the next death. The basic assumptions are that, if $Y_t = i$, then for some $\lambda(i)$ and $p(i)$

$$(6.1) \qquad P\{V_t > u, W_t > u \,|\, Y_s; s \leq t\} = e^{-\lambda(i)u},$$

and

$$(6.2) \qquad P\{V_t \leq W_t \,|\, Y_s; s \leq t\} = p(i).$$

These imply that $Y = \{Y_t; t \geq 0\}$ is a Markov process.

It is convenient to define $a_i = p(i)\lambda(i)$ and $b_i = (1 - p(i))\lambda(i)$, $i \in E$, so that a_i and b_i can be thought as the respective time rates of births and deaths at an instant at which the population size is i. The parameters a_i and b_i are arbitrary non-negative numbers, except that $b_0 = 0$ and no pair (a_i, b_i) is equal to $(0, 0)$. It follows from (4.6) that the generator of the process Y is

$$(6.3) \qquad A = \begin{bmatrix} -a_0 & a_0 & & & 0 \\ b_1 & -a_1 - b_1 & a_1 & & \\ & b_2 & -a_2 - b_2 & a_2 & \\ & & \cdot & \cdot & \cdot \\ 0 & & & \cdot & \cdot & \cdot \end{bmatrix}.$$

The underlying Markov chain X has the transition probabilities $Q(i, i - 1) = 1 - Q(i, i + 1) = b_i/(a_i + b_i)$ for $i \geq 1$, and $Q(0, 1) = 1$. Let $R(i, j)$ be the expected number of visits to j starting from i. Since each visit to j lasts $1/\lambda(j)$ units of time on the average, the expected time spent in state j is equal to, starting with j,

$$(6.4) \qquad U(j, j) = E_j\left[\int_0^\infty 1_j(Y_s)\,ds\right] = \frac{R(j, j)}{\lambda(j)}.$$

If the sum

(6.5)
$$\sum_j \frac{R(j, j)}{\lambda(j)} < \infty,$$

then the process is *not* regular. It turns out that in the converse direction, the divergence of the sum in (6.5) is also sufficient for Y to be regular.

If Y is not regular, then the population size reaches infinity in a finite time with a positive probability. Otherwise, if Y is regular, then the population size remains finite with probability one. Even then, it is possible that the size grows in time to have an infinite limit. In particular, this is the case when Y is regular but transient.

(6.6) EXAMPLE. Suppose $a_i = ia, b_i = ib$ for some $a, b > 0$. This corresponds to the case where each individual of the population has an exponential lifetime with parameter b, and each individual can give birth to another after an exponential time with parameter a. Then the underlying Markov chain X has the transition matrix Q with $Q(0, 0) = 1$, $Q(i, i + 1) = p$, $Q(i, i - 1) = 1 - p$ for $i \geq 1$, where $p = a/(a + b)$. Of course, $\lambda(0) = 0$, $\lambda(i) = i(a + b)$ for $i \geq 1$.

In this case, the $\lambda(i)$ are not bounded. If $a \leq b$, then the X chain is absorbing and therefore Y is regular by Theorem (3.25) (cf. Example (5.3.31) for this chain). Even if $a > b$, the sum in (6.5) becomes

$$\sum_j \frac{R(j, j)}{j(a + b)} \geq \frac{1}{a + b} \sum_{j=0}^{\infty} \frac{1}{j} = +\infty,$$

which means that the process is regular. Hence, $P\{Y_t < \infty\} = 1$ for any t; but when $a > b$ there is a tendency for growth, and

$$\lim_{t \to \infty} P\{Y_t \leq j\} = 0$$

for all j, that is, in the limit as $t \to \infty$, the population size becomes infinite. $\square$

Going back to the general birth and death process, by Corollary (5.13), a limiting distribution exists if and only if $vA = 0$ has a solution satisfying $v1 = 1$. For A given by (6.3), the equation $vA = 0$ becomes, with $v = (1, v_1, v_2, \ldots)$,

$$b_1 v_1 = a_0,$$
$$a_0 + b_2 v_2 = (a_1 + b_1)v_1,$$
$$a_1 v_1 + b_3 v_3 = (a_2 + b_2)v_2, \ldots.$$

Solving these consecutively, we get

$$(6.7) \qquad v_i = \frac{a_0 \cdots a_{i-1}}{b_1 \cdots b_i}, \quad i = 1, 2, \dots .$$

Hence, a limiting distribution exists if and only if

$$(6.8) \qquad c = \sum_{i=0}^{\infty} v_i = 1 + \sum_{i=1}^{\infty} \frac{a_0 \cdots a_{i-1}}{b_1 \cdots b_i} < \infty .$$

If $c < \infty$, then the limiting distribution is

$$(6.9) \qquad \pi(j) = \begin{cases} \dfrac{1}{c} & \text{if } j = 0, \\[2ex] \dfrac{a_0 \cdots a_{j-1}}{c b_1 \cdots b_j} & \text{if } j \geq 1. \end{cases}$$

Otherwise, if $c = +\infty$, then

$$(6.10) \qquad \lim_{t \to \infty} P_i\{Y_t = j\} = 0, \quad j = 0, 1, \dots .$$

In the remainder of this section we discuss applications of this general model to the theory of Markovian queues.

(6.11) EXAMPLE. $M/M/1/\infty$ queue. This is the process discussed in Examples (1.10) and (3.12). Customers arrive according to a Poisson process with rate a, service times are exponential with mean $1/b$, there is a single server, and infinite queues are permissible. Here, Y_t is the number of customers in the system at time t. Y is a special birth and death process where

$$a_0 = a_1 = \cdots = a; \quad b_1 = b_2 = \cdots = b.$$

Writing $r = a/b$, the sum c of (6.8) is

$$c = \sum_{n=0}^{\infty} r^n = \begin{cases} +\infty & \text{if } a \geq b, \\ 1/(1 - r) & \text{if } a < b. \end{cases}$$

Hence, if $r = a/b < 1$, then there is a limiting distribution, and using (6.9), that limiting distribution is

$$(6.12) \qquad \pi(j) = (1 - r)r^j, \quad j = 0, 1, \dots .$$

Otherwise, if $r \geq 1$, then $P\{Y_t = j\} \to 0$ for all j as $t \to \infty$. The ratio r is called the *traffic intensity* for the system. If $r < 1$, the arrival rate is less than the service "rate" possible, and the queue size does not grow to infinity. If $r \geq 1$, arrivals occur faster than the server can handle. □

(6.13) EXAMPLE. *M/M/1/m queue.* This is a queueing system exactly as the preceding one except that the waiting room has finite capacity $m - 1$. Hence, starting with less than m customers, the total number in the system cannot exceed m; that is, a customer arriving to find m or more customers in the system leaves and never returns (we do allow initial sizes of greater than m, however).

The generator for the queue size process is of shape (6.3) with

$$a_0 = \cdots = a_{m-1} = a; \quad a_m = a_{m+1} = \cdots = 0; \quad b_1 = b_2 = \cdots = b.$$

The states $m + 1, m + 2, \ldots$ are all transient; $C = \{0, 1, \ldots, m\}$ is a recurrent irreducible set. We leave the computation of the limiting distribution as an exercise. □

(6.14) EXAMPLE. *M/M/s/∞ queue.* Arrivals form a Poisson process with rate a, service times are exponential with mean $1/b$, there are s servers, and the waiting room is of infinite size. If there are $i < s$ customers in the system, then i servers are busy working independently of each other, and thus

(6.15) $\beta = P\{V_t > u, W_t > v \,|\, Y_t = i\} = e^{-au}e^{-ibv}.$

If $i \geq s$, then all s servers are busy, and

$$\beta = e^{-au}e^{-sbv}.$$

Hence, the queue size process is a birth and death process with

$$a_0 = a_1 = \cdots = a; \quad b_1 = b, \quad b_2 = 2b, \quad \ldots, \quad b_s = sb, \quad b_{s+1} = sb, \ldots.$$

It is easy to see that a limiting distribution exists if and only if $r = a/sb < 1$. □

(6.16) EXAMPLE. *M/M/s/∞ queue with non-identical servers.* This system is just like the preceding one except that the service times (though still exponential) have different means depending on the server. Suppose for simplicity that $s = 2$. Suppose the mean service time is $1/b'$ for the first server and $1/b''$ for the second; and suppose the policy is to work the second server only when there are two or more customers in the system, so that if the first server becomes idle while the second is busy, then the customer switches to the first server immediately. Arguing as in (6.14), we see that

$$a_0 = a_1 = \cdots = a; \quad b_1 = b', \quad b_2 = b_3 = \cdots = b' + b'' = b.$$

This process has a limiting distribution if and only if $a/b < 1$. □

(6.17) EXAMPLE. $M/M/\infty$ *queue.* Arrivals form a Poisson process with rate a, service times are exponential with mean $1/b$, and there are infinitely many servers so that no customer ever waits. In this case (6.15) holds for all $i \geq 0$. Thus, the constant c of (6.8) is, writing $r = a/b$,

$$c = \sum_{n=0}^{\infty} \frac{r^n}{n!} = e^r < \infty,$$

no matter what $a, b > 0$ are. Therefore, we see from (6.9) that the limiting distribution is

$$(6.18) \qquad \pi(j) = \frac{e^{-r}r^j}{j!}, \quad j = 0, 1, \ldots,$$

which we recognize as the Poisson distribution with parameter r.

This model applies to very large parking lots; then the customers are the arriving vehicles, servers are the stalls, and the service times are the times vehicles remain parked.

Another application is to the kinetic theory of gases. Then the arrivals are gas molecules entering a fixed region, and service times are the times spent in that region by the molecules. If the temperature is low, molecules do not interact very much and the model fits well.

This model also fits the short-run behavior of the size (measured in terms of the number of families living in them) of larger cities. Then arrivals are the families moving into the city, and a service time is the amount of time spent there by a family before it moves out. The national experience in the United States is that $1/b$ is about 4 years. If the city has a stable population size of about 1,000,000 families (note that the mean and variance of the Poisson distribution in (6.18) are both $r = a/b$), then the arrival rate must be $a = 250,000$ families per year. Hence, on the average a quarter of a million new families move in every year and about the same number move out. Yet, the size of the city remains fairly constant: We can approximate the distribution (6.18) by the normal distribution with mean r and variance r, which means that for large t,

$$P\{997,000 \leq Y_t \leq 1,003,000\} \cong 0.9974;$$

that is, the number of families in the city is about one million plus or minus three thousand. Of course, this model is rather crude: it does not take into account possible interactions between families, and the term "family" itself is not defined. ∎

(6.19) EXAMPLE. $M/M/s/\infty$ *queue with balking.* Arrivals form a Poisson process with rate a, services are exponential with mean $1/b$, there are s servers, and there is no limit on the number of customers allowed in. Furthermore, an arriving customer is permitted to balk: if he finds the system too crowded, he may leave; but once he joins the system he cannot change his mind later. Suppose the probability of joining the queue is p_i if there are i customers in the system at the time of arrival. Thus, if the queue size at time t is $Y_t = i$, and if there were no service completions during $[t, t + u]$, then the probability that there are no additions to the queue during $[t, t + u]$ is

$$\sum_{n=0}^{\infty} \frac{e^{-au}(au)^n}{n!}(1 - p_i)^n = e^{-ap_iu}.$$

Hence, the process Y is a Markov process with

$$P\{V_t > u, W_t > u \,|\, Y_t = i\} = e^{-ap_iu}e^{-b_iu};$$

that is, $\lambda(i) = a_i + b_i$ with $a_i = ap_i$, and the generator is the same as (6.3) but with

$$a_0 = ap_0, \quad a_1 = ap_1, \ldots;$$
$$b_1 = b, \quad b_2 = 2b, \ldots, \quad b_s = sb, \quad b_{s+1} = b_{s+2} = \cdots = sb.$$

The model fits barbershop-type situations well. If the queue size has a negative effect upon joining probabilities, then p_i decreases as i increases. Sometimes, a large queue size will attract further customers, which means that p_i increases as i increases. $\qquad\square$

(6.20)   EXAMPLE.   *Machine repair problem.* Suppose there are m machines serviced by one repairman. Each machine runs without failure, independent of all others, an exponential time with mean $1/a$. When it fails, it waits until the repairman can come to repair it, and the repair itself takes an exponentially distributed amount of time with mean $1/b$. Once repaired, the machine is as good as new.

If the number of failed machines at time t is $Y_t = i$, then there are $m - i$ machines working, and the time until the next failure is exponential with parameter $(m - i)a$ if no machines are repaired in the meantime. Hence, Y has the generator A of (6.3) but with

$$a_0 = ma, \quad a_1 = (m - 1)a, \quad \ldots, a_{m-1} = a, \quad a_m = a_{m+1} = \cdots = 0$$

and

$$b_1 = b_2 = \cdots = b.$$

Since $Y_0 \in \{0, \ldots, m\}$, so is Y_t for all t. Restricted to $\{0, \ldots, m\}$, then, Y is a recurrent Markov process. The limiting distribution is

$$\pi(j) = pm(m - 1) \cdots (m - j)(a/b)^j, \quad j = 0, 1, \ldots, m,$$

where

$$\frac{1}{p} = \sum_{i=0}^{m} m(m-1) \cdots (m-i)\left(\frac{a}{b}\right)^i.$$ □

7. Exercises

(7.1) Show that

$$P_t = \begin{bmatrix} 0.6 + 0.4e^{-5t} & 0.4 - 0.4e^{-5t} \\ 0.6 - 0.6e^{-5t} & 0.4 + 0.6e^{-5t} \end{bmatrix}$$

is a transition function. For a Markov process Y with this transition function on the state space $E = \{a, b\}$, compute

$$P_a\{Y_{2.4} = b, \ Y_{3.8} = a, \ Y_{4.2} = a\}.$$

(7.2) Suppose the arrivals at a counter form a Poisson process with rate λ, and suppose that each arrival is of type a or type b with respective probabilities p and q ($p, q > 0, p + q = 1$), independent of all past data. Let Y_t be the type of the last arrival before t. Show that Y is a Markov process with state space $E = \{a, b\}$ and transition function

$$P_t = \begin{bmatrix} p + qe^{-\lambda t} & q - qe^{-\lambda t} \\ p - pe^{-\lambda t} & q + pe^{-\lambda t} \end{bmatrix}.$$

(7.3) Let W be the time until the first jump of the process Y of Exercise (7.1).
   (a) Use Proposition (1.6) to justify writing

$$P_a\{W > t\} = \lim_{n \to \infty} [P_{t/n}(a, a)]^n.$$

   (b) Use this to compute $P_a\{W > t\}$. (Hint: $e^{-x} \cong 1 - x$ for small x, and $(1 + (x/n))^n \longrightarrow e^x$ as $n \longrightarrow \infty$ for any x.)
   (c) Check your result against Theorem (2.9).

(7.4) Use the method of the preceding exercise to compute the distributions of the sojourn times in states a and b for the Markov process Y of Exercise (7.2). Check your result against that obtained directly by using the decomposition arguments for Poisson processes.

(7.5) Classify (as stable or instantaneous) the states of the Markov process of Exercise (7.2) by using the results of the preceding exercise. Draw the picture of the set $G_a(\omega) = \{t \geq 0: Y_t(\omega) = a\}$ for a typical realization ω.

(7.6) Show that the transition matrix of the imbedded Markov chain X of the Markov process of (7.1) is

$$Q = \begin{bmatrix} 0 & 1 \\ 1 & 0 \end{bmatrix}.$$

(7.7) Consider a system whose successive states form a Markov chain X with state space $E = \{a, b, c\}$ and transition matrix

$$Q = \begin{bmatrix} 0 & 1 & 0 \\ 0.4 & 0 & 0.6 \\ 0 & 1 & 0 \end{bmatrix}.$$

Suppose the lengths of sojourns in states a, b, and c are all independent of each other and have exponential distributions with parameters $\lambda(a) = 2$, $\lambda(b) = 5$, $\lambda(c) = 3$. Compute the generator A of the Markov process Y describing the evolution of this system.

(7.8) Let Y be a two-state Markov process whose sojourn distributions have the parameters a and b (a and b are positive).

(a) Show that the generator A must be

$$A = \begin{bmatrix} -a & a \\ b & -b \end{bmatrix}.$$

(b) In Theorem (4.31) choose $c = a + b$ to obtain

$$K = \begin{bmatrix} p & q \\ p & q \end{bmatrix}$$

with $p = b/(a + b)$, $q = a/(a + b)$ for the transition matrix of the chain $\hat{X}$.

(c) Use this to compute P_t. Conclude from the form of this P_t that any two-state Markov process can be thought of as if it were obtained by the mechanism of Exercise (7.2).

(d) Use the result of part (b) to compute the α-potential matrix U^α directly by (4.26).

(7.9) For the Markov process Y of Exercise (7.1), compute the α-potential matrix U^α directly by using (4.14).

(7.10) Let Y be the Markov process of Exercise (7.7). Compute the α-potential matrix U^α.

(7.11) Let Y be a Markov process with generator

$$A = \begin{bmatrix} -6 & 5 & 1 \\ 0 & -4 & 4 \\ 2 & 1 & -3 \end{bmatrix},$$

and let g be the function $g = (15, 32, 27)$; the state space is $E = \{a, b, c\}$.

(a) Compute $U^\alpha g(a) = E_a\left[\int_0^\infty e^{-\alpha t} g(Y_t)\, dt\right]$.

(b) For the first entrance time $T = \inf\{t \geq 0: Y_t = c\}$, compute $E_a[e^{-\alpha T}]$ for $\alpha > 0$.

(c)   Compute

$$E_a\left[\int_0^T e^{-\alpha t} g(Y_t)\, dt\right].$$

(d)   Compute

$$E_a\left[\int_0^T g(Y_t)\, dt\right]$$

by letting $\alpha \downarrow 0$ in part (c).

(7.12)   Let Y be the same process as in (7.11), and let g be as before. Let X be the underlying Markov chain of Y, and $T_1, T_2, \ldots$ the jump times.
   (a)   For the same first entrance time T, compute

$$E_a\left[\int_0^T g(Y_t)\, dt \,\middle|\, X_0, X_1, X_2, \ldots\right]$$

by considering the values of T and $\int_0^T g(Y_t)\, dt$ in terms of the X_n and T_n.

   (b)   Compute $E_a\left[\int_0^T g(Y_t)\, dt\right]$ by taking the expected value of the result obtained in (a).

(7.13)   Consider the queueing system $M/M/1$ of Example (1.10) with the following modification: the capacity of the system is two, so an arriving customer who finds two customers in the system must leave immediately and does not return.
   (a)   Show that, then, the queue size process Y is a Markov process with state space $E = \{0, 1, 2\}$ and generator

$$A = \begin{bmatrix} -a & a & 0 \\ b & -(a+b) & a \\ 0 & b & -b \end{bmatrix}.$$

   (b)   Compute the α-potential matrix U^α for the case $a = 2$, $b = 8$ by using (4.24).
   (c)   Invert the Laplace transforms $U^\alpha(i, j)$ to obtain the transition functions $P_t(i, j)$.

(7.14)   Suppose that a penalty cost of one dollar per unit time is applicable for each customer present in the system of Exercise (7.13). Suppose the discount rate is α. Show that the expected total discounted cost is, starting at i, $U^\alpha g(i)$, where $g = (0, 1, 2)$. Compute this for every i.

(7.15)   *Continuation.* Suppose a penalty of one dollar is paid at time t if a customer arrived at instant t to find the system full (and therefore was turned away). If $S_1, S_2, \ldots$ are the times of arrivals, then the total discounted value of this penalty is

$$W = \sum_{n=1}^{\infty} e^{-\alpha S(n)} I_{\{2\}}(Y_{S_n-}).$$

(a) Show that

$$E[W \mid Y_t; \, t \geq 0] = a \int_0^\infty e^{-\alpha t} I_{\{2\}}(Y_t) \, dt.$$

(*Hint*: Over each sojourn interval $[T_n, T_{n+1})$ on which Y is equal to 2, the arrivals of customers to be rejected form a Poisson process with rate a.)

(b) Show that

$$E_i[W] = a U^\alpha(i, 2).$$

(7.16) Consider the Markov process Y of Exercise (7.11), and let $g = (15, 32, 27)$ as before.

(a) Find the limiting distribution $\pi(j) = \lim_{t \to \infty} P_i\{Y_t = j\}$ for all $i, j \in E$.

(b) Compute the time average $\lim_{t \to \infty} B_t / t$ for

$$B_t = \int_0^t g(Y_s) \, ds.$$

(7.17) Consider a Markov process Y with state space $E = \{a, b, c, d\}$ and generator

$$A = \begin{bmatrix} -1 & 1 & 0 & 0 \\ 0 & -5 & 2 & 3 \\ 1 & 2 & -4 & 1 \\ 3 & 0 & 0 & -3 \end{bmatrix}.$$

(a) Find the limiting distribution π.

(b) Compute the average expectation

$$\lim_{t \to \infty} \frac{1}{t} E_i \left[\int_0^t f(Y_s) \, ds \right]$$

for $f = (1, 2, -1, -2)$.

(c) Compute

$$\lim_{t \to \infty} \frac{\int_0^t f(Y_s) \, ds}{\int_0^t g(Y_s) \, ds}$$

for f as in part (b) and $g = (-1, 3, -1, 4)$.

(7.18) Let Y be a Markov process which moves through the states a, b, and c in the cyclic order $a, b, c, a, b, c, \ldots$ with sojourns in states a, b, c having the respective expectations 1, 3, and 5. Compute the limiting distribution for this process.

(7.19) Let Y be a Markov process with state space $E = \{r, w, b\}$ and suppose every jump from i to j brings a reward $f(i, j)$. The generator A and this function f are given as

$$A = \begin{bmatrix} -3 & 2 & 1 \\ 1 & -1 & 0 \\ 1 & 3 & -4 \end{bmatrix}, \qquad f = \begin{bmatrix} 0 & 1 & 2 \\ 2 & 0 & 8 \\ 3 & 5 & 0 \end{bmatrix}.$$

(a)   Compute the limiting distribution for Y.
(b)   Compute the limit, as $t \to \infty$, of the average reward B_t/t where

$$B_t = \sum_{s \leq t} f(Y_{s-}, Y_s).$$

(**7.20**)   Compute the limiting distribution of the queue size in Example (6.13).

(**7.21**)   Compute the limiting distribution of the queue size in Example (6.14) when $a = 8$, $b = 3.5$, $s = 3$.

(**7.22**)   Compute the limiting distribution of the queue size in Example (6.16) for the case where $a = 8$, $b' = 4$, $b'' = 6$.

(**7.23**)   In Example (6.19) let $a = 3$, $b = 2$, $s = 2$, $p_0 = p_1 = 1$, $p_2 = \frac{1}{2}$, $p_3 = \frac{1}{4}$, $p_4 = p_5 = \cdots = 0$. Compute the limiting distribution.

(**7.24**)   *Excessive functions.* Let Y be a Markov process with state space E and transition function P_t. A function $f \geq 0$ on E is said to be *excessive* provided that $f \geq P_t f$ for all $t > 0$ and $\lim_{t \downarrow 0} P_t f = f$.

When E is finite, show using (2.2) that a function $f \geq 0$ on E is excessive if and only if $f \geq P_t f$ for all $t > 0$.

(**7.25**)   *Continuation.* Suppose E is finite. Show that a function $f \geq 0$ is excessive if and only if

$$f \geq \alpha U^\alpha f$$

for all $\alpha > 0$; here U^α is the α-potential matrix of Y.

(**7.26**)   *Continuation.* Show that for any $g \geq 0$, the potential $f = Ug$ is excessive.

(**7.27**)   *Continuation.* Show that $f \geq 0$ is excessive if and only if

$$-Af \geq 0.$$

(**7.28**)   Show that $f \geq 0$ is excessive if and only if

$$f(i) \geq E_i[f(Y_T)]$$

for all stopping times T.

(**7.29**)   Let Q be the transition matrix of the underlying Markov chain X of Y. Show that $f \geq 0$ is excessive for Y if and only if

$$f \geq Qf.$$

(**7.30**)   *Optimal stopping.* Let f be an arbitrary function on E (assume E is finite), and define

$$v(i) = \sup_T E_i[f(Y_T)],$$

where the supremum is taken over all stopping times T.
(a)   Show that v is excessive and $v \geq f$.

(b)   Show that if h is another excessive function with $h \geq f$, then $h \geq v$ as well.

(c)   Show that

$$T_0 = \inf\{t \colon f(Y_t) = v(Y_t)\}$$

is an optimal stopping time in the sense that

$$E_i[f(Y_{T_0})] = v(i), \quad i \in E.$$

Compared with Markov chains, the continuous time version of this chapter needs a much higher level of sophistication in the tools required. In Section 2 we made an attempt to describe the sample path behavior of Markov processes. This is properly handled through measurability and separability arguments which, however, we could not use in this elementary book. The interested reader should consult CHUNG [1] *for an almost complete account of the theory of discrete state-space Markov processes.*

The treatment given here for right continuous processes, on the other hand, is fairly complete. By considering the jump times and the successive states visited in an explicit fashion, we were able to obtain many results which are difficult to get by purely analytic techniques. Incidentally, this technique is just the explicit version of that which lurks behind the analytical doings of FELLER *in his famous paper* [1].

Concerning the ergodic limit theorems of Section 5, we should mention that the methods used are those of Markov renewal theory, an account of which will be found in Chapter 10. Of the somewhat lesser-known results put in here, Theorem (5.21) is due to NAGASAWA *and* SATO [1], *and Theorems (5.29) and (5.34) are due to* AZEMA, DUFLO, *and* REVUZ [1].

If the reader enjoyed this chapter, then he might want to ask what lies further ahead and how he might prepare himself for a life of randomness. If so, I would advise him first to rebuild the foundations of his knowledge of probabilities by studying a book such as NEVEU [1]; *then study the general theory of stochastic processes through a book such as* DELLACHERIE [1]; *then read the theory of Markov processes in a general setting, say through* BLUMENTHAL *and* GETOOR [1]. *And this is a most straightforward Christian piece of advice.*

CHAPTER 9

Renewal Theory

In all the stochastic processes we have studied so far, an important property has been the existence of times, usually random, from which onward the future of the process is a probabilistic replica of the original process. In Markov chains and Markov processes, for example, if the initial state is i, then the times of successive entrances to that state i play this role; and this fact in turn enables us to obtain many of the limiting results we listed before. This "regeneration" property may hold in much more general situations, and when it holds, surprisingly sharp results can be obtained by the methods to be developed here.

In the first section we introduce renewal processes; in the second section the regeneration idea mentioned above is made precise, and the tools needed to take advantage of it are introduced. The tools are results concerning the solution, and the limiting properties of the solution, of the so-called renewal equation. Then, in Section 3, some generalizations are introduced to allow the regeneration times to start later than the time origin. A number of exercises in Section 4 provide more examples of situations where renewal theory has been useful.

1. Renewal Processes

Consider a fixed phenomenon, and let $W_1, W_2, \ldots$ be the times between its successive occurrences. Then

$$(1.1) \qquad S_0 = 0; \qquad S_{n+1} = S_n + W_{n+1}, \quad n \in \mathbb{N},$$

define the times of occurrence, assuming that the time origin is taken to be an instant of such an occurrence.

(1.2) DEFINITION. The sequence $S = \{S_n; n \in \mathbb{N}\}$ is called a *renewal process* provided that $W_1, W_2, \ldots$ be independent and identically distributed non-negative random variables. Then the S_n are called *renewal times*.

(1.3) EXAMPLE. Let $S_0, S_1, \ldots$ be the successive instants at which vehicles cross a certain fixed point on a highway. If the successive headways W_1, $W_2, \ldots$ are independent of each other, and if they all have the same distribution, then $S = \{S_n; n \in \mathbb{N}\}$ is a renewal process. In particular, if the headway distribution is exponential, then the renewal process S becomes the sequence of arrival times in a Poisson process. □

(1.4) EXAMPLE. Consider an item installed at time $S_0 = 0$. When it fails it is replaced by an identical item; when that item fails, it in turn is replaced by a new item; and so on. Suppose the lifetimes of the successive items are $U_1, U_2, \ldots$ and that the replacements take $V_1, V_2, \ldots$ units of time. Hence, the successive new items start working at times $S_0 = 0$, $S_1 = U_1 + V_1$, $S_2 = S_1 + U_2 + V_2$, $S_3 = S_2 + U_3 + V_3, \ldots$.   It is reasonable to assume that $U_1, U_2, \ldots$ are independent and identically distributed, and further that $V_1, V_2, \ldots$ are independent and identically distributed and are independent of the lifetimes. Then $W_1 = U_1 + V_1$, $W_2 = U_2 + V_2, \ldots$ are also independent and identically distributed, and hence $S_0, S_1, \ldots$ form a renewal process. □

It follows from the definition that the joint distribution of any number of the S_n can be determined once the distribution function of the times between renewals is specified. We will take this computation up after we introduce the following definition. We are writing $\varphi(dx)$ for $d\varphi(x)$ throughout.

(1.5) DEFINITION. Let φ be a distribution function on $\mathbb{R}_+$, and let f be a non-negative function defined on $\mathbb{R}_+$ which is bounded over any finite interval $[0, t]$. Then the function $\varphi * f$ defined by

$$\varphi * f(t) = \int_{[0,t]} \varphi(dx) f(t - x), \qquad t \geq 0,$$

is called the *convolution of φ and f*. □

It follows that $\varphi * f$ is non-negative and bounded over finite intervals, so the operation can be repeated: we write $\varphi * (\varphi * f)$ as $\varphi^2 * f$ and write $\varphi^n * f$ for the nth iterate.

In particular, when f is replaced by a distribution function ψ, the resulting convolution $\varphi * \psi$ is again a distribution function. To see the significance of $\varphi * \psi$, suppose X and Y are two independent random variables taking values in $\mathbb{R}_+$, and let φ and ψ be their respective distributions. Then using Proposition (2.2.16), we have

$$(1.6) \qquad P\{X + Y \le t\} = E[P\{X + Y \le t \,|\, X\}]$$

$$= E[\psi(t - X)] = \int_{[0,t]} \varphi(dx)\psi(t - x)$$

for any $t \ge 0$. That is, $\varphi * \psi$ is the distribution of the sum $X + Y$ of the independent random variables X and Y with distributions φ and ψ. Since the roles of X and Y can be reversed, we see that for any two distributions φ and ψ,

$$(1.7) \qquad\qquad\qquad \varphi * \psi = \psi * \varphi.$$

When φ and ψ are the same, we write φ^2 for $\varphi * \varphi$, and further define $\varphi^3 = \varphi * \varphi^2 = \varphi * \varphi * \varphi$, $\varphi^4 = \varphi * \varphi^3$, etc. Defining φ^0 to be the distribution function 1 defined by

$$(1.8) \qquad\qquad\qquad 1(t) = \begin{cases} 0 & t < 0, \\ 1 & t \ge 0, \end{cases}$$

we can then write

$$\varphi^{n+1} = \varphi * \varphi^n$$

for all $n \in \mathbb{N}$. The distribution φ^n so defined is called the *n-fold convolution* of φ with itself. From the interpretation (1.6) we see that φ^n is the distribution of the sum of n independent random variables each of which has the same distribution φ.

Throughout the remainder of this section, $S = \{S_n; n \in \mathbb{N}\}$ is a renewal process, and F is the distribution of the interrenewal times. It follows from the preceding discussion that for any m, the distribution of $S_{n+m} - S_n$ is the m-fold convolution F^m of F with itself (independent of n). Along with the independence of the increments, this implies that

$$(1.9) \qquad\qquad P\{S_{n+m} - S_n \le t \,|\, S_0, \ldots, S_n\} = F^m(t), \quad t \ge 0.$$

Next we consider the number of renewals N_t in the interval $[0, t]$; this is

$$(1.10) \qquad\qquad N_t(\omega) = \sum_{n=0}^{\infty} I_{[0,t]}(S_n(\omega)), \quad t \ge 0, \omega \in \Omega,$$

where $I_A(x) = 1$ or 0 according as $x \in A$ or $x \notin A$. Note that $N_0(\omega) \ge 1$ always, and that $N_t(\omega) = \inf\{n \in \mathbb{N}: S_n(\omega) > t\}$. Thus, the event $\{N_t = k\}$

is equal to the event $\{S_{k-1} \leq t;\ S_k > t\} = \{S_{k-1} \leq t\} \cap \{S_k \leq t\}^c$, and $\{S_k \leq t\} \subset \{S_{k-1} \leq t\}$, since $S_k > S_{k-1}$. Thus, for any $k = 1, 2, \ldots$,

$$(1.11) \qquad P\{N_t = k\} = P\{S_{k-1} \leq t\} - P\{S_k \leq t\} = F^{k-1}(t) - F^k(t).$$

One can compute the expected number of renewals in $[0, t]$ by using this distribution; however, it is easier to use (1.10) directly:

$$(1.12) \qquad R(t) = E[N_t] = \sum_{n=0}^{\infty} E[I_{[0,t]}(S_n)]$$

$$= \sum_{n=0}^{\infty} P\{S_n \leq t\} = \sum_{n=0}^{\infty} F^n(t).$$

The function

$$(1.13) \qquad R = 1 + F + F^2 + F^3 + \cdots$$

is called the *renewal function* corresponding to the distribution F.

The renewal function R plays the same role in renewal theory as the potential matrix R does in the theory of Markov chains. In fact, the sequence S is a Markov chain with state space $E = [0, \infty]$, since the conditional distribution of S_{n+1} given $S_0, \ldots, S_n$ depends only on S_n. Moreover, the definition (1.13) shows R to be an operator similar to the potential R defined by (5.2.16), and analogous to the potential of a function in the sense of (7.1.2) we have the following result. Here and throughout this chapter, it is important to note that $R(0) \geq 1$. Therefore, in the integral (1.15) for instance, one must not forget the term $R(0)f(0)$ which is to be added to the integral over $(0, \infty)$.

(1.14) PROPOSITION. Suppose $F(0) < 1$. Then R is right continuous, nondecreasing, and finite. The number Rf defined by

$$Rf = E\left[\sum_{n=0}^{\infty} f(S_n)\right]$$

is finite for any bounded function f defined on $\mathbb{R}_+$ and vanishing outside a finite interval. For any $f \geq 0$ on $\mathbb{R}_+$,

$$(1.15) \qquad Rf = \int_{[0,\infty)} R(du)f(u).$$

Proof. We will show the finiteness of $R(t)$ in Lemma (1.16) below. We now complete the proof assuming that $R(t) < \infty$ for all t.

First, observe that $t \to N_t$ is non-decreasing and right continuous by the definition (1.10). Thus, R is non-decreasing. To show right continuity, let $t_0, t_1, \ldots$ be a sequence decreasing to t. By the right continuity of the process N, the random variables N_{t_n} decrease to N_t. Moreover, $N_{t_n} \leq N_{t_0}$ for all n, and $E[N_{t_0}] = R(t_0) < \infty$ by hypothesis. Thus, the bounded convergence theorem applies, and we have

$$\lim_{n \to \infty} R(t_n) = \lim_{n \to \infty} E[N_{t_n}] = E[N_t] = R(t).$$

Hence, R is right continuous.

If f is bounded by c and vanishes outside $[0, t]$, then $\sum f(S_n) \leq cN_t$ and $Rf \leq cR(t) < \infty$. To prove the formula (1.15), first note that for $f = I_{(s,t]}$ we have $Rf = E[N_t - N_s] = R(t) - R(s)$, so (1.15) holds. If f is a linear combination of finitely many such indicator functions, that is, if f is simple, then (1.15) holds again, since the expectation of a sum is the sum of the expectations and the integral of a sum is the sum of the integrals. For arbitrary $f \geq 0$ we can find a sequence f_n increasing to f such that each f_n is simple. Then the monotone convergence theorem shows that $Rf_n \to Rf$ and $\int R(ds)f_n(s) \to \int R(ds)f(s)$. This completes the proof. $\qquad\square$

(1.16) LEMMA. If $F(0) < 1$, then $R(t) < \infty$ for every $t < \infty$.

Proof. Let t be fixed. Since $F(0) < 1$ and F is right continuous, there is some $b > 0$ such that $F(b) < 1$. Choose $k \in \mathbb{N}$ such that $t \leq kb$. The event $\{S_k \leq t\}$ implies the event $\{S_k \leq kb\}$, which in turn implies that the event $\{W_1 > b, \ldots, W_k > b\}$ has not occurred. Thus,

$$P\{S_k \leq t\} \leq 1 - P\{W_1 > b, \ldots, W_k > b\} = 1 - [1 - F(b)]^k = 1 - \beta,$$

where $\beta > 0$. Similarly, the event $\{S_{mk} \leq t\}$ implies the event $\{S_k - S_0 \leq t, \ldots, S_{mk} - S_{mk-k} \leq t\}$; thus,

$$(1.17) \qquad P\{S_{mk} \leq t\} \leq (P\{S_k \leq t\})^m \leq (1 - \beta)^m.$$

Finally, since $\{S_{mk+j} \leq t\}$ implies $\{S_{mk} \leq t\}$,

$$\sum_{n=mk}^{mk+k-1} P\{S_n \leq t\} \leq kP\{S_{mk} \leq t\}.$$

Putting this into (1.12) and noting (1.17), we obtain

$$R(t) \leq k \sum_{m=0}^{\infty} (1 - \beta)^m = \frac{k}{\beta},$$

and this is finite, since $\beta > 0$. $\qquad\square$

(1.18) EXAMPLE. Suppose $F(t) = 1 - e^{-\lambda t}$, $t \geq 0$. Then, the renewal process is actually a Poisson process with an arrival at time 0. Hence, $R(t) = 1 + \lambda t$ for all $t \geq 0$. $\qquad\square$

(1.19) EXAMPLE. Suppose $F(t) = 1 - e^{-\lambda(t-b)}$, $t \geq b$, in Example (1.3). This means that a headway (measured in time units) is equal to the sum of a constant b and an exponentially distributed random variable with mean $1/\lambda$ (this tends to hold under moderate traffic conditions where the drivers experience a slight amount of restriction in their movements; here b is the reaction time of a driver and no headway can be less than that).

Now F^n is the distribution of the sum of the constant nb and n independent exponentially distributed random variables; thus,

$$F^n(t) = \begin{cases} 0 & \text{if } t < nb, \\ \sum_{k=n}^{\infty} \dfrac{e^{-\lambda(t-nb)}[\lambda(t-nb)]^k}{k!} & \text{if } t \geq nb; \end{cases}$$

and

$$R(t) = \sum_{n \leq t/b} \left[1 - \sum_{k=0}^{n-1} \frac{e^{-\lambda(t-nb)}[\lambda(t-nb)]^k}{k!} \right], \quad t \geq 0. \qquad \square$$

(1.20) EXAMPLE. Let $X = \{X_n; n \in \mathbb{N}\}$ be a Markov chain with a discrete state space, and let j be a fixed state. Let $S_1, S_2, \ldots$ be the successive step numbers at which state j is visited. If the initial state is j, then the times S_1, $S_2 - S_1, \ldots$ between returns are independent and identically distributed, by Proposition (5.2.1). Hence, if $X_0 = j$, then $S = \{S_n; n \in \mathbb{N}\}$ is a renewal process. (This fact can be used to prove Theorem (5.3.2) by using the renewal theory to be developed; see Example (2.22) for this.)

Consider the renewal function R for this process S. Since the number of visits to j during $[0, t]$ is

$$\sum_{n=0}^{\infty} I_{[0,t]}(S_n) = \sum_{n \leq t} 1_j(X_n),$$

we have

$$R(t) = E_j \left[\sum_{n \leq t} 1_j(X_n) \right] = \sum_{n \leq t} P^n(j, j),$$

where $P^n(i, j)$ are the n-step transition probabilities. Note that in this case all the S_n are integer-valued, so R is a step function whose jumps are restricted to the times $0, 1, 2, \ldots$. $\qquad \square$

Returning to an arbitrary renewal process, we note that the case $F(0) = 1$ is the trivial case where $S_0 = S_1 = S_2 = \cdots = 0$. We will exclude this case from further consideration and always assume that $F(0) < 1$ (then $R(t) < \infty$ for all t).

(1.21) DEFINITION. A renewal process S is said to be *recurrent* if $W_n < \infty$ almost surely for every n; otherwise, S is called *transient*. S is said to be *periodic with period* δ if the random variables $W_1, W_2, \ldots$ take values in a discrete set $\{0, \delta, 2\delta, \ldots\}$ and δ is the largest such number. Otherwise, if there is no such $\delta > 0$, S is said to be *aperiodic*.

S is recurrent or transient according as $F(\infty) = \lim_{t \to \infty} F(t) = 1$ or < 1. If S is periodic with period δ, then the distribution F is constant over intervals of form $[n\delta, n\delta + \delta)$, and all its jumps occur at points of the form $n\delta, n \in \mathbb{N}$; clearly, δ is the greatest common divisor of the set of points x at which $F(x) > F(x-) = \lim_{y \uparrow x} F(y)$. A distribution function all of whose jumps occur at points of the form $n\delta$ for some $\delta > 0$ is said to be *arithmetic with span* δ. For example, if F jumps only at 0.2, 0.6, 0.65, then F is arithmetic with span 0.05. If F jumps only at $\sqrt{2}$ and 3, then F is *not* arithmetic (since there is no number δ such that both $\sqrt{2}$ and 3 are integer multiples of δ). If F has a density it is not arithmetic.

If S is recurrent, then $F(+\infty) = \lim_{t \to \infty} F(t) = 1$ and vice versa. When S is recurrent, all the S_n are finite and therefore $N_\infty = \sup_t N_t = +\infty$ with probability one. Of course, as a result, $R(+\infty) = \lim_{t \to \infty} R(t) = +\infty$ as well. In this case the following is the main limit theorem on the behavior of the renewal function. We define

$$(1.22) \qquad m = \int_0^\infty x F(dx) = \int_0^\infty [1 - F(t)] \, dt,$$

that is, m is the expected value of the times between successive renewals. In interpreting the next theorem, and others to come, $1/m$ is equal to zero whenever $m = +\infty$.

(1.23) THEOREM. If S is recurrent aperiodic, then

$$\lim_{t \to \infty} [R(t + \tau) - R(t)] = \tau/m$$

for any $\tau > 0$. If S is recurrent and periodic with period δ, then the same result holds provided that τ be an integer multiple of δ.

Proof is omitted. As an immediate corollary we have the following elementary result.

(1.24) COROLLARY. If S is recurrent, then

$$\lim_{t \to \infty} \frac{R(t)}{t} = \frac{1}{m}.$$

A somewhat finer result will be given later. The following strong law of large numbers can be proved by carrying out the steps indicated in Exercise (3.5.21).

(1.25) PROPOSITION. If S is recurrent, then

$$\lim_{t\to\infty} \frac{N_t(\omega)}{t} = \frac{1}{m}$$

for almost all $\omega \in \Omega$.

Next suppose S is transient. This happens if and only if

$$(1.26) \qquad\qquad F(\infty) = \lim_{t\to\infty} F(t) < 1.$$

For example, the renewal process formed by the times of successive entrances to a fixed state j of a Markov process is transient if and only if j is transient.

If S is transient, then there is a positive probability $1 - F(\infty)$ that an interval W_n is of infinite length. Then $S_{k-1} < \infty$ and $S_k = S_{k+1} = \cdots = +\infty$ if and only if $\{W_1 < \infty, \ldots, W_{k-1} < \infty$ and $W_k = +\infty\}$ occurs; the probability of this event is $F(\infty)^{k-1}(1 - F(\infty))$. Of course, this is also the probability that the total number of renewals in $[0, \infty)$ is equal to k; that is, if

$$(1.27) \qquad\qquad N = \sup_{t<\infty} N_t,$$

then

$$(1.28) \qquad P\{N = k\} = [1 - F(\infty)]\, F(\infty)^{k-1}, \qquad k = 1, 2, \ldots.$$

This implies in particular that the total number of renewals N is finite with probability one and that its expected value is

$$(1.29) \qquad R(\infty) = \lim_{t\to\infty} R(t) = E[N] = \frac{1}{1 - F(\infty)} < \infty.$$

Still assuming that S is transient, let

$$(1.30) \qquad\qquad L = \sup\{S_n : S_n < \infty\};$$

then L is the time of the last renewal and is also called the *lifetime* of the renewal process S. Clearly, $L = S_{N-1} < \infty$. The following proof provides a good example of the "renewal-theoretic reasoning" to be made formal in the next section.

(1.31) PROPOSITION. If S is transient, then

$$P\{L \le t\} = (1 - F(\infty))R(t); \quad t \ge 0.$$

Proof. Let $f(t) = P\{L > t\}$. If the event $\{t < S_1 < \infty\}$ occurs, then

$$P\{L > t \,|\, S_1\} = 1.$$

If, on the other hand, the event $\{S_1 \leq t\}$ occurs, then the lifetime L is equal to the sum of S_1 and the lifetime of the renewal process $\hat{S}$ with $\hat{S}_0 = 0$, $\hat{S}_1 = S_2 - S_1$, $\hat{S}_2 = S_3 - S_1$, ... This latter renewal process has the same probability law as the former, and hence,

$$P\{L > t \mid S_1\} = f(t - S_1) \quad \text{on } \{S_1 \leq t\}.$$

Therefore,

$$
\begin{aligned}
f(t) &= P\{t < S_1 < \infty\} + E[f(t - S_1)I_{\{S_1 \leq t\}}] \\
&= F(\infty) - F(t) + \int_{[0,t]} F(ds)f(t - s).
\end{aligned}
$$

By Theorem (2.3) below the solution is

$$
\begin{aligned}
f(t) &= \int_{[0,t]} R(ds)[F(\infty) - F(t - s)] \\
&= 1 - [1 - F(\infty)]R(t);
\end{aligned}
$$

(here we used (1.13) to obtain $R * F = R - 1$). $\qquad \square$

It is clear from (1.31) that L is finite with probability one. The expected value of L may be obtained directly from its distribution; however, it is easier to use the arguments of the preceding proof: If $S_1 = +\infty$, then $L = 0$; if $S_1 < \infty$, then L is the sum of S_1 and the lifetime of the renewal process whose time origin is at S_1 and whose successive intervals are $S_2 - S_1$, $S_3 - S_2$, ... Hence,

$$
\begin{aligned}
E[L] &= E[S_1 I_{\{S_1 < \infty\}}] + E[L]E[I_{\{S_1 < \infty\}}] \\
&= \int_{[0,\infty)} t F(dt) + E[L]F(\infty) \\
&= \int_0^\infty [F(\infty) - F(t)] \, dt + E[L]\, F(\infty),
\end{aligned}
$$

from which we solve for $E[L]$ to obtain

$$(1.32) \qquad E[L] = \frac{1}{1 - F(\infty)} \int_0^\infty [F(\infty) - F(t)] \, dt.$$

(1.33) EXAMPLE. *Pedestrian Delay Problem.* Consider the traffic model of Example (1.3). Suppose, at time $t = 0$, a pedestrian (or a vehicle on a side street) arrives at that fixed point and wants to cross the highway. To cross it, the pedestrian needs τ units of time; therefore he starts crossing at $L = S_n$ if and only if $W_1 \leq \tau, \ldots, W_n \leq \tau$, and $W_{n+1} > \tau$. We see that L is the lifetime of the renewal process $\hat{S}$ whose nth interval is given by

$$\widehat{W}_n(\omega) = \begin{cases} W_n(\omega) & \text{if } W_n(\omega) \leq \tau, \\ +\infty & \text{otherwise.} \end{cases}$$

If the distribution of W_n is φ, then that of $\widehat{W}_n$ is

$$F(t) = \begin{cases} \varphi(t) & \text{if } t \leq \tau, \\ \varphi(\tau) & \text{if } t > \tau. \end{cases}$$

Hence, the delay the pedestrian experiences has the distribution

$$P\{L \leq t\} = (1 - F(\infty))R(t) = (1 - \varphi(\tau))R(t)$$

where $R = \sum F^n$. The expected delay is, from (1.32),

$$E[L] = \frac{1}{1 - \varphi(\tau)} \int_0^\tau [\varphi(\tau) - \varphi(t)] \, dt.$$

In particular, if the traffic flow is a Poisson process with rate λ, then $\varphi(t) = 1 - e^{-\lambda t}$, $t \geq 0$, and

$$E[L] = \frac{1}{\lambda} (e^{\lambda \tau} - 1) - \tau. \qquad \square$$

(1.34) EXAMPLE. *Geiger Counters of type II.* Particles arrive at the counter according to a renewal process. At time $t = 0$, a particle arrives, is registered, and locks the counter for a fixed time τ. If there are no arrivals during $[0, \tau]$, then the counter gets unlocked at time τ, the next particle to arrive gets registered, and the counter is locked again for a time of length τ. If an arrival occurs during a locked period, it does not get registered, but it extends the locked period so that the counter remains locked until τ units of time passes after that arrival.

If $W_1, W_2, \ldots$ are the successive interarrival times, then the counter gets unlocked for the first time at $S_n + \tau = W_1 + \cdots + W_n + \tau$ if and only if $W_1 \leq \tau, \ldots, W_n \leq \tau, W_{n+1} > \tau$. Thus, the length of the first locked period is equal to $L + \tau$, where L is the lifetime of the renewal process $\hat{S}$ with intervals defined by

$$\widehat{W}_n(\omega) = \begin{cases} W_n(\omega) & \text{if } W_n(\omega) \leq \tau, \\ +\infty & \text{otherwise.} \end{cases}$$

The distribution of the length of a locked period can be obtained just as in the preceding example once the distribution φ of the W_n is known. $\square$

2. Regenerative Processes and Renewal Theory

Consider a stochastic process $Z = \{Z_t; t \geq 0\}$ with state space E. Suppose that every time a certain phenomenon occurs, the future of the process Z after that time becomes a probabilistic replica of the future after time zero. Such times (usually random) are called regeneration times of Z, and the process Z is then said to be regenerative. For example, if Z is a Markov chain or a Markov process with a countable state space E and if j is a fixed state, then every time at which state j is entered is a time of regeneration for Z starting at j. For another example, let Z_t be the queue size at time t for a single-server queueing system subject to a Poisson process of arrivals and independent and identically distributed service times. Suppose the time origin is taken to be an instant of departure which left behind exactly j customers. Then every time a departure occurs leaving behind j customers, the future of Z after such a time has exactly the same probability law as the process Z had starting at time zero.

We will formalize the concept of regeneration later, in Definition (2.18), after developing the theory needed to exploit it. For the present, to motivate this theory, let Z be a regenerative process with a discrete state space, and consider the probability $f(t)$ that $Z_t = i$ for some fixed state i. As in the proof of (1.31), we condition the event $\{Z_t = i\}$ on the time S_1 of first regeneration, and argue as follows. The process Z regenerates itself at S_1, and the future process $\hat{Z}$ defined by $\hat{Z}_u = Z_{S_1+u}$ has the same probability law as Z itself. Given S_1, if $S_1 = s \leq t$, then $Z_t = \hat{Z}_{t-s}$, and therefore,

$$P\{Z_t = i \,|\, S_1\} = P\{\hat{Z}_{t-s} = i\} = f(t-s) \quad \text{on } \{S_1 = s \leq t\}.$$

Hence, if we define

$$g(t) = P\{Z_t = i, S_1 > t\},$$

then we have

(2.1)
$$f(t) = g(t) + \int_{[0,t]} F(ds) f(t-s).$$

This equation is called a *renewal equation*. The results we will be obtaining in this section will enable us to solve such equations for f and to study the behavior of the solution f for large t. Hence, by means of the theory we will develop, studying the behavior of Z is reduced to computing the function g.

This theory is called *renewal theory*, and it is the main tool for studying regenerative processes in the absence of further properties. Conversely, it is this applicability to regenerative processes which makes renewal theory the most important tool in elementary probability theory.

Renewal theory is the study of the so-called *renewal equation*

$$(2.2) \qquad\qquad f = g + F * f,$$

where F is a distribution function on $\mathbb{R}_+$ and f and g are functions which are bounded over finite intervals. It is assumed that F and g are known, and the problem is to solve for f and examine its behavior at infinity. Concerning the solution, we have

(2.3) THEOREM. The renewal equation (2.2) has one and only one solution; it is

$$f = R * g$$

where $R = \sum F^n$ is the renewal function corresponding to F.

Proof. It follows from (1.13) that

$$F * (R * g) = F * g + F^2 * g + \cdots = R * g - g,$$

which shows that $R * g$ is a solution to (2.2). If f is another solution, then $h = f - R * g$ must satisfy $h = F * h$, which implies that

$$h = F^n * h$$

for all n. It follows from the finiteness of $R(t)$ that $F^n(t) \rightarrow 0$ as $n \rightarrow \infty$ for any fixed t. Hence, as $n \rightarrow \infty$, $F^n * h(t) \rightarrow 0$, which implies that $h(t) = 0$. $\square$

In the case where the renewal process S associated with the distribution F is transient, the limiting behavior of the solution $R * g$ is easy to see:

(2.4) PROPOSITION. If $F(\infty) < 1$, then

$$\lim_{t \to \infty} R * g(t) = R(\infty)g(\infty)$$

provided that $g(\infty) = \lim_{t \to \infty} g(t)$ exists.

In the recurrent case the key result follows from Theorem (1.23) and provides the limiting behavior of $R * g(t)$ for large t for a certain class of functions g which we introduce next.

Let $g \geq 0$ be a function which is bounded over finite intervals, and for $b > 0$ define

$$\gamma_b'(n) = \inf\{g(t): nb \leq t < nb + b\}, \quad n \in \mathbb{N}$$
$$\gamma_b''(n) = \sup\{g(t): nb \leq t < nb + b\}, \quad n \in \mathbb{N}$$

and put

(2.5) $$\sigma'_b = b \sum_n \gamma'_b(n), \qquad \sigma''_b = b \sum_n \gamma''_b(n).$$

(2.6) DEFINITION.   The function g is said to be *directly Riemann integrable* (notation: $g \in \mathbb{D}$) provided that the sums (2.5) both converge for every $b > 0$ and that

$$\lim_{b \to 0} (\sigma''_b - \sigma'_b) = 0. \qquad \square$$

If $g \in \mathbb{D}$, then

(2.7) $$\lim_{b \to 0} \sigma'_b = \lim_{b \to 0} \sigma''_b = \int_0^\infty g(t)\, dt,$$

where the last integral is the usual Riemann integral of g over $[0, \infty)$. Hence, direct Riemann integrability implies Riemann integrability. Some results in the converse direction will be put after the following theorem. This is the main limit theorem of this section and is sometimes referred to as the *key renewal theorem*. It implies Theorem (1.23) and, as the proof will show, is implied by (1.23). As before, the number m appearing below is the expected value defined by (1.22), and $1/m$ is to be taken as zero when m is infinite.

(2.8) THEOREM.   If $g \in \mathbb{D}$, $F(\infty) = 1$, and F is not arithmetic, then

(2.9) $$\lim_{t \to \infty} R * g(t) = \frac{1}{m} \int_0^\infty g(y)\, dy.$$

If $F(\infty) = 1$, F is arithmetic with span δ, and $\sum_{k=0}^\infty g(x + k\delta)$ converges, then

(2.10) $$\lim_{n \to \infty} R * g(x + n\delta) = \frac{\delta}{m} \sum_{k=0}^\infty g(x + k\delta).$$

Proof. (a) By (1.13), $R * F = R - 1$, so $R * (1 - F) = 1$. Since the function $1 - F$ is monotone,

$$1 = \int_0^t R(ds)[1 - F(t - s)] \geq [R(t) - R(t - b)][1 - F(b)],$$

and choosing b so that $F(b) < 1$ we see that

(2.11) $$\beta_b = \sup_t [R(t) - R(t - b)] < \infty.$$

Since any interval can be partitioned into a finite number of intervals of smaller length, (2.11) holds for any $b > 0$.

(b) Suppose F is not arithmetic. For fixed $b > 0$ and $n \in \mathbb{N}$ let i_n be the indicator function of the interval $[nb, nb + b)$, that is, $i_n(x) = 1$ or 0 according as $nb \leq x < nb + b$ or not. Then $R * i_n(t) = R(t - nb) - R(t - nb - b)$ for $t > nb + b$ and hence, by Theorem (1.23),

$$(2.12) \qquad\qquad \lim_{t \to \infty} R * i_n(t) = \frac{b}{m}.$$

(c) Next let $h = \sum_n c_n i_n$, where the constants $c_n \geq 0$ are such that $\sum c_n < \infty$. By (2.11), $R * i_n(t) \leq \beta_b$, and hence

$$\sum_{n=0}^{k} c_n R * i_n(t) \leq R * h(t) \leq \sum_{n=0}^{k} c_n R * i_n(t) + \beta_b \sum_{n=k+1}^{\infty} c_n.$$

Taking limits as $t \to \infty$ by using (2.12), and then letting k go to infinity, we obtain

$$(2.13) \qquad\qquad \lim_{t \to \infty} R * h(t) = \frac{b}{m} \sum c_n.$$

(d) For $g \in \mathbb{D}$, let $h_b'(t) = \sum_n \gamma_b'(n) i_n(t)$ and $h_b''(t) = \sum_n \gamma_b''(n) i_n(t)$ in the notations preceding (2.6). Then

$$(2.14) \qquad\qquad R * h_b' \leq R * g \leq R * h_b'',$$

and (2.13) applies to h_b' and h_b'' to yield

$$(2.15) \qquad \lim_{t \to \infty} R * h_b'(t) = \frac{1}{m} \sigma_b', \qquad \lim_{t \to \infty} R * h_b''(t) = \frac{1}{m} \sigma_b''.$$

The proof is completed by using (2.14) and (2.15), letting $b \to 0$, and using the definition of direct integrability.

(e) For the arithmetic case the proof needs to be modified slightly by taking $b = \delta$ and passing to the limit only over lattices of form $\{x + n\delta; n \in \mathbb{N}\}$. $\qquad\square$

The following puts together some results on direct integrability which are encountered quite often.

(2.16) PROPOSITION. (a) Let $g \geq 0$ be continuous and vanishing outside a finite interval; then $g \in \mathbb{D}$.

(b) Let $g \geq 0$ be bounded and continuous; then $g \in \mathbb{D}$ if and only if $\sigma_b'' < \infty$ for some $b > 0$.

(c) Let $g \geq 0$ be monotone non-increasing; then $g \in \mathbb{D}$ if and only if g is Riemann integrable.

(d)   Let $g \geq 0$ and let φ be a distribution function on $\mathbb{R}_+$. If $g \in \mathbb{D}$, then $\varphi * g \in \mathbb{D}$ as well.

Proof. Proofs of (a), (b), and (c) are immediate from the definition. To prove (d), first we observe that it is sufficient to show it for $g \in \mathbb{D}$ of form $g = \sum c_n i_n$, where i_n is the indicator function of $[nb, nb + b)$ for some fixed $b > 0$. For such a g and for $t \in [nb, nb + b) = A_n$,

$$\varphi * g(t) = \sum_{k=0}^{n-1} c_k [\varphi(t - kb) - \varphi(t - kb - b)] + c_n \varphi(t \qquad nb).$$

Noting that

$$\sup_{t \in A_n} [\varphi(t - kb) - \varphi(t - kb - b)] \leq \int_{(n-k)b-b}^{(n-k)b+b} \varphi(dx)$$

and that

$$\sup_{t \in A_n} \varphi(t - nb) = \varphi(b),$$

we obtain

$$a_n = \sup_{t \in A_n} \varphi * g(t) \leq \sum_{k=0}^{n} c_{n-k} \int_{kb-b}^{kb+b} \varphi(dx).$$

Hence,

$$\sum_n a_n \leq \sum_{k=0}^{\infty} \sum_{n=k}^{\infty} c_{n-k} \int_{kb-b}^{kb+b} \varphi(dx) \leq 2 \sum_n c_n < \infty$$

as needed. $\square$

As a first application of the renewal theory developed above, we now give a refinement of Corollary (1.24), which stated that for large t the renewal function varies as t/m. To estimate the behavior of the difference

$$f(t) = R(t) - \frac{t}{m},$$

we first note that f satisfies the renewal equation $f = g + F * f$ with

$$g(t) = \frac{1}{m} \int_t^{\infty} (1 - F(x)) \, dx.$$

This function g is monotone, and

$$\int_0^{\infty} g(t) \, dt = \frac{1}{2m} \int_0^{\infty} t^2 F(dt) = \frac{m^2 + v^2}{2m}$$

where v^2 is the variance of an interrenewal time. Thus, applying Theorem (2.8) yields the proof of the following

(2.17) PROPOSITION.   If S is aperiodic and $v^2 < \infty$, then

$$\lim_{t \to \infty} \left[R(t) - \frac{t}{m} \right] = \frac{m^2 + v^2}{2m^2}. \qquad \square$$

The remainder of this section is devoted to the applications of renewal theory to regenerative processes. We start by making the concept of regeneration more precise.

Consider a stochastic process $Z = \{Z_t ; t \geq 0\}$ with state space E. For technical reasons, we assume that E has a topology and that the paths $t \to Z_t(\omega)$ are right continuous for almost all $\omega \in \Omega$. We recall that a random time T is called a stopping time of Z provided that for every t, whether or not the event $\{T \leq t\}$ has occurred can be determined once the history $\{Z_u ; u \leq t\}$ before t is known.

(2.18) DEFINITION.   The process $Z = \{Z_t ; t \geq 0\}$ is said to be *regenerative* provided that there exists a sequence $S_0, S_1, \ldots$ of stopping times such that

(a) $S = \{S_n ; n \in \mathbb{N}\}$ is a renewal process,

(b)   for any $n, m \in \mathbb{N}$, $t_1, \ldots, t_n \in \mathbb{R}_+$, and any bounded function f defined on E^n,

$$(2.19) \qquad E[f(Z_{S_m + t_1}, \ldots, Z_{S_m + t_n}) \,|\, Z_u ; u \leq S_m] = E[f(Z_{t_1}, \ldots, Z_{t_n})]. \qquad \square$$

Property (2.18b) is called the regeneration property applied at S_m, and the times $S_0, S_1, \ldots$ are called the regeneration times. To understand (2.19) better, let $W_t = f(Z_{t+t_1}, \ldots, Z_{t+t_n})$; and let $\hat{Z}$ be the future process obtained from Z by taking $T = S_m$ as the time origin, that is, let $\hat{Z}_u = Z_{T+u}$ for all $u \geq 0$. First note that

$$W_T = f(Z_{T+t_1}, \ldots, Z_{T+t_n}) = f(\hat{Z}_{t_1}, \ldots, \hat{Z}_{t_n}) = \hat{W}_0.$$

Then, observe that (2.19) contains two statements:

$$(2.20) \qquad\qquad\qquad E[\hat{W}_0 \,|\, Z_u ; u \leq T] = E[\hat{W}_0],$$
$$(2.21) \qquad\qquad\qquad\qquad E[\hat{W}_0] = E[W_0].$$

Of these, (2.20) states that the future process $\hat{Z}$ is independent of the past history $\{Z_u ; u \leq T\}$ before T, and (2.21) states that the probability law of $\hat{Z}$ is the same as that of Z.

Let $Z = \{Z_t ; t \geq 0\}$ be a regenerative process with regeneration points $S = \{S_n ; n \in \mathbb{N}\}$, and let F be the distribution function of the times between regenerations. Define, for an open subset $A \subset E$,

(2.22) $$K(t, A) = P\{S_1 > t, Z_t \in A\}, \quad t \geq 0,$$

and let

$$f(t) = P\{Z_t \in A\}, \quad t \geq 0.$$

By the regeneration property applied at S_1,

$$P\{Z_t \in A \,|\, S_1\} = f(t - S_1) \quad \text{on } \{S_1 \leq t\}.$$

Hence, the function f satisfies

(2.23) $$f(t) = K(t, A) + \int_{[0, t]} F(ds)f(t - s).$$

This is a renewal equation; by Theorem (2.3) its solution is $f = R * g$ with $g(t) = K(t, A)$. We put this next.

(2.24) PROPOSITION. Let Z be a regenerative process with state space E. Then for any $t \geq 0$ and open $A \subset E$,

$$P\{Z_t \in A\} = \int_{[0, t]} R(ds)K(t - s, A),$$

where K is as defined by (2.22) and R is the renewal function of the imbedded renewal process S.

Following is the main limit theorem for regenerative processes.

(2.25) THEOREM. Let A be such that $t \longrightarrow K(t, A)$ is Riemann integrable in the ordinary sense.
  (a) If S is recurrent aperiodic and $m < \infty$, then

$$\lim_{t \to \infty} P\{Z_t \in A\} = \frac{1}{m} \int_0^\infty K(s, A) \, ds.$$

  (b) If S is recurrent periodic with period δ and $m < \infty$, then

$$\lim_{n \to \infty} P\{Z_{x+n\delta} \in A\} = \frac{\delta}{m} \sum_{k=0}^\infty K(x + k\delta, A).$$

  (c) If S is recurrent and $m = \infty$, then

$$\lim_{t \to \infty} P\{Z_t \in A\} = 0.$$

Proof follows easily from Proposition (2.24) and Theorem (2.8). We should only note that, in case (a), the function $t \rightarrow K(t, A)$ is directly integrable. This follows from observing that it is Riemann integrable by hypothesis and that it is bounded by the function $1 - F$ which is directly integrable, by Proposition (2.16c), since $1 - F$ is monotone and its integral is $m < \infty$. ☐

(2.26) EXAMPLE.   Let Z_t be the time from t until the instant of next renewal in a renewal process S; that is,

$$Z_t(\omega) = \begin{cases} S_{n+1}(\omega) - t & \text{if } S_n(\omega) \leq t < S_{n+1}(\omega) \text{ for some } n \in \mathbb{N}, \\ +\infty & \text{otherwise.} \end{cases}$$

Then Z is a regenerative process with state space $E = [0, \infty]$. For $A = (y, \infty]$,

$$K(t, A) = P\{S_1 > t, Z_t > y\} = P\{S_1 > t + y\} = 1 - F(t + y).$$

By Proposition (2.24), then,

$$P\{Z_t > y\} = \int_{[0, t]} R(ds)[1 - F(t + y - s)].$$

If S is recurrent aperiodic and $m < \infty$, then the monotone function $t \rightarrow g(t)$ $= 1 - F(t + y)$ is in $\mathbb{D}$, since

$$\int_0^\infty [1 - F(t + y)]\, dt = \int_y^\infty [1 - F(x)]\, dx \leq m < \infty.$$

Hence, by Theorem (2.8),

$$\lim_{t \to \infty} P\{Z_t > y\} = \frac{1}{m} \int_y^\infty (1 - F(x))\, dx. \qquad \qquad ☐$$

(2.27) EXAMPLE. In Example (1.4), let $Z_t(\omega) = 1$ or 0 according as the item is working or being replaced at time t for the outcome $\omega \in \Omega$. Then $Z = \{Z_t; t \geq 0\}$ is a regenerative process with state space $E = \{0, 1\}$. If φ and ψ are the respective distributions of the lifetime and replacement time and $A = \{1\}$, then

$$K(t, A) = P\{Z_t = 1, S_1 > t\} = P\{U_1 > t\} = 1 - \varphi(t),$$

and Proposition (2.24) yields, with $R = \sum F^n$, $F = \varphi * \psi$,

$$P\{Z_t = 1\} = \int_{[0, t]} R(ds)[1 - \varphi(t - s)].$$

If the mean lifetime is

$$a = \int_0^\infty (1 - \varphi(t))\, dt$$

and the mean replacement time b, then $m = a + b$, and in the recurrent aperiodic case we have

$$\lim_{t \to \infty} P\{Z_t = 1\} = \frac{a}{a + b}$$

by Theorem (2.25).   □

(2.28) EXAMPLE. *Markov chains.* Let X be a Markov chain with initial state j. If we define $Z_t = X_n$ for all $t \in [n, n + 1)$, we obtain a regenerative process $Z = \{Z_t; t \geq 0\}$ with regeneration times S_n being the times of successive visits to the fixed state j. In this case,

$$K(t, \{j\}) = P\{Z_t = j, S_1 > t\} = \begin{cases} 1 & \text{if } t < 1, \\ 0 & \text{otherwise.} \end{cases}$$

If j is a transient state, then S is transient, and by Proposition (2.4) we have

$$\lim_{t \to \infty} P\{Z_t = j\} = \lim_{n \to \infty} P\{X_n = j\} = 0.$$

If j is recurrent null, the same is true by (2.25c). Suppose next that j is recurrent non-null. If state j is aperiodic, then S is periodic with $\delta = 1$; if state j is periodic with period δ, then S is periodic with period δ. By Theorem (2.25b), then, we have

$$\lim_{n \to \infty} P\{Z_{n\delta} = j\} = \lim_{n \to \infty} P\{X_{n\delta} = j\} = \frac{\delta}{m}$$

where m is the mean recurrence time of state j. This constitutes the proof of Theorem (5.3.2).   □

(2.29) EXAMPLE. *Markov Processes.* Let Z be a Markov process with initial state j. Then Z is a regenerative process with regeneration times S_n being the instants of successive entrances to state j. Now,

$$K(t, \{j\}) = P_j\{Z_t = j, S_1 > t\} = e^{-\lambda(j)t}$$

by Theorem (8.3.3). If state j is transient, then by Proposition (2.4), we have

$$\lim_{t \to \infty} P\{Z_t = j\} = 0.$$

Otherwise, if state j is recurrent (the corresponding renewal process S is automatically aperiodic), Theorem (2.25a, c) gives

$$\lim_{t \to \infty} P\{Z_t = j\} = \frac{1}{m(j)} \int_0^\infty e^{-\lambda(j)x} \, dx = \frac{1}{m(j)\lambda(j)},$$

where $m(j)$ is the mean recurrence time of j.   □

3.   Delayed and Stationary Processes

In certain applications the time S_0 of the first renewal is not necessarily the time origin. For example, in the pedestrian delay problem of Example (1.33), the instant $t = 0$ at which the pedestrian arrives does not necessarily coincide with a time of arrival of a vehicle. To handle such situations we introduce the following

(3.1)   DEFINITION.   A sequence $S = \{S_n; n \in \mathbb{N}\}$ is called a *delayed renewal process* provided that $\hat{S} = \{S_n - S_0; n \in \mathbb{N}\}$ be a renewal process and that $S_0 \geq 0$ be independent of $\hat{S}$.   □

In other words, if S is a delayed renewal process, then at the time S_0 of the first renewal there starts a renewal process $\hat{S}$. Similarly, we define a *delayed regenerative process* to be a process Z having a sequence S of stopping times which form a delayed renewal process and is such that Definition (2.18b) holds. For example, for any initial state i, the times of successive entrances to a fixed state j in a Markov process form a delayed renewal process. Any Markov process is a delayed regenerative process; any M/G/1 queueing process is a delayed regenerative process; etc.

There are no new tools needed. In handling a delayed regenerative process, first we condition the event or expectation in question on the time S_0 of first regeneration, and then we use the fact that at S_0 there starts an ordinary regenerative process. The following treatment of delayed renewal processes will provide several examples of this.

Let G be the distribution of S_0, and let F denote the common distribution of the interrenewal times $S_1 - S_0$, $S_2 - S_1, \ldots$. Note that a delayed renewal process becomes an ordinary renewal process when $G(t) = 1$ for all $t \geq 0$, that is, when $S_0 = 0$ almost surely.

Writing $S_n = S_0 + (S_n - S_0)$ and noting the independence of S_0 from $S_n - S_0$ shows that the distribution of S_n is the convolution of those of S_0 and $S_n - S_0$ which are G and F^n respectively; namely,

$$P\{S_n \leq t\} = G * F^n(t), \quad t \geq 0.$$

As before, let N_t be the number of renewals occurring in the interval

$[0, t]$; we still have (1.10) holding: $N_t = \sum_{n=0}^{\infty} I_{[0, t]}(S_n)$. Thus, just as in (1.12), we have

(3.2) $$\hat{R}(t) = E[N_t] = \sum_{n=0}^{\infty} P\{S_n \le t\} = \sum_{n=0}^{\infty} G * F^n(t).$$

Defining

$$R = 1 + F + F^2 + \cdots$$

as before in (1.13), we now have

(3.3) $$\hat{R}(t) = E[N_t] = G * R(t) = R * G(t).$$

The limiting behavior of this expectation is similar to that of R; this, of course, would be expected, since the influence of the initial distribution should disappear in the long run, provided that the first renewal ever occurs.

(3.4) THEOREM. (a) If $F(\infty) < 1$, then

$$\hat{R}(\infty) = G(\infty)R(\infty) = \frac{G(\infty)}{1 - F(\infty)}.$$

(b) If $F(\infty) = 1$ and F is not arithmetic, then

$$\lim_{t \to \infty} [\hat{R}(t) - \hat{R}(t - \tau)] = \frac{1}{m} G(\infty)\tau,$$

for any $\tau > 0$. If $F(\infty) = 1$ and F is arithmetic with span δ, the same holds for τ an integer multiple of δ, as $t \to \infty$ over $0, \delta, 2\delta, \ldots$.

Proof. If $F(\infty) < 1$, the result claimed is immediate from Proposition (2.4) applied to (3.3).

For the recurrent case $F(\infty) = 1$, first note that $\hat{R}(t) - \hat{R}(t - \tau) = R * g(t)$ for $t > \tau$, where

$$g(t) = G * I_{[0, \tau)}(t) = \int_{[0, t]} G(dx)I_{[0, \tau)}(t - x), \quad t \ge 0.$$

It is clear that $I_{[0, \tau)} \in \mathbb{D}$; therefore, by Proposition (2.16d), $g \in \mathbb{D}$, and

$$\int_0^{\infty} g(t) \, dt = G(\infty)\tau.$$

The proof in the non-arithmetic case now follows from Theorem (2.8). In the arithmetic case, when $\tau = j\delta$ for some integer j,

$$\delta \sum_{k=0}^{\infty} g(k\delta) = \delta \sum_{k=0}^{\infty} \int_0^{k\delta} G(dx)I_{[0, \tau)}(k\delta - x)$$

$$= \delta \int_0^{\infty} G(dx) \sum_{k=0}^{\infty} I_{[0, \tau)}(k\delta - x) = \delta G(\infty)j = G(\infty)\tau;$$

thus, from Theorem (2.8) again, we get

$$\lim_{n \to \infty} [\hat{R}(n\delta) - \hat{R}(n\delta - \tau)] = \frac{1}{m} G(\infty)\tau$$

as desired.   □

Next consider the time V_t from t until the instant of next renewal; the time of the first renewal after t is S_{N_t}, so

$$(3.5) \qquad\qquad V_t = S_{N_t} - t, \quad t \geq 0.$$

The process $\{V_t; t \geq 0\}$ is a delayed regenerative process. In (2.26) we had computed the distribution of V_t for the regenerative case. That result now applies to the regenerative process $\{V_{S_0 + u}; u \geq 0\}$, and we have

$$(3.6) \qquad\qquad P\{V_{S_0 + u} > y\} = \int_{[0, u]} R(dx)[1 - F(u + y - x)]$$

for any $y \geq 0$. To obtain the distribution of V_t, we first observe that $V_t = S_0 - t$ if $\{S_0 > t\}$ occurs, and that $V_t = V_{S_0 + u}$ with $u = t - s$ if $\{S_0 = s \leq t\}$ occurs; this argument yields

$$P\{V_t > y\} = P\{V_t > y; S_0 > t\} + P\{V_t > y; S_0 \leq t\}$$
$$= P\{S_0 > t + y\} + P\{S_0 \leq t, V_{S_0 + (t - S_0)} > y\}.$$

Evaluating this and using (3.6), we obtain

$$P\{V_t > y\} = 1 - G(t + y) + \int_{[0, t]} G(ds) \int_{[0, t - s]} R(dx)[1 - F(t - s + y - x)].$$

The second term on the right is $G * R * g(t)$ with $g(t) = 1 - F(t + y)$. Replacing $G * R$ by $\hat{R}$, we obtain the following

(3.7) PROPOSITION. For any $t \geq 0$ and $y \geq 0$,

$$P\{V_t > y\} = 1 - G(t + y) + \int_{[0, t]} \hat{R}(dx)[1 - F(t + y - x)].$$

A case of special interest occurs when G is the limiting distribution of V_t as $t \to \infty$ in a renewal process; that is, when

$$(3.8) \qquad\qquad G(t) = \frac{1}{m} \int_0^t (1 - F(x))\, dx, \quad t \geq 0.$$

Intuitively, this case corresponds to a renewal process that had been going on for a long time before $t = 0$. As would be expected, then, $\hat{R}(t) = t/m$, and the distribution of V_t as given in (3.7) becomes independent of t. For these and other reasons to be listed below, a delayed renewal process S is called a *stationary renewal process* if G is as given by (3.8).

(3.9) PROPOSITION. If S is stationary, then

$$\hat{R}(t) = \frac{1}{m}\, t, \quad t \geq 0.$$

Proof. From (3.3) we have $\hat{R} = R * G$, which now becomes

$$\hat{R}(t) = \int_{[0,t]} R(dy) \frac{1}{m} \int_0^{t-y} [1 - F(x)]\, dx, \quad t \geq 0.$$

Making a change of variable $z = x + y$, and then changing the order of integration, we obtain

$$\hat{R}(t) = \frac{1}{m} \int_{[0,t]} R(dy) \int_y^t [1 - F(z - y)]\, dz$$

$$= \frac{1}{m} \int_0^t dz \int_{[0,z]} R(dy)[1 - F(z - y)] = \frac{1}{m}\, t$$

since $R * (1 - F) = 1$ identically. □

(3.10) PROPOSITION. If S is stationary, then

$$P\{V_t > y\} = \frac{1}{m} \int_y^\infty [1 - F(x)]\, dx = 1 - G(y), \quad y \geq 0,$$

independent of t.

Proof. Putting $\hat{R}(t) = t/m$ in the formula given by (3.7), we obtain

$$P\{V_t > y\} = 1 - G(t + y) + \frac{1}{m} \int_0^t [1 - F(t + y - x)]\, dx$$

$$= \frac{1}{m} \int_{t+y}^\infty [1 - F(z)]\, dz + \frac{1}{m} \int_y^{t+y} [1 - F(z)]\, dz$$

$$= \frac{1}{m} \int_y^\infty [1 - F(z)]\, dz,$$

as claimed. □

The process $V = \{V_t; t \geq 0\}$ is an example of a Markov process with a non-discrete state space; that is, V is a stochastic process with state space $E = [0, \infty]$, and it satisfies

(3.11) $P\{V_{t+s} \in B \mid V_u; u \leq t\} = P\{V_{t+s} \in B \mid V_t\}$

for any $t, s \geq 0$ and interval subset B of E. In fact, the right-hand side above is equal to $P_s(V_t, B)$, where

$$P_t(x, B) = \begin{cases} I_B(x - t), & t < x, \\ H_{t-x}(B), & t \geq x, \end{cases}$$

where $H_s(B) = P\{V_s \in B\}$ is as computed in Proposition (3.7). Note now that $V_0 = S_0$, so that G is the initial distribution of the Markov process V. What Proposition (3.10) states is that (3.8) is the invariant distribution of V. The process V here is the continuous-time analog of the Markov chain of Example (5.1.17). What we showed above are the analogs to (6.2.18).

Proposition (3.10) shows that at any time t, the interval from t until the next renewal has the distribution G independent of t. Of course, once that first renewal after t occurs, there starts a new renewal process. Thus, the future of S from any time t onward has the same probability law as S itself. The following are two ways of expressing this; we omit the formal proofs.

(3.12) THEOREM. S is a stationary renewal process if and only if the joint distribution of $V_{t+t_0}, \ldots, V_{t+t_n}$ is the same as that of $V_{t_0}, \ldots, V_{t_n}$ for any $t \geq 0, n \in \mathbb{N}$, and $t_0, \ldots, t_n \in \mathbb{R}_+$.

(3.13) THEOREM. S is a stationary renewal process if and only if the joint distribution of $N_{t_1} - N_{t_0}, \ldots, N_{t_n} - N_{t_{n-1}}$ is the same as that of $N_{t+t_1} - N_{t+t_0}, \ldots, N_{t+t_n} - N_{t+t_{n-1}}$ for any $t \geq 0, n \in \mathbb{N}$, and $t_0, \ldots, t_n \in \mathbb{R}_+$.

The simplest stationary renewal process is the Poisson process. That the Poisson process is stationary follows from the fact that

$$P\{V_t > y\} = P\{N_{t+y} - N_t = 0\} = e^{-\lambda y}$$

independent of t. We see, in particular, that if $F(t) = 1 - e^{-\lambda t}, t \geq 0$, then

(3.14) $\dfrac{1}{m} \displaystyle\int_0^t [1 - F(x)] \, dx = F(t), \quad t \geq 0.$

Now the question arises as to whether there are other renewal processes where the distribution of V_t is the same as that of an interrenewal time; in other words, are there distribution functions F satisfying (3.14) which are not exponential? The answer is, no. For if F satisfies (3.14), then it must be continuous (since the left side is an integral of a right-continuous function) and therefore differentiable (since then the left side is the integral of a continuous function). Taking derivatives on both sides, we see that

$$(3.15) \qquad \frac{F'(t)}{1 - F(t)} = \frac{1}{m}, \quad t \geq 0.$$

The left-hand side of (3.15) is the derivative of $-\log[1 - F(t)]$; hence the only solution of (3.15) and therefore of (3.14) is

$$1 - F(t) = e^{-t/m}.$$

4. Exercises

(4.1) Let F be a distribution, F^n its n-fold convolution with itself, and $R = \sum F^n$ the corresponding renewal function. Define the Laplace transforms

$$F_\alpha = \int_{[0,\infty)} e^{-\alpha t} F(dt), \qquad R_\alpha = \int_{[0,\infty)} e^{-\alpha t} R(dt), \quad \alpha \geq 0.$$

    (a) Show that the Laplace transform of F^n is $(F_\alpha)^n$. (*Hint*: F^n is the distribution of the sum $W_1 + \cdots + W_n$.)

    (b) Use this to show that

$$R_\alpha = (1 - F_\alpha)^{-1}; \qquad F_\alpha = 1 - 1/R_\alpha, \quad \alpha > 0.$$

(4.2) Let $F(t) = 1 - e^{-3t} - 3te^{-3t}, \ t \geq 0.$

    (a) Show that

$$F_\alpha = \left(\frac{3}{3+\alpha}\right)^2, \qquad R_\alpha = 1 + \frac{3}{2\alpha} - \frac{1}{4} \cdot \frac{6}{\alpha + 6}.$$

    (b) Invert R_α to obtain

$$R(t) = 1 + \frac{3}{2}t - \frac{1}{4}(1 - e^{-6t}), \quad t \geq 0.$$

    (c) Compute the limits indicated in Theorem (1.23), Corollary (1.24), and Proposition (2.17) directly from (b).

(4.3) Prove Proposition (1.25).

(4.4) *Wald's lemma.* Let N_t be the number of renewals in $[0, t]$ of a renewal process S, and consider the time S_{N_t} of the first renewal after t.

    (a) Show that S_{N_t} can be written as

$$S_{N_t} = \sum_{n=0}^{\infty} \sum_{k=0}^{n} W_{k+1} I_{[S_n \leq t < S_{n+1}]}.$$

(b)   Change the order of summation to obtain

$$S_{N_t} = \sum_{k=0}^{\infty} W_{k+1} I_{\{S_k \leq t\}}.$$

(c)   Use the independence of S_k and W_{k+1} to compute

$$E[S_{N_t}] = mR(t).$$

(4.5)   Use Wald's lemma to compute the expected value of the time V_t from t until the next renewal. Use this to compute $\lim_{t \to \infty} E[V_t]$ via the formula given in Proposition (2.17).

(4.6)   Let U_t be the time since the last renewal before t in a renewal process S, that is, let $U_t = t - S_{N_t-1}$.
   (a)   Show that $f(t) = P\{U_t > x\}$ satisfies, for fixed x, the renewal equation

$$f(t) = (1 - F(t))I_{(x,\infty)}(t) + \int_{[0,t]} F(ds)f(t - s).$$

(b)   Show that for $t > x$

$$P\{U_t > x\} = \int_{[0,t-x)} R(ds)(1 - F(t - s)).$$

(c)   Show that the limit as $t \to \infty$ of $P\{U_t > x\}$ is the same as the limit given for $V_t = S_{N_t} - t$ in Example (2.26).

(4.7)   Let $U_t = t - S_{N_t-1}$ and $V_t = S_{N_t} - t$, that is, U_t and V_t are, respectively, the time since the last renewal and time to the next renewal at t.
   (a)   For fixed $x, y \geq 0$ show that $f(t) = P\{U_t > x, V_t > y\}$ satisfies the renewal equation $f = g + F * f$ with

$$g(t) = I_{(x,\infty)}(t)[1 - F(t + y)].$$

(b)   Show that when $m < \infty$,

$$\lim_{t \to \infty} P\{U_t > x, V_t > y\} = \frac{1}{m} \int_{x+y}^{\infty} (1 - F(u)) \, du.$$

(4.8)   Let S be a renewal process with interval distribution F and renewal function R, and let U and V be as defined in the preceding exercise. Show that the following are equivalent.
   (a) S forms a Poisson process.
   (b) $R(t) = 1 + \lambda t$ for some constant $\lambda > 0$ for all $t \geq 0$.
   (c) $E[V_t] = c$ for some constant $c > 0$ independent of t.
   (d) $E[U_t] = \int_0^t (1 - F(u)) \, du$ for all $t \geq 0$.

(4.9)   *Counters of type II.* In Example (1.34), let $T_0 = 0, T_1, T_2, \ldots$ be the times of successive registrations. Show that T is a renewal process and compute the distribution G of the times between registrations. Compute $\lim_{t \to \infty} M_t/t$ for the number M_t of registrations during $[0, t]$.

(4.10) *Continuation.* Suppose the locked period effected by an arriving particle is a random variable (rather than a constant τ). Let V_n be the length of the locked period caused by the nth particle. Suppose $V_1, V_2, \ldots$ are independent and have the same distribution ψ.

(a) Compute the distribution of the length of a locked period.

(b) Compute the distribution G of the time between two registrations.

(c) Compute the limit of the average M_t/t of registrations per unit time.

(4.11) *Counters of type I.* Arrivals at a counter form a renewal process with interarrival distribution F. An arriving particle which finds the counter free gets registered and locks it for a time of length τ. Arrivals during a locked period have no effects whatsoever. Suppose $t = 0$ is a time of registration.

(a) Show that the successive times of registrations form a renewal process, and compute the distribution G of the times between registrations.

(b) Find the probability that the counter is locked at time t.

(c) Compute the ratio M_t/N_t of the number of registered particles to the number of arrivals during $[0, t]$ in the limit as $t \to \infty$.

(4.12) *Continuation.* In the preceding exercise, suppose that the locked period effected by a registered particle is a random variable with distribution ψ independent of the arrival times and of the other locked periods. Do (a), (b), and (c) of (4.11) for this more general setup.

(4.13) In the problem of Examples (1.4) and (2.27), let W_t be the cumulative time during $(0, t]$ which is spent working, that is, let

$$W_t = \int_0^t Z_s \, ds,$$

with Z as defined in (2.27). Compute $E[W_t]$. Show that $W_t/t \to a/(a + b)$ almost surely as $t \to \infty$. (*Hint*: Use (1.25) and the fact that $\hat{W}_n = W_{S_n}$ is a sum of n independent identically distributed random variables.)

(4.14) *Interval reliability.* For the problem of Examples (1.4) and (2.27) let $f_x(t)$ be the probability that the item is in working condition at time t and will remain so at least x more units of time.

(a) Write the renewal equation which f_x satisfies; solve it to show that

$$f_x(t) = \int_{[0, t]} R(ds)[1 - \varphi(t + x - u)].$$

(b) Compute the limit of $f_x(t)$ for fixed x as $t \to \infty$.

(4.15) *Failure rate function.* Let the lifetime W of an item have distribution φ, and assume that φ has a right-hand derivative φ'. Then

$$r(t) = \frac{\varphi'(t)}{1 - \varphi(t)}, \quad t \geq 0,$$

is called the failure rate function for W. To explain this, show that

$$r(t) = \lim_{s \downarrow 0} \frac{1}{s} P\{W \le t + s \mid W > t\}.$$

A failure rate function is non-negative and finite. Show that it defines the distribution φ by

$$\varphi(t) = 1 - \exp[-\int_0^t r(s)\, ds].$$

(4.16) *Continuation.* Show that W has an exponential distribution if and only if its failure rate is constant. (This is the case where the item is the same as new as long as it lasts.) Show that the *Weibull distribution*

$$\varphi(t) = 1 - \exp(-at^b), \qquad t \ge 0$$

has increasing failure rate if $b > 1$ and decreasing failure rate if $b < 1$.

(4.17) *Age replacement.* In Example (1.4), suppose the replacements take no time, and instead of replacing only the failed items, suppose the rule is to replace whenever the item in use fails or has been in use τ units of time. We now distinguish between replacements and failures. Let φ be the distribution of the lifetimes.

(a)   Show that successive replacement times S_n form a renewal process, and compute the distribution F of the time between two replacements.

(b)   Show that if N_t is the number of replacements during $[0, t]$,

$$\lim_{t \to \infty} \frac{N_t}{t} = 1 \Big/ \int_0^\tau [1 - \varphi(u)]\, du.$$

(4.18) *Continuation.* Show that the successive failure times $T_0, T_1, \dots$ form a renewal process with the distribution G between two failures being given by

$$1 - G(t) = [1 - \varphi(\tau)]^k [1 - \varphi(t - k\tau)], \quad k\tau \le t < k\tau + \tau.$$

Show that the average number of failures M_t/t (per unit time) has the limit, as $t \longrightarrow \infty$,

$$\varphi(\tau) \Big/ \int_0^\tau [1 - \varphi(u)]\, du.$$

(4.19) *Continuation.* It is of interest to know if reducing τ improves reliability. The answer is yes if φ has an increasing failure rate. Following are the steps leading up to this assertion. For fixed t, let $f(\tau) = 1 - G(t)$, and assume φ is differentiable and has an increasing failure rate.

(a)   Show that for any integer $k \ge 1$, the derivative $f'(\tau) \le 0$ for τ in the interval $(t/(k + 1), t/k)$. Show the same for $\tau > t$.

(b)   Show that f is continuous at points of the form $\tau = t/k$. This together with (a) shows that f is non-increasing.

(c)   Conclude from this that $1 - G(t) \ge 1 - \varphi(t)$.

(4.20) *Randomized age replacement.* In Exercise (4.17), suppose that the nth item is replaced whenever it fails or has been in use Y_n units of time, whichever comes first; here $Y_1, Y_2, \ldots$ are (instead of all being equal to a constant τ) independent and identically distributed random variables independent of the lifetimes. Let φ be the distribution of lifetimes and ψ the distribution of Y_n.

(a) Show that the times S_n of successive replacements form a renewal process, and compute the distribution F of the times between replacements.

(b) Show that the times of failures T_n form a renewal process with distribution G of the time between two failures satisfying

$$1 - G(t) = 1 - F(t) + \int_{[0,t]} \psi(dx)[1 - \varphi(x)][1 - G(t - x)].$$

(c) Compute the expected value m_f of the time between two failures directly from (b) to find

$$m_f = \int_0^\infty (1 - \varphi(x))(1 - \psi(x)) \, dx \Big/ \int_{[0,\infty)} \psi(dx)\varphi(x).$$

(4.21) *Continuation: Reliability at t for x units of time.* This is the probability that the item in use at time t will last at least x more units of time without replacement; let this be denoted by $f_x(t)$.

(a) Show that for fixed x, the function f_x satisfies the renewal equation $f = g + F * f$ with

$$g(t) = [1 - \psi(t)][1 - \varphi(t + x)].$$

(b) Compute the limit of $f_x(t)$ as $t \longrightarrow \infty$.

(4.22) *Optimal replacement policy.* In the problem of Exercise (4.20), suppose that a replacement costs a units if the item being replaced had failed and b units otherwise; take $b = 1$, $a > b$, and put $c = a - b$. Let M_t and N_t be respectively the numbers of failures and replacements during $[0, t]$.

(a) Show that the average cost per unit time is, in the long run,

$$C_\psi = \lim_{t \to \infty} [aM_t + b(N_t - M_t)]/t = \frac{1 + c \int_{[0,\infty)} \psi(dx)\varphi(x)}{\int_0^\infty (1 - \psi(x))(1 - \varphi(x)) \, dx}.$$

(b) To minimize this cost by choosing ψ properly, show first that we can write

$$C_\psi = \frac{\int \psi(dx)f(x)}{\int \psi(dx)g(x)}$$

by defining $f(x) = 1 + c\varphi(x)$, $g(x) = \int_0^x (1 - \varphi(y)) \, dy$.

(c) Suppose φ is continuous, let $h = f/g$, and choose the number

$\tau \in [0, +\infty]$ so that $h(\tau) \leq h(x)$ for all x. Show that then

$$C_\psi \geq h(\tau)$$

for any distribution ψ. Conclude from this that the optimal replacement policy is to replace the item whenever it fails or has been used for τ units of time.

(4.23) *Continuation.* Show that if the lifetimes have the exponential distribution, the optimal policy is to replace only when failures occur. If, on the other hand, the lifetimes have a strictly increasing failure rate function r, then τ is the only point satisfying

$$r(\tau) \int_0^\tau [1 - \varphi(x)] \, dx - \varphi(\tau) = 1/c.$$

(4.24) *Decomposition of renewal processes.* Let S be a renewal process with interval distribution F, and let $T_1, T_2, \ldots$ be the successive trial numbers at which a "success" has occurred in a Bernoulli process with probability of success p at any one trial; as usual, we set $T_0 = 0$. Show that $\hat{S}_n = S_{T_n}$ form a renewal process with interval distribution

$$\hat{F} = \sum_{n=1}^\infty pq^{n-1} F^n.$$

(4.25) *Superposition of renewal processes.* Let S^1 and S^2 be two independent renewal processes with interval distributions F and G which have the respective densities F' and G'. Let $S = S^1 \vee S^2$ be the set of all points which belong to either S^1 or S^2 or both. Then S is called the *superposition* of S^1 and S^2. Show that S is a renewal process if and only if both S^1 and S^2 are Poisson processes.

$$* \quad * \quad *$$

Renewal theory is one of the main tools in the elementary theory of probability. A renewal process, by itself, does not have a rich enough structure to be of much interest. The main importance of renewal theory is, therefore, largely due to its applications to regenerative processes and to the elegant formalism of the renewal equation. In this light, the importance of the renewal function itself is due to its use as a potential operator and not to its connection with the expectations of the numbers of renewals.

The present treatment follows FELLER [7] fairly closely; for omitted proofs and extensions we refer the reader there. For refinements we refer the reader to the stimulating account given by SMITH [2].

CHAPTER 10

Markov Renewal Theory

In this chapter we will introduce a class of processes which includes as special cases all the processes which we have examined so far. Its theory combines renewal theory with the theory of Markov chains to create tools that are more powerful than those which either could provide. For purposes of picturing such a process, we think of a particle (or system) which moves from one state to another with random sojourn times in between; the successive states visited form a Markov chain, and a sojourn time has a distribution which depends on the state being visited as well as the next state to be entered. Such a process becomes a Markov process if the distributions of the sojourn times are all exponential independent of the next state; it becomes a Markov chain if the sojourn times are all equal to one; and it becomes a renewal process if there is only one state.

Throughout the chapter Ω is a fixed sample space and P a probability measure on it.

1. Markov Renewal Processes

Suppose we have defined, for each $n \in \mathbb{N}$, a random variable X_n taking values in a countable set E and a random variable T_n taking values in $\mathbb{R}_+ = [0, +\infty)$ such that $0 = T_0 \leq T_1 \leq T_2 \leq \cdots$.

(1.1) DEFINITION. The stochastic process $(X, T) = \{X_n, T_n; n \in \mathbb{N}\}$ is said to be a *Markov renewal process with state space E* provided that

(1.2) $$P\{X_{n+1} = j, T_{n+1} - T_n \leq t \,|\, X_0, \ldots, X_n; T_0, \ldots, T_n\}$$
$$= P\{X_{n+1} = j, T_{n+1} - T_n \leq t \,|\, X_n\}$$

for all $n \in \mathbb{N}$, $j \in E$, and $t \in \mathbb{R}_+$. $\qquad\square$

We will always assume that (X, T) is *time-homogeneous:* that is, for any $i, j \in E, t \in \mathbb{R}_+$,

(1.3) $$P\{X_{n+1} = j, T_{n+1} - T_n \leq t \,|\, X_n = i\} = Q(i, j, t)$$

independent of n. The family of probabilities $Q = \{Q(i, j, t): i, j \in E, t \in \mathbb{R}_+\}$ is called a *semi-Markov kernel over E.*

For each pair (i, j), the function $t \longrightarrow Q(i, j, t)$ has all the properties of a distribution function except that

(1.4) $$P(i, j) = \lim_{t \to \infty} Q(i, j, t)$$

is not necessarily one. Indeed, it is easy to see from (1.3) that

(1.5) $$P(i, j) \geq 0, \qquad \sum_{j \in E} P(i, j) = 1;$$

that is, the $P(i, j)$ are the transition probabilities for some Markov chain with state space E. It follows from (1.2) and (1.4) that

(1.6) $$P\{X_{n+1} = j \,|\, X_0, \ldots, X_n; T_0, \ldots, T_n\} = P(X_n, j)$$

for all $n \in \mathbb{N}$ and $j \in E$. This implies, in particular, the following

(1.7) PROPOSITION. $X = \{X_n; n \in \mathbb{N}\}$ is a Markov chain with state space E and transition matrix P.

If $P(i, j) = 0$ for some pair (i, j), then $Q(i, j, t) = 0$ for all t; we then define $Q(i, j, t)/P(i, j) = 1$. With this convention, we define

(1.8) $$G(i, j, t) = \frac{Q(i, j, t)}{P(i, j)}, \qquad i, j \in E, t \in \mathbb{R}_+.$$

Then for each pair (i, j) the function $t \longrightarrow G(i, j, t)$ is a distribution function. From (1.3), (1.6), and (1.8) we deduce the following interpretation:

(1.9) $$G(i, j, t) = P\{T_{n+1} - T_n \leq t \,|\, X_n = i, X_{n+1} = j\}.$$

Using this result together with (1.2), we obtain the following important statement on the structure of (X, T).

(1.10) PROPOSITION. For any integer $n \geq 1$ and numbers $u_1, \ldots, u_n \in \mathbb{R}_+$,

$$P\{T_1 - T_0 \leq u_1, \ldots, T_n - T_{n-1} \leq u_n \,|\, X_0, \ldots, X_n\}$$
$$= G(X_0, X_1, u_1)G(X_1, X_2, u_2) \cdots G(X_{n-1}, X_n, u_n);$$

in words, the increments $T_1 - T_0, T_2 - T_1, \ldots$ are conditionally independent given the Markov chain $X_0, X_1, \ldots$, with the distribution of $T_{n+1} - T_n$ depending only on X_n and X_{n+1}.

In particular, if the state space E consists of a single point, then the increments are independent and identically distributed: namely, we have

(1.11) COROLLARY. If E consists of a single point, then $\{T_n; n \in \mathbb{N}\}$ is a renewal process.

This result, together with Proposition (1.7), justifies the term "Markov renewal process" somewhat. The full justification, however, is contained in Proposition (1.7) and the following important

(1.12) PROPOSITION. Let $j \in E$ be fixed, and define $S_0^j, S_1^j, \ldots$ as the successive T_n for which $X_n = j$. Then $S^j = \{S_n^j; n \in \mathbb{N}\}$ is a (possibly delayed) renewal process.

Proof follows from Corollary (1.11) above and the theorem to be stated next. Before we can state it, however, we need some conventions. First, we append a new point Δ to E, and set $X_\infty(\omega) = \Delta, T_\infty(\omega) = +\infty$ for all $\omega \in \Omega$. Then we extend the definition of a Markov renewal process slightly to allow for $T_n(\omega)$ to take on the value $+\infty$, but only if $X_n(\omega) = \Delta$.

(1.13) THEOREM. Let D be a fixed subset of E, and let $N_0, N_1, N_2, \ldots$ be the successive indices $n \geq 0$ such that $X_n \in D$. For each $n \in \mathbb{N}$, define

$$\hat{X}_n = X_{N_n}, \qquad \hat{T}_n = T_{N_n}.$$

Then $(\hat{X}, \hat{T}) = \{\hat{X}_n, \hat{T}_n; n \in \mathbb{N}\}$ is a Markov renewal process with state space $D \cup \{\Delta\}$.

Proof will be omitted; we merely remark that (X, T) is a Markov chain with a two-dimensional state space $E \times \mathbb{R}_+$, and the present theorem follows from applying the strong Markov property at the stopping times N_n to this chain (X, T).

To see the reasons for the conventions introduced in the paragraph preceding the theorem, consider a finite set D of transient states in E. Then the chain X enters D only finitely many times with probability one. Therefore, for almost all $\omega \in \Omega$, there is a finite integer $n = n(\omega)$ such that $N_n(\omega) = N_{n+1}(\omega) = \cdots = +\infty$. Then by the conventions introduced, we have $\hat{X}_n(\omega) = \hat{X}_{n+1}(\omega) = \cdots = \Delta$ and $\hat{T}_n(\omega) = \hat{T}_{n+1}(\omega) = \cdots = +\infty$.

Going back to Proposition (1.12), we see that to each state $j \in E$ there corresponds a renewal process S^j; the superposition of all these renewal processes gives the points T_n, $n \in \mathbb{N}$; the renewal process which contributed the point T_n is the jth one if and only if $X_n = j$; the types of the successive points, namely $X_0, X_1, \ldots$, form a Markov chain.

Another convenient picture in describing a Markov renewal process is provided by the process $Y = \{Y_t; t \geq 0\}$ defined by putting, for each $t \geq 0$ and $\omega \in \Omega$,

$$(1.14) \qquad Y_t(\omega) = \begin{cases} X_n(\omega) & \text{if } T_n(\omega) \leq t < T_{n+1}(\omega) \\ \Delta & \text{if } t \geq \sup_n T_n(\omega), \end{cases}$$

where Δ is a point not in E.

(1.15) DEFINITION. The stochastic process $Y = \{Y_t; t \geq 0\}$ defined by (1.14) is called the *minimal semi-Markov process* associated with (X, T).   □

We may think of Y_t as the state at time t of some system or particle which moves from one state to another with random sojourn times in between. The length of a sojourn interval $[T_n, T_{n+1})$ is a random variable whose distribution depends on both the state X_n being visited and the state X_{n+1} to be visited next. The successive states visited form a Markov chain; given that sequence, the successive sojourn times are conditionally independent.

(1.16) EXAMPLE. *Markov Processes.* It follows from Theorem (8.3.3) that if Q has the form

$$Q(i, j, t) = P(i, j)[1 - e^{-\lambda(i)t}]$$

for all i, j, t for some function $\lambda(i)$, $i \in E$, then Y is a Markov process. Conversely, every regular Markov process is a semi-Markov process.   □

(1.17) EXAMPLE. *Traffic Theory.* This model of traffic flow is supposed to hold well when the interactions between different vehicles are limited to nearest-neighbor interactions. Let $T_0 = 0, T_1, T_2, \ldots$ be the positions (in time or space) of the vehicles along a highway; and let X_n denote the "type" of the nth vehicle (for example, whether the nth vehicle is a car or a truck, whether it is a "leader" or a "follower" in a platoon, its make, weight, or speed). Usually $X = \{X_n\}$ is a Markov chain, and the headway $T_{n+1} - T_n$ depends only on the types X_n and X_{n+1}. Then (X, T) is a Markov renewal process.   □

(1.18) EXAMPLE. *Counters of Type I:* Arrivals at a particle counter form a Poisson process with rate λ. An arriving particle which finds the counter free gets registered and locks it for a random duration with distribution function ψ. Arrivals during a locked period have no effect. Let $T_0 = 0, T_1, T_2, \ldots$ be the successive instants of changes in the state of the counter, and write $X_n = 1$

or 0 according as the nth change locks or frees the counter. Then $\{X_n, T_n; n \in \mathbb{N}\}$ is a Markov renewal process with state space $E = \{0, 1\}$. In this case, if $Q(t)$ denotes the matrix whose (i, j)th entry is $Q(i, j, t)$, then

$$Q(t) = \begin{bmatrix} 0 & 1 - e^{-\lambda t} \\ \psi(t) & 0 \end{bmatrix}.$$

Here, the semi-Markov process Y associated with (X, T) indicates the state of the counter. $\qquad\qquad\qquad\qquad\qquad\qquad\qquad\qquad\qquad\qquad$ $\square$

(1.19) EXAMPLE. *M/G/1 queue.* We refer the reader to the sections on queues at the end of Chapter 6 for a more detailed description of this queueing system as well as of the G/M/1 queue to be discussed below. This is a single-server queueing system subject to a Poisson process of arrivals with rate a and independent service times with the common distribution φ. Let $T_0 = 0$, $T_1, T_2, \ldots$ be the instants of successive departures, and let X_n be the number of customers left behind by the nth departure. Then $(X, T) = \{X_n, T_n; n \in \mathbb{N}\}$ is a Markov renewal process. If $Q(t)$ is the matrix whose (i, j)th entry is $Q(i, j, t)$, then we have

$$Q(t) = \begin{bmatrix} p_0(t) & p_1(t) & p_2(t) & \cdot & \cdot & \cdot \\ q_0(t) & q_1(t) & q_2(t) & \cdot & \cdot & \cdot \\ & q_0(t) & q_1(t) & \cdot & \cdot & \cdot \\ & & & \cdot & \cdot & \cdot \\ 0 & & & & \cdot & \cdot \\ & & & & & \cdot \end{bmatrix},$$

where

$$q_n(t) = \int_0^t \varphi(ds) \frac{e^{-as}(as)^n}{n!}, \qquad n = 0, 1, \ldots,$$

and

$$p_n(t) = \int_0^t ae^{-as} q_n(t - s)\, ds, \qquad n = 0, 1, \ldots.$$

Note that the $P(i, j) = Q(i, j, \infty)$ are as given before in Chapter 6 for this Markov chain X. There we obtained the limiting behavior of X_n; here, by the methods we will develop, we will obtain the behavior of the queue size at any time t. $\qquad\qquad\qquad\qquad\qquad\qquad\qquad\qquad\qquad\qquad\qquad\qquad$ $\square$

(1.20) EXAMPLE. *G/M/1 queue.* Let $T_0 = 0, T_1, T_2, \ldots$ be the instants of successive arrivals, and let X_n be the number of customers in the system just before the nth arrival. Then $(X, T) = \{X_n, T_n; n \in \mathbb{N}\}$ is a Markov renewal process. If φ is the distribution of the interarrival times and a is the parameter of the service distribution, then the semi-Markov kernel Q^* is given by

$$Q^*(t) = \begin{bmatrix} r_0(t) & q_0(t) & & & 0 \\ r_1(t) & q_1(t) & q_0(t) & & \\ r_2(t) & q_2(t) & q_1(t) & \cdot & \\ \cdot & \cdot & \cdot & \cdot & \cdot \\ \cdot & \cdot & \cdot & & \cdot & \cdot \\ \cdot & \cdot & \cdot & & & \cdot \end{bmatrix},$$

where the $q_n(t)$ are as defined above in (1.19), and

$$r_n(t) = \varphi(t) - \sum_{i=0}^{n} q_i(t), \quad n \in \mathbb{N}.$$

Again, the methods to be developed below will enable us to study the time-dependent behavior of the queue size. ☐

2. Markov Renewal Functions and Classification of States

Let $(X, T) = \{X_n, T_n; n \in \mathbb{N}\}$ be a Markov renewal process with a semi-Markov kernel Q over a countable state space E. Throughout this section we will write $P_i\{A\}$ for the conditional probability $P\{A \mid X_0 = i\}$ and, similarly, E_i for the conditional expectations given $\{X_0 = i\}$. Throughout this chapter, all integrals of form $\int_0^t$ are integrals over the closed set $[0, t]$. Finally, to avoid trivialities, we assume throughout that for all $i \in E$

(2.1) $P_i\{T_0 = T_1 = T_2 = \cdots = 0\} = 0.$

Let us define

(2.2) $Q^n(i, j, t) = P_i\{X_n = j, T_n \le t\}, \quad i, j \in E, t \in \mathbb{R}_+,$

for all $n \in \mathbb{N}$. Then

(2.3) $Q^0(i, j, t) = I(i, j) = \begin{cases} 1 & \text{if } i = j, \\ 0 & \text{if } i \ne j, \end{cases}$

for all $t \ge 0$; and for $n \ge 0$ we have the recursive relation

(2.4) $Q^{n+1}(i, k, t) = \sum_{j \in E} \int_0^t Q(i, j, ds) Q^n(j, k, t - s),$

where the integration is over $[0, t]$. The probability argument for (2.4) is as follows: in order that $X_{n+1} = k$ and $T_{n+1} \le t$, the first transition must occur at some time s smaller than t into some state j, and then, starting from j, the remaining n transitions should take no longer than $t - s$ units of time and end with state k.

For each fixed state $j \in E$, the instants T_n for which $X_n = j$ form a possibly delayed renewal process S^j by Proposition (1.12). The number of renewals during $[0, t]$ of this renewal process is

$$\sum_{n=0}^{\infty} I_{[0,t]}(S_n^j) = \sum_{n=0}^{\infty} 1_j(X_n)I_{[0,t]}(T_n).$$

By assumption (2.1), the probability that all the S_n^j, $n \in \mathbb{N}$, are equal to S_0^j is zero. Therefore, by Proposition (9.1.14), the expected number of renewals in a finite interval is finite. Hence,

$$(2.5) \qquad R(i, j, t) = E_i\left[\sum_{n=0}^{\infty} 1_j(X_n)I_{[0,t]}(T_n)\right]$$
$$= \sum_{n=0}^{\infty} P_i\{X_n = j, T_n \leq t\} = \sum_{n=0}^{\infty} Q^n(i, j, t)$$

is finite for any $i, j \in E$ and $t < \infty$.

The functions $t \rightarrow R(i, j, t)$ are called *Markov renewal functions* and the collection $R = \{R(i, j, \cdot): i, j \in E\}$ of these functions is called a *Markov renewal kernel*. We mention once more the fact that $R(i, j, \cdot)$ is a renewal function, namely the one corresponding to the renewal process S^j.

Let $F(i, j, t)$ be the distribution of the first passage time from state i to state j, that is, let

$$(2.6) \qquad F(i, j, t) = P_i\{S_0^j \leq t\}, \quad i \neq j,$$

and let $F(j, j, t)$ be the distribution of the time between two successive occurrences of j, that is, let

$$(2.7) \qquad F(j, j, t) = P_j\{S_1^j \leq t\}$$

(note that $P_j\{S_0^j = 0\} = 1$). Then it follows from (9.1.13) and (9.3.3) that

$$(2.8) \qquad R(j, j, t) = \sum_{n=0}^{\infty} F^n(j, j, t)$$

and

$$(2.9) \qquad R(i, j, t) = \int_0^t F(i, j, ds)R(j, j, t - s), \quad i \neq j,$$

where $F^n(j, j, \cdot)$ is the n-fold convolution of the distribution $F(j, j, \cdot)$ with itself.

Once the $R(i, j, t)$ are obtained from (2.5), the preceding two expressions can be used to solve for the first-passage distributions $F(i, j, t)$. Computationally, especially if the state space E is finite, a feasible method is to use Laplace transforms. For $\alpha \geq 0$, we define

$$(2.10) \qquad Q_\alpha(i, j) = \int_0^\infty e^{-\alpha t} Q(i, j, dt), \qquad F_\alpha(i, j) = \int_0^\infty e^{-\alpha t} F(i, j, dt),$$

and for $\alpha > 0$,

$$(2.11) \qquad R_\alpha(i, j) = \int_0^\infty e^{-\alpha t} R(i, j, dt).$$

It follows from (2.8) and (2.9) that

$$(2.12) \qquad R_\alpha(i, j) = F_\alpha(i, j) R_\alpha(j, j), \quad i \neq j,$$

$$(2.13) \qquad R_\alpha(j, j) = \sum_{n=0}^\infty [F_\alpha(j, j)]^n = \frac{1}{1 - F_\alpha(j, j)},$$

and conversely,

$$(2.14) \qquad F_\alpha(j, j) = 1 - \frac{1}{R_\alpha(j, j)},$$

$$(2.15) \qquad F_\alpha(i, j) = R_\alpha(i, j)/R_\alpha(j, j), \quad i \neq j.$$

Next we concentrate on the computation of R_α. It follows from (2.4) that the Laplace transform $Q_\alpha^n(i, j)$ of $Q^n(i, j, t)$ satisfies

$$(2.16) \qquad Q_\alpha^{n+1}(i, k) = \sum_j Q_\alpha(i, j) Q_\alpha^n(j, k), \quad i, k \in E,$$

for all $n \in \mathbb{N}$. Hence Q_α^n is precisely the nth power of the matrix Q_α; that is,

$$Q_\alpha^n = (Q_\alpha)^n, \quad \alpha \geq 0, n \in \mathbb{N},$$

and this holds for $n = 0$ with the usual convention that the 0th power of any matrix is the identity matrix.

In matrix notation, (2.5) implies that

$$(2.17) \qquad R_\alpha = I + Q_\alpha + Q_\alpha^2 + \cdots, \quad \alpha > 0,$$

which implies that

$$(2.18) \qquad R_\alpha = I + Q_\alpha R_\alpha, \qquad R_\alpha = I + R_\alpha Q_\alpha,$$

or equivalently,

$$(2.19) \qquad R_\alpha(I - Q_\alpha) = I; \qquad (I - Q_\alpha) R_\alpha = I.$$

(2.20) PROPOSITION. If E is finite, then

$$R_\alpha = (I - Q_\alpha)^{-1}.$$

If E is not finite, then R_α is the minimal solution of

$$(2.21) \qquad\qquad (I - Q_\alpha)M = I, \qquad M \geq 0.$$

Proof. If E is finite, the conclusion is immediate from (2.19). Suppose E is infinite: then (2.19) shows that R_α is a solution of (2.21); to show that R_α is the minimal solution, let M be another solution. Then

$$M = I + Q_\alpha M,$$

and repeatedly replacing M on the right by $I + Q_\alpha M$, we obtain

$$M = I + Q_\alpha + Q_\alpha^2 + \cdots + Q_\alpha^n + Q_\alpha^{n+1}M.$$

Taking limits as $n \longrightarrow \infty$ we see that $M \geq R_\alpha$. ∎

This proposition is the generalization of both Corollary (7.1.8) and Theorem (7.2.5). For example, since a Markov chain may be looked upon as a Markov renewal process with $T_n = n$, we have $Q_\alpha = e^{-\alpha} P$, and then the preceding proposition is exactly the same as (7.1.8) with $\alpha < 1$ replaced by $e^{-\alpha}$. In (7.1.8) and (9.2.3) we had shown further that there is only one solution. From the computational point of view, it is important to know when the solution is unique. The question is tied in with the possibility of having infinitely many transitions in a finite time; this was encountered earlier with Markov processes as well, and we refer the reader to Section 3 of Chapter 8 for examples. We will treat this matter more fully in the next section. For the present, we merely mention that R_α is the unique solution of the equation (2.21) having bounded columns (note that $R_\alpha(i, j) = F_\alpha(i, j)R_\alpha(j, j) \leq R_\alpha(j, j)$, so every column of R_α is bounded by some number) if and only if the only bounded solution of $Q_\alpha h = h$ is $h = 0$. The latter is true, in particular, when E is finite, or when every state is recurrent, or when there are only finitely many transient states.

We end this section by giving a classification of states.

(2.22) DEFINITION. State $j \in E$ is said to be *recurrent*, or *transient*, or *aperiodic*, or *periodic with period δ* respectively if the corresponding renewal process $\hat{S}^j = \{S_n^j - S_0^j; n \in \mathbb{N}\}$ is recurrent, or transient, or aperiodic, or periodic with period δ. ∎

If state j is recurrent, then it is also recurrent for the Markov chain X; similarly, j is transient if and only if j is transient for the Markov chain X. The recurrence question therefore reduces to that in the Markov chain X, and we refer the reader to Chapter 5 for a full treatment. Periodicity, however, has nothing to do with periodicity in the Markov chain X. A state j can be periodic in (X, T) without being periodic for the Markov chain X, and conversely, j can be periodic for X and aperiodic for (X, T). The periodicity question is as with a renewal process: that is, j is periodic in (X, T) with period δ if and only if the distribution $F(j, j, t)$ is arithmetic with span δ. The following are some results reducing this to tests concerning the elementary data, namely, the $Q(i, j, t)$.

(2.23) PROPOSITION. If i and j can be reached from each other, then either they are both aperiodic or they are both periodic. In the latter case they have the same period.

Before proving this we list the following as an immediate

(2.24) COROLLARY. If X is irreducible, then either all states are aperiodic in (X, T) or they are all periodic. In the latter case they all have the same period.

Proof of (2.23). Let i and j in question be fixed, let $D = \{i, j\}$, and consider the Markov renewal process $(\hat{X}, \hat{T})$ defined in Theorem (1.13). It is clear that the renewal times S_n^i and S_n^j are the same whether they are obtained from (X, T) or $(\hat{X}, \hat{T})$. Hence, it is sufficient to consider the periodicity question in $(\hat{X}, \hat{T})$. This is a three-state Markov renewal process if $X_0 \in D$; and if the states are ordered as i, j, Δ, then the semi-Markov kernel corresponding to it is simply

$$\hat{Q} = \begin{bmatrix} H & K & A \\ L & M & B \\ 0 & 0 & C \end{bmatrix}$$

for some H, K, L, and M with none vanishing. For the return distributions we easily have

(2.25) $$F(i, i, \cdot) = H + K * L + K * M * L + \cdots,$$

(2.26) $$F(j, j, \cdot) = M + L * K + L * H * K + \cdots,$$

where $*$ denotes convolution.

Suppose i is periodic with period δ. Then all the points of increase of $F(i, i, \cdot)$ belong to $B = \{0, \delta, 2\delta, \dots\}$. Now (2.25) shows that the points of increase of H, $K * L$, $K * M * L, \dots$ all are in B. A point of increase $c \in B$ of $K * L * M$ is of the form $c = a + b$, where $a \in B$ is a point of increase for $K * L$ and b is one for M. Since a, $c \in B$, we must have $b \in B$ as well. Hence, all points of increase of H, $K * L$, M are in B. This implies, through (2.26), that the points of increase of $F(j, j, \cdot)$ are also in B, that is, j is periodic with a period δ', which is an integer multiple of δ. Reversing the argument, i must be periodic with period δ equal to an integer multiple of δ'. Thus, $\delta = \delta'$ as well. $\qquad\square$

The following shows how to recognize periodicity by examining the semi-Markov kernel Q.

(2.27) PROPOSITION. If X is irreducible and the states are periodic with period δ, then every $Q(i, j, t)$ is a step function with jumps in the set

$$\{\delta_{ij}, \delta_{ij} + \delta, \delta_{ij} + 2\delta, \ldots \}$$

for some $\delta_{ij} \geq 0$. Moreover, if $P(i_0, i_1)$, $P(i_1, i_2), \ldots, P(i_{n-1}, i_n) > 0$ and $i_0 = i_n$, then $\delta_{i_0 i_1} + \delta_{i_1 i_2} + \cdots + \delta_{i_{n-1} i_n}$ is equal to an integer multiple of δ.

Proof. If it is possible to go from i to j in one step by taking a units of time and b units of time, and c is such that it is possible to go from j to i in c units of time, then both $a + c$ and $b + c$ must be multiples of δ. Therefore, their difference $b - a$ is an integer multiple of δ. This shows that both a and b are of the form $\delta_{ij} + n\delta$ with $n \in \mathbb{N}$. The reasoning behind the second statement is similar. $\qquad\square$

3. Markov Renewal Equations

Just as the real usefulness of renewal theory lies in its applicability to regenerative processes, the real power of Markov renewal theory lies in its usefulness in dealing with semi-regenerative processes. We will now introduce the system of equations which generalizes the renewal equation in a natural manner, and leave the applications to semi-regenerative processes aside until we get to Sections 5 and 6.

Throughout this section $(X, T) = \{X_n, T_n; n \in \mathbb{N}\}$ is a Markov renewal process with a semi-Markov kernel Q, Q^n is the nth iterate of Q, and $R = \sum Q^n$ is the Markov renewal kernel corresponding to Q. The class of functions which we will be working with will be denoted by $\mathbb{B}$: these are functions $f: E \times \mathbb{R}_+ \to \mathbb{R}$ such that for every $i \in E$, the function $t \to f(i, t)$ is bounded over finite intervals, and for every fixed $t \in \mathbb{R}_+$, the function $i \to f(i, t)$ is bounded. In particular, we note that for any fixed $j \in E$, the functions $(i, t) \to Q(i, j, t)$ and $(i, t) \to R(i, j, t)$ both belong to $\mathbb{B}$.

For any function $f \in \mathbb{B}$, the function $Q * f$ defined by

$$(3.1) \qquad\qquad Q * f(i, t) = \sum_{j \in E} \int_0^t Q(i, j, ds) f(j, t - s)$$

is well defined, and $Q * f$ belongs to $\mathbb{B}$ again. Hence, the operation can be repeated, and the nth iterate $Q^n * f$ is given by

$$(3.2) \qquad Q^n * f(i, t) = \sum_{j \in E} \int_0^t Q^n(i, j, ds) f(j, t - s).$$

If $f \geq 0$, we can replace Q on the right side of (3.1) by R and still have a well-defined function which we will denote by $R * f$; that is, for $f \in \mathbb{B}$, $f \geq 0$

$$(3.3) \qquad R * f(i, t) = \sum_{j \in E} \int_0^t R(i, j, ds) f(j, t - s).$$

A function $f \in \mathbb{B}$ is said to satisfy a *Markov renewal equation* if for all $i \in E$ and $t \in \mathbb{R}_+$,

$$(3.4) \qquad f(i, t) = g(i, t) + \sum_{j \in E} \int_0^t Q(i, j, ds) f(j, t - s)$$

for some function $g \in \mathbb{B}$. Here the point of view is that g and Q are known and the problem is to solve for f and study the limiting behavior of $t \to f(i, t)$ as $t \to \infty$.

To avoid certain unpleasant difficulties, we will limit ourselves to functions $f, g \in \mathbb{B}$ which are non-negative; we denote this set by $\mathbb{B}_+$. Using the notation introduced in (3.1), the Markov renewal equation (3.4) of interest now becomes

$$(3.5) \qquad f = g + Q * f, \qquad f, g \in \mathbb{B}_+.$$

Formally, (3.5) is the same as a renewal equation. Indeed, it becomes one if E consists of a single state. Just as in that simpler case, $R * g$ is a solution, but we no longer have uniqueness without further conditions. The following is the result; it generalizes both (7.2.5) and (9.2.3).

(3.6) Theorem. The Markov renewal equation (3.5) has a solution $R * g$. Every solution f is of the form

$$(3.7) \qquad f = R * g + h$$

where h satisfies

$$(3.8) \qquad h = Q * h, \qquad h \in \mathbb{B}_+.$$

Proof. Replacing f on the right side of (3.5) by $g + Q * f$ repeatedly, we obtain

$$f = (g + Q * g + \cdots + Q^n * g) + Q^{n+1} * f.$$

Since $g \geq 0$, the term in parenthesis increases to $R * g$, whereas $Q^{n+1} * f$ decreases to some function h. Thus, $f = R * g + h$. Since h is the limit of the monotone sequence f, $Q * f$, $Q^2 * f, \ldots$, we have

$$h = \lim_n Q^{n+1} * f = Q * (\lim_n Q^n * f) = Q * h. \qquad \square$$

Since $R(i, j, t) = \sum_n Q^n(i, j, t)$ is finite,

$$\lim_{n \to \infty} Q^n(i, j, t) = 0$$

for any $i, j \in E$ and $t < \infty$. If there are only finitely many states, this implies that

$$\lim_{n \to \infty} Q^n * h = 0$$

for any $h \in \mathbb{B}_+$. Therefore, if E is finite, $h = Q * h$ implies $h = Q^n * h$, which implies that $h = 0$, and we have the following

(3.9)   COROLLARY.   If E is finite, then

$$f = R * g$$

is the only solution of (3.5).

The following are some criteria for deciding when $R * g$ is the only solution. First we show that a non-zero solution to (3.8) exists if and only if there is a non-zero bounded solution, and then give several criteria for the latter.

(3.10)   PROPOSITION.   There is a solution $h \neq 0$ to (3.8) if and only if

(3.11) $$h = Q * h, \qquad 0 \leq h \leq 1$$

has a solution $h \neq 0$.

Proof. If (3.11) has a non-trivial solution, the solution is also a non-trivial solution to (3.8). Suppose next that (3.8) has a non-trivial solution. Then there is some $b > 0$ such that $h(t) \neq 0$ for some $t < b$, and then $\beta = \sup\{h(i, t) : i \in E, t \leq b\}$ is positive. Now define

$$k(i, t) = \begin{cases} \dfrac{h(i, t)}{\beta} & \text{if } t < b, \\ 1 & \text{if } t \geq b. \end{cases}$$

Then

(3.12) $$Q * k(i, t) = k(i, t) \qquad \text{if } t < b,$$

and since $k \leq 1$,

$$(3.13) \qquad Q * k(i, t) \leq \sum_j Q(i, j, t) \leq 1 = k(i, t) \quad \text{if } t \geq b.$$

It follows from (3.12) and (3.13) that $k \geq Q * k$, which by iteration gives

$$1 \geq k \geq Q * k \geq Q^2 * k \geq \cdots.$$

Hence, $\hat{h} = \lim_n Q^n * k$ exists, and by the monotonicity of the convergence, it satisfies $\hat{h} = Q * \hat{h}$. This and the obvious fact that $0 \leq \hat{h} \leq 1$ shows that $\hat{h}$ is a solution to (3.11). Furthermore, $\hat{h}$ is not trivial, since (3.12) implies

$$\hat{h}(i, t) = k(i, t) \quad \text{if } t < b.$$

Thus $\hat{h}(i, t) \neq 0$ for some i and $t < b$. $\qquad \square$

The following is a qualitative condition which is equivalent to the statement that the only solution of (3.11) is $h = 0$.

(3.14) PROPOSITION.   The only solution of (3.11) is $h = 0$ if and only if

$$(3.15) \qquad\qquad L = \sup_n T_n$$

is almost surely infinite.

Proof. Since $T_0 \leq T_1 \leq T_2 \leq \cdots$, the probability

$$f_n(i, t) = P_i\{T_n \leq t\} = \sum_{j \in E} Q^n(i, j, t)$$

decreases as $n \to \infty$, and the limit is

$$f(i, t) = \lim_n f_n(i, t) = P_i\{L \leq t\}.$$

Since the convergence to f is monotone, and since $f_{n+1} = Q * f_n$, we must have

$$f = Q * f, \quad 0 \leq f \leq 1.$$

Hence, f is a solution of (3.11). Moreover, if h is any other solution, $h \leq 1$ implies that

$$h = Q^n * h \leq Q^n * 1 = f_n$$

for all n, which in turn shows that $h \leq f$.

If the only solution of (3.11) is $h = 0$, then f, being a solution, must be zero and thus $L = +\infty$ almost surely. Conversely, if $L = +\infty$ almost surely, then $f = 0$, and f being the maximal solution of (3.11), we must have $h = 0$ as the only solution of (3.11).   □

The following are two easy criteria which imply that the *lifetime L* defined by (3.15) is infinite almost surely.

(3.16) PROPOSITION. If all states are recurrent, then $L = +\infty$ almost surely.

Proof. If state i is recurrent, then starting at i, the renewal process $\{S_n^i; n \in \mathbb{N}\}$ has infinite lifetime. Since $\{S_n^i; n \in \mathbb{N}\}$ is a subset of $\{T_n; n \in \mathbb{N}\}$, the supremum of $\{T_n, n \in \mathbb{N}\}$ is at least as great as that of $\{S_n^i, n \in \mathbb{N}\}$. Hence, $P_i\{L = +\infty\} \geq P_i\{\sup_n S_n^i = +\infty\} = 1$. This being true for every $i \in E$, we must have $L = +\infty$ almost surely.   □

We see that the argument of the proof goes through if the set of recurrent states is reached with probability one no matter what the initial state is. We thus have

(3.17) COROLLARY. If the probability of staying in the set of transient states forever is equal to zero, then $L = +\infty$ almost surely.

In particular, the condition of the preceding corollary holds if there are only finitely many transient states:

(3.18) COROLLARY. If there are only finitely many transient states, in particular if there are only finitely many states, then $L = \infty$ almost surely.

The following is a condition on the sojourn times which makes $L = \infty$ almost surely.

(3.19) PROPOSITION. If for some $b > 0$

$$\sum_j Q(i, j, b) \leq c$$

for all $i \in E$ and for some constant $c < 1$, then $L = +\infty$ almost surely.

Proof. The condition implies that

$$\sum_{j \in E} Q(i, j, t) \leq \varphi(t), \quad t \geq 0,$$

for all $i \in E$ if φ is defined as

$$\varphi(t) = \begin{cases} c & \text{if } t < b, \\ 1 & \text{if } t \geq b. \end{cases}$$

This in turn implies that

$$f_n(i, t) = P_i\{T_n \leq t\} = \sum_j Q^n(i, j, t) \leq \varphi^n(t)$$

for all $i \in E$, $t \geq 0$, if φ^n is the n-fold convolution of φ with itself. Since $\varphi(0) = c < 1$, $\lim_n \varphi^n(t) = 0$ for all $t \in \mathbb{R}_+$, and this implies that

$$f(i, t) = P_i\{L \leq t\} = \lim_n f_n(i, t) = 0$$

for every $i \in E$ and $t \in \mathbb{R}_+$. Hence, $L = +\infty$ almost surely. $\qquad\square$

It follows from (3.18) that in order for the lifetime L to be finite, there must be infinitely many transient states within which the chain X can remain forever. Furthermore, the sojourn times in these states must be short enough for the sum to be finite. The following is a sufficient condition for L to be finite with a positive probability.

(3.20) PROPOSITION. If there exists a set A of transient states such that

$$\sum_{j \in A} R(i, j, \infty)m(j) < \infty$$

for some $i \in A$ (here $m(j) = E_j[T_1]$), then $P_i\{L < \infty\} > 0$.

We omit the proof and only note that for j transient, the renewal function $R(j, j, t)$ approaches a finite limit as $t \to \infty$, and $R(i, j) = R(i, j, \infty) = F(i, j, \infty) R(j, j, \infty) = F(i, j) R(j, j)$, where the quantities $R(i, j)$, $F(i, j)$, and $R(j, j)$ are as computed by considering only the Markov chain X.

A simple Markov renewal process with finite lifetime is the one where X moves from i to $i + 1$, from $i + 1$ to $i + 2$, and so on, and the sojourn time in state i has mean $1/2^i$. Then X is transient, we can take A to be the whole space $E = \{0, 1, \ldots\}$, and then

$$\sum_j R(i, j, +\infty)m(j) = \sum_{j=i+1}^{\infty} \frac{1}{2^j} = \frac{1}{2^i} < \infty.$$

So all the transitions of this process will fall in a finite interval. For other examples in the case of Markov processes we refer the reader to Section 3 of Chapter 8. In particular, note that Theorem (8.3.23), Corollary (8.3.24), Theorem (8.3.25), Corollary (8.3.26), and Theorem (8.3.27) are the specialized versions of the results of this section.

4.   Limit Theorems

For fixed $i, j \in E$, the function $R(i, j, \cdot)$ is a possibly delayed renewal function. Its limiting behavior, therefore, can be obtained from Theorem

(9.3.4): If state j is transient and $i \neq j$, then

(4.1) $\lim\limits_{t \to \infty} R(i, j, t) = F(i, j, +\infty) \, R(j, j, +\infty) = F(i, j)R(j, j),$

which is exactly the expected number of visits to j by the Markov chain X starting at i. For the computation of $F(i, j)$ and $R(j, j)$ we refer the reader back to the first section of Chapter 6.

If state j is recurrent aperiodic, then Theorem (9.3.4) implies that

(4.2) $\lim\limits_{t \to \infty} [R(i, j, t) - R(i, j, t - \tau)] = \tau \, F(i, j)\eta(j),$

where $\eta(j)$ is the inverse of the expected value of the time between two returns to state j. In the case where j is recurrent periodic with period δ, the same holds when τ is a multiple of δ.

The behavior of the individual renewal functions $R(i, j, \cdot)$ can thus be obtained directly from renewal theory. In this section we concentrate on the computation of quantities such as the $\eta(j)$ appearing in (4.2), and more importantly, on the investigation of the limiting behavior of the solutions of Markov renewal equations. The main results are (4.3), (4.9), and (4.17).

As far as computing $\eta(i)$ is considered, it is sufficient to consider only the irreducible recurrent class which contains i. Therefore, the following theorem should be applied to each recurrent class separately to obtain the $\eta(i)$ for all i in that recurrent class.

(4.3) THEOREM. Suppose X is irreducible recurrent, and let v be an invariant measure for X, that is, let v be a solution of

(4.4) $$\sum_{i \in E} v(i)P(i, j) = v(j), \quad j \in E.$$

Then the inverse of the mean recurrence time of state i in (X, T) is

(4.5) $$\eta(i) = \frac{v(i)}{vm}, \quad i \in E,$$

where

$$m(j) = E_j[T_1] = \int_0^\infty [1 - \sum_k Q(j, k, t)] \, dt.$$

(4.6) REMARK. It follows from Theorem (6.2.25) that the system of linear equations (4.4) has only one (up to multiplication by a constant) solution. In matrix notation, (4.4) becomes $v = vP$; and in interpreting vm in (4.5) we recall that v is to be thought of as a row vector and m as a column vector, so that

$$vm = \sum_j v(j)m(j).$$

Using the interpretation of $v(j)/v(i)$ as the expected number of visits to j between two visits to i, (4.5) yields the following explanation. The mean recurrence time of i is the sum, over all states j, of the expected time spent in j between two returns to i; the term corresponding to j is equal to the product of $v(j)/v(i)$ and the expected sojourn time $m(j)$ in each visit to j.

Proof of (4.3). Let a be the mean recurrence time of i; then $\eta(i) = 1/a$. Define

(4.7) $$m(j, k) = E[T_{n+1} - T_n | X_n = j, X_{n+1} = k], \quad j, k \in E.$$

Then

$$a = P(i, i)m(i, i) + \sum_{j \in D} P(i, j)P(j, i)[m(i, j) + m(j, i)]$$

$$+ \sum_{j \in D} \sum_{k \in D} P(i, j)P(j, k)P(k, i)[m(i, j) + m(j, k) + m(k, i)] + \cdots$$

where $D = E - \{i\}$. Changing the order of summation this becomes, with $N = \inf\{n \geq 1 : X_n = i\}$,

(4.8) $$a = P(i, i)m(i, i) + \sum_{j \in D} P(i, j)m(i, j) + \sum_{n=1}^{\infty} P_i\{N = n\}$$

$$+ \sum_{j \in D} \sum_{n=1}^{\infty} P_i\{N > n, X_n = j\}P(j, i)m(j, i)$$

$$+ \sum_{j \in D} \sum_{k \in D} \sum_{r=2}^{\infty} \sum_{n=1}^{r-1} P_i\{N > n, X_n = j\}P(j, k)m(j, k) P_k\{N = r - n\}$$

$$= P(i, i)m(i, i) + \sum_{j \in D} P(i, j)m(i, j)F(j, i)$$

$$+ \sum_{j \in D} \mu(i, j)P(j, i)m(j, i)$$

$$+ \sum_{j \in D} \sum_{k \in D} \mu(i, j)P(j, k)m(j, k)F(k, i),$$

where we have put

$$\mu(i, j) = \sum_{n=1}^{\infty} P_i\{N > n, X_n = j\}$$

and $F(k, i)$ is the probability of ever reaching i from k. Since X is irreducible recurrent, $F(k, j) = 1$ for all k, j. Concerning $\mu(i, j)$, we note that it is the expected number of visits to j starting at i, before returning to i; hence $\mu(i, j) = v(j)/v(i)$. Putting these in the last expression of (4.8), taking the sums indicated, and noting that

$$m(j) = \sum_{k \in E} P(j, k) m(j, k),$$

we arrive at

$$a = \frac{1}{v(i)} \sum_{j \in E} v(j)m(j).$$

Since $\eta(i) = 1/a$, this completes the proof.   □

One pleasant consequence of the preceding theorem is that the mean recurrence times can be computed directly from the elementary data, without computing the distributions $F(i, i, t)$. Next we consider the limiting behavior of the solutions of the Markov renewal equation. In the case of finitely many states this reduces immediately to renewal theory:

(4.9) PROPOSITION. Let (X, T) be an irreducible aperiodic Markov renewal process with a finite state space E. Then

$$(4.10) \quad \lim_{t \to \infty} \sum_{j \in E} \int_0^t R(i, j, ds)g(j, t - s) = \frac{1}{vm} \sum_j v(j) \int_0^\infty g(j, s)\, ds$$

provided that each one of the functions $g(j, \cdot)$ be directly Riemann integrable. Here v is a solution of $v = vP$ and $m(i)$ is the mean sojourn time in state i.

Proof. Since E is finite, the limit in question is equal to

$$(4.11) \qquad \sum_{j \in E} \lim_{t \to \infty} \int_0^t R(i, j, ds)g(j, t - s).$$

To obtain the limit as $t \to \infty$ of the jth term, we note that $R(i, j, \cdot)$ is a delayed renewal function, and we can write it as the convolution of $F(i, j, \cdot)$ and $R(j, j, \cdot)$. Since $g(j, \cdot)$ is directly Riemann integrable, by Proposition (9.2.16), so is the convolution of $F(i, j, \cdot)$ with $g(j, \cdot)$. Thus, applying the renewal theorem (9.2.8) with $R = R(j, j, \cdot)$ and $g = F(i, j, \cdot) * g(j, \cdot)$, we obtain

$$\lim_{t \to \infty} \int_0^t R(i, j, ds)g(j, t - s) = \eta(j) \int_0^\infty ds \int_0^s F(i, j, du)g(j, t - u)$$

$$= \eta(j) \int_0^\infty g(j, s)\, ds,$$

since $F(i, j, +\infty) = 1$ by the irreducibility and recurrence of (X, T). Putting this result in (4.11) and replacing $\eta(j)$ by its value (4.5), we obtain (4.10).   □

In the case of infinitely many states, however, we can no longer pass the limit in (4.10) inside the summation without some justification. The class of functions g which permit this turns out to be the following analog of the class of directly Riemann integrable functions.

Consider the class $\mathbb{B}$ of functions introduced in Section 3, and write $g \in \mathbb{B}_+$ if $g \in \mathbb{B}$ and $g \geq 0$. Let π be a positive measure on E. Then the function $g \in \mathbb{B}_+$ is said to be *directly integrable with respect to* π provided that the sums

$$(4.12) \qquad \sigma'_b = b \sum_n \sum_j \pi(j) \sup \{g(j, t) : nb \leq t < nb + b\}$$

$$(4.13) \qquad \sigma''_b = b \sum_n \sum_j \pi(j) \inf \{g(j, t) : nb \leq t < nb + b\}$$

both be finite for any $b > 0$ and $(\sigma'_b - \sigma''_b) \to 0$ as $b \to 0$. We then write $g \in \mathbb{D}_\pi$.

The following summarizes some of the useful facts about $\mathbb{D}_\pi$. Here we write $\mathbb{D}$ for the class of directly Riemann integrable functions introduced in Definition (9.2.6), and we put πg for the function defined by

$$(4.14) \qquad \pi g(t) = \sum_{j \in E} \pi(j) g(j, t), \quad t \geq 0.$$

(4.15) PROPOSITION. (a) If $g \in \mathbb{D}_\pi$, then $\pi g \in \mathbb{D}$, and the sums (4.12) and (4.13) both converge, as $b \to 0$, to the value

$$(4.16) \qquad \gamma = \int_0^\infty \pi g(t)\, dt.$$

(b) Suppose $g \in \mathbb{B}_+$ and $\int \pi g(t)\, dt < \infty$. Then $g \in \mathbb{D}_\pi$ if and only if the sum in (4.12) is finite for some $b > 0$.

(c) Suppose $g \in \mathbb{B}_+$ is such that $t \to g(j, t)$ is monotone decreasing for each j. Then $g \in \mathbb{D}_\pi$ provided that $\pi g(0) < \infty$ and $\int \pi g(t)\, dt < \infty$.

(d) Suppose $g \in \mathbb{D}_\pi$ and $f \in \mathbb{B}_+$ and $0 \leq f \leq g$. Then $f \in \mathbb{D}_\pi$ provided that each $f(i, \cdot)$ be Riemann integrable.

(e) Suppose $g \in \mathbb{D}_\pi$ and let, for each j,

$$f(j, t) = \int_0^t \varphi(j, ds) g(j, t - s)$$

for some distribution $\varphi(j, \cdot)$. Then $f \in \mathbb{D}_\pi$ as well.

Proof. (a) For any interval $A \subset \mathbb{R}_+$,

$$\sum_j \pi(j) \inf_{t \in A} g(j, t) \leq \inf_{t \in A} \sum_j \pi(j) g(j, t)$$

$$\leq \sup_{t \in A} \sum_j \pi(j) g(j, t) \leq \sum_j \pi(j) \sup_{t \in A} g(j, t).$$

In view of the definitions of $\mathbb{D}_\pi$ and $\mathbb{D}$, this is sufficient to show that $\pi g \in \mathbb{D}$. Since $\pi g \in \mathbb{D}$, πg is Riemann integrable in the ordinary sense and $\gamma < \infty$ is well defined. The above approximation shows that this number γ is also the limit of the sums in (4.12) and (4.13).

(b) Suppose $g \in \mathbb{B}_+$, $\int \pi g < \infty$, and $\sigma'_b < \infty$ for $b = 1$. Then (4.12) and (4.13) are finite for all $b > 0$, and all we need to show is the convergence of their difference to zero as $b \to 0$. Pick $\varepsilon > 0$; choose N such that

$$\sum_{n \geq N} \sum_{j \in E} \pi(j) \sup\{g(j, t): n \leq t < n + 1\} < \varepsilon.$$

Next, choose a finite subset D of E, possibly depending on N, such that

$$\sum_{n \leq N} \sum_{j \in E-D} \pi(j) \sup\{g(j, t); n \leq t < n + 1\} < \varepsilon.$$

Finally, choose $K \in \mathbb{N}$ so large that, for $b = 1/K$, with $A_n = [nb, nb + b)$,

$$b \sum_{n \leq KN} \sum_{j \in D} \pi(j)[\sup\{g(j, t): t \in A_n\} - \inf\{g(j, t): t \in A_n\}] < \varepsilon.$$

That N and D can be so selected follows from the finiteness of the sum (4.12) for $b = 1$. That K can be so chosen follows from the integrability of πg and the finiteness of D. Now, for $b = 1/K$, with $A_n = [nb, nb + b)$

$$\sigma'_b - \sigma''_b \leq b \sum_{n \leq KN} \sum_{j \in D} \pi(j)[\sup\{g(j, t): t \in A_n\} - \inf\{g(j, t): t \in A_n\}]$$

$$+ b \sum_{n \leq KN} \sum_{j \in E-D} \pi(j) \sup\{g(j, t): t \in A_n\}$$

$$+ b \sum_{n > KN} \sum_{j \in E} \pi(j) \sup\{g(j, t): t \in A_n\}$$

$$\leq \varepsilon + \sum_{n \leq N} \sum_{j \in E-D} \pi(j) \sup\{g(j, t): n \leq t < n + 1\}$$

$$+ \sum_{n > N} \sum_{j \in E} \pi(j) \sup\{g(j, t): n \leq t < n + 1\} \leq 3\varepsilon.$$

Since $\varepsilon > 0$ is arbitrary, this proves (b).

(c) When each $g(j, \cdot)$ is monotone decreasing,

$$\sigma'_b = b \sum_j \sum_n \pi(j)g(j, nb) \leq b \sum_j \pi(j)g(j, 0) + \int \sum_j \pi(j) g(j, t) \, dt.$$

Thus, $\pi g(0) < \infty$ and $\int \pi g < \infty$ imply that $\sigma'_b < \infty$, and this implies, in view of part (b), that $g \in \mathbb{D}_\pi$.

(d) Proof is immediate from part (b).

(e) Let $A_n = [nb, nb + b)$. From the proof of Proposition (9.2.16d) we have that

$$\sup_{t \in A_n} f(j, t) \leq \sum_{k=0}^{\infty} \sup_{t \in A_{n-k}} g(j, t) \int_{kb-b}^{kb+b} \varphi(j, ds).$$

Thus,

$$b \sum_n \sum_j \pi(j) \sup_{t \in A_n} f(j, t) \leq b \sum_j \pi(j) \sum_{k=0}^{\infty} \sum_{n=k}^{\infty} \sup_{t \in A_{n-k}} g(j, t) \int_{kb-b}^{kb+b} \varphi(j, ds)$$

$$\leq 2b \sum_j \pi(j) \sum_m \sup_{t \in A_m} g(j, t).$$

The last term is finite since $g \in \mathbb{D}_\pi$, and in view of (b) part above, this completes the proof since

$$\int_0^{\infty} \pi f(t) dt = \sum_j \pi(j) \int_0^{\infty} dt \int_0^t \varphi(j, ds) g(j, t - s)$$

$$= \sum_j \pi(j) \int_0^{\infty} \varphi(j, ds) \int_s^{\infty} g(j, t - s) \, dt = \int_0^{\infty} \pi g(t) \, dt.$$

This completes the proof of Proposition (4.15). $\square$

The following is the main limit theorem of Markov renewal theory.

(4.17)   THEOREM.   Suppose (X, T) is irreducible recurrent, let v be a positive solution of $v = vP$, and let $m(i)$ denote the mean sojourn time in state i as before in (4.3). If $g \in \mathbb{D}_v$, then

$$(4.18) \qquad \lim_{t \to \infty} \sum_{j \in E} \int_0^t R(i, j, ds) g(j, t - s) = \frac{1}{vm} \int_0^{\infty} vg(s) \, ds$$

in the aperiodic case, and

$$(4.19) \quad \lim_{n \to \infty} \sum_{j \in E} \int_0^{x+n\delta} R(i, j, ds) g(j, t - s) = \frac{\delta}{vm} \sum_{k=0}^{\infty} vg(x - \delta_{ij} + k\delta)$$

in the case where the states are periodic with period δ; here δ_{ij} is the first jump point of the distribution $F(i, j, \cdot)$ of the first passage time from i to j.

Proof. Define a new kernel $\tilde{Q}$ by

$$(4.20) \qquad\qquad v(i)\tilde{Q}(i, j, t) = v(j)Q(j, i, t).$$

Then $t \to \tilde{Q}(i, j, t)$ is non-decreasing and right continuous, and if $\tilde{P}(i, j) = \tilde{Q}(i, j, +\infty)$, we have $\tilde{P}(i, j) \geq 0$ and $\sum_j \tilde{P}(i, j) = 1$. So $\tilde{Q}$ is a semi-Markovian kernel. It is easy to see that the iterates $\tilde{Q}^n$ of $\tilde{Q}$ are related to Q^n by the same formula: $v(i) Q^n(i, j, t) = v(j) \tilde{Q}^n(j, i, t)$, so for the Markov renewal functions $\tilde{R}(i, j, t)$ corresponding to $\tilde{Q}$ we have

$$(4.21) \qquad\qquad v(i)R(i, j, t) = v(j)\tilde{R}(j, i, t).$$

We see that in particular, $R(i, i, t) = \tilde{R}(i, i, t)$ and that we have, by (2.9),

$$(4.22) \qquad v(i)R(i, j, t) = v(j) \int_0^t R(i, i, ds)\tilde{F}(j, i, t - s)$$

for $i \neq j$, where $\tilde{F}(j, i, \cdot)$ is the distribution of the first passage time from j to i in the Markov renewal process corresponding to $\tilde{Q}$.

Using (4.22) in computing $f = R * g$, we obtain

$$(4.23) \qquad v(i)f(i, t) = \sum_{j \in E} v(i) \int_0^t R(i, j, ds)g(j, t - s)$$
$$= \int_0^t R(i, i, ds)\hat{f}(i, t - s),$$

where

$$(4.24) \qquad \hat{f}(i, t) = \sum_{j \neq i} v(j) \int_0^t \tilde{F}(j, i, ds)g(j, t - s) + v(i)g(i, t).$$

For fixed i, since $g \in \mathbb{D}_v$, Proposition (4.15e) applies to show that the function $\hat{g}$ defined as $\hat{g}(j, t) = \int_0^t \tilde{F}(j, i, ds) g(j, t - s)$ is in $\mathbb{D}_v$, and this in turn implies, by Proposition (4.15a), that $\hat{f}$ is in $\mathbb{D}$. There remains to apply the key renewal theorem (9.2.8) to (4.23). In the aperiodic case this yields, through Theorem (4.3),

$$\lim_{t \to \infty} v(i)f(i, t) = \eta(i) \int_0^\infty \hat{f}(i, s)\, ds$$
$$= \frac{1}{vm} v(i) \sum_j v(j) \int_0^\infty g(j, s)\, ds.$$

as desired.

We omit the proof in the periodic case. $\square$

The heart of the preceding theorem is that the condition of direct integrability of g with respect to v implies the direct Riemann integrability of $\hat{f}$ defined by (4.24). If more is known about the process, for example, if the $\tilde{F}(j, i, t)$ can be calculated explicitly, then the condition on g can be relaxed.

The preceding proof introduced a dual process which can be used further to obtain limits of certain important ratios. For this purpose we need the following technical result.

(4.25) LEMMA. Let φ be a non-decreasing function defined on $\mathbb{R}_+$ and with $\varphi(t + b) - \varphi(t) \leq c$ for all t for some constant c and $b > 0$. Then for any non-decreasing right continuous function g we have

$$\lim_{t \to \infty} \frac{\varphi * g(t)}{\varphi(t)} = g(+\infty).$$

Proof will be omitted. We note that, being a renewal function, each Markov renewal function $R(i, j, \cdot)$ satisfies the hypothesis of (4.25) regarding φ; see (9.2.11) for this.

(4.26) PROPOSITION. Let (X, T) be an irreducible recurrent Markov renewal process, and let $g(k, \cdot)$ be a right continuous non-decreasing function. Then for any $h, i, j, k \in E$,

$$(4.27) \qquad \lim_{t \to \infty} \frac{1}{R(h, i, t)} \int_0^t R(j, k, ds)g(k, t - s) = \frac{1}{v(i)} v(k)g(k, \infty),$$

where v is a solution of $v = vP$. In particular, for any $h, i, j, k \in E$,

$$(4.28) \qquad \lim_{t \to \infty} \frac{R(j, k, t)}{R(h, i, t)} = \frac{v(k)}{v(i)}.$$

Proof. Using Lemma (4.25) with $\varphi = R(j, j, \cdot)$ and $g = F(i, j, \cdot)$, we obtain, since $F(i, j, +\infty) = 1$ by hypothesis,

$$(4.29) \qquad \lim_{t \to \infty} \frac{R(i, j, t)}{R(j, j, t)} = 1, \quad i, j \in E.$$

On the other hand, applying Lemma (4.25) with $\varphi = R(i, i, \cdot)$ and $g = \tilde{F}(j, i, \cdot)$ to the right-hand side of (4.22), we have

$$(4.30) \qquad \lim_{t \to \infty} \frac{R(i, j, t)}{R(i, i, t)} = \frac{v(j)}{v(i)}, \quad i, j \in E.$$

In particular, (4.29) and (4.30) together imply that

$$(4.31) \qquad \lim_{t \to \infty} \frac{R(j, j, t)}{R(i, i, t)} = \frac{v(j)}{v(i)}.$$

To obtain (4.28), we write

$$\frac{R(j, k, t)}{R(h, i, t)} = \frac{R(j, k, t)}{R(j, j, t)} \cdot \frac{R(j, j, t)}{R(i, i, t)} \cdot \frac{R(i, i, t)}{R(h, i, t)},$$

and apply (4.30), (4.31), and (4.29) respectively to the first, second, and third factors on the right. Finally, (4.27) follows from (4.28) and Lemma (4.25) applied to the left-hand side of (4.27) with $\varphi = R(j, k, \cdot)$ and $g = g(k, \cdot)$. $\square$

It is interesting to note that the limit of the ratio (4.28), for example, does not depend on the distributions of the sojourn times.

5. Semi-Markov Processes

Let (X, T) be a Markov renewal process with state space E and semi-Markov kernel Q. Define, as before in (3.14), $L = \sup_n T_n$; then L is the lifetime of (X, T). If E is finite or if X is irreducible recurrent, then $L = +\infty$ almost surely. But it is possible, if there are infinitely many transient states which can be reached from each other with very small sojourn times in between, for the lifetime L to be finite with a positive probability. We refer to Proposition (3.14) and the propositions following it for a complete picture.

In this section we will be interested in the process $Y = \{Y_t; t \geq 0\}$ defined by (1.14), that is,

$$
(5.1) \qquad Y_t = \begin{cases} X_n & \text{if } T_n \leq t < T_{n+1}, \\ \Delta & \text{if } t \geq L, \end{cases}
$$

where Δ is a point not in E. The continuous time parameter process so defined is called the *minimal semi-Markov process* associated with (X, T). In terms of Y, the times $T_1, T_2, \ldots$ are the successive times of transitions for Y, and $X_0, X_1, X_2, \ldots$ are the successive states visited (this picture should be taken with some reserve: we allow transitions from a state into itself in X, and such transitions are not noticeable to an observer of the path of Y).

If the semi-Markov kernel Q has the same form as in Example (1.16), then Y is a Markov process. Otherwise Y is not Markovian. The word *semi-Markov* comes from the somewhat limited Markov property which Y enjoys, namely, that the future of Y is independent of its past given the present state provided the "present" is a time of jump. We will study the process Y by the techniques developed in this chapter so far. This treatment is parallel to, and in fact extends, that of Markov processes which we put in Chapter 8.

We start with a consideration of the transition probabilities. For $i, j \in E, t \geq 0$, let

$$
(5.2) \qquad P_t(i, j) = P_i\{Y_t = j\},
$$

where, as before, we write P_i for the conditional probability given $\{X_0 = i\}$. In the proposition below, R is the Markov renewal kernel corresponding to Q, and h is defined by

$$
(5.3) \qquad h(j, t) = 1 - \sum_{k \in E} Q(j, k, t), \quad j \in E, t \geq 0.
$$

(5.4) PROPOSITION. For any $i, j \in E$ and $t \geq 0$

$$
P_t(i, j) = \int_0^t R(i, j, ds) h(j, t - s).
$$

Proof. (a) We first show that $P_t(i, j)$ satisfies

(5.5) $$P_t(i, j) = I(i, j)h(i, t) + \sum_{k \in E} \int_0^t Q(i, k, ds)P_{t-s}(k, j).$$

We have

$$P_t(i, j) = P_i\{Y_t = j, T_1 > t\} + P_i\{Y_t = j, T_1 \leq t\}$$
$$= P_i\{T_1 > t\}P_i\{Y_t = j \mid T_1 > t\}$$
$$+ E_i[P_i\{Y_t = j, T_1 \leq t \mid X_1, T_1\}].$$

On the set $\{T_1 > t\}$ we have $Y_t = X_0$, and clearly $P_i\{T_1 > t\} = h(i, t)$; hence, the first term on the right is $I(i, j) h(i, t)$. To obtain the second term, first observe that it is equal to

(5.6) $$E_i[I_{[0, t]}(T_1)P_i\{Y_t = j \mid X_1, T_1\}].$$

Consider the Markov renewal process $(\hat{X}, \hat{T})$ obtained by setting $\hat{X}_n = X_{n+1}$ and $\hat{T}_n = T_{n+1} - T_1$ for all $n \in \mathbb{N}$. The semi-Markov process $\hat{Y}$ associated with $(\hat{X}, \hat{T})$ is the same as Y except that the time origin of $\hat{Y}$ is the time T_1 of first transition for Y; therefore $Y_t = \hat{Y}_{t-s}$ if $T_1 = s$. On the other hand, $(\hat{X}, \hat{T})$ has the same probability law as (X, T), and therefore $P\{\hat{Y}_{t-s} = j \mid \hat{Y}_0 = k\} = P_k\{Y_{t-s} = j\}$. Thus

$$P_i\{Y_t = j \mid X_1, T_1\} = P_{t-T_1}(X_1, j) \quad \text{on } \{T_1 \leq t\}.$$

Putting this into (5.6), the latter expression becomes

$$E_i[I_{[0, t]}(T_1)P_{t-T_1}(X_1, j)] = \sum_{k \in E} \int_0^t Q(i, k, ds)P_{t-s}(k, j),$$

which is the second term on the right of (5.5).

(b)   For fixed j, if we put $f(i, t) = P_t(i, j)$, we see that f satisfies a Markov renewal equation $f = g + Q * f$ with $g(i, t) = 0$ for all i except $i = j$, for which it is $h(j, t)$. It follows from Theorem (3.6) that

$$P_t(i, j) = \int_0^t R(i, j, ds)h(j, t - s)$$

is a solution, and to complete the proof we need to show that this is the desired solution.

If $L = \infty$ almost surely, then by (3.6), (3.10), and (3.14), this is the only solution, and we are finished. Otherwise, we need to show that

(5.7) $$\sum_{j \in E} \int_0^t R(i, j, ds)h(j, t - s) = P_i\{L > t\}$$

since, by the definition (5.1) of Y,

$$\sum_j P_t(i, j) = P_i\{Y_t \in E\} = 1 - P_i\{Y_t = \Delta\} = P_i\{L > t\}.$$

The left side of (5.7) is $R * h(i, t)$; and noting that $h = 1 - Q * 1$, we have

$$R * h = \lim_{n \to \infty} \sum_{m=0}^{n} Q^m * h$$

$$= \lim_{n \to \infty} \left[\sum_{m=0}^{n} (Q^m * 1 - Q^{m+1} * 1) \right]$$

$$= \lim_{n \to \infty} [1 - Q^{n+1} * 1] = 1 - \lim_n Q^n * 1.$$

This completes the proof since, by the proof of (3.14),

$$\lim_n Q^n * 1(i, t) = P_i\{L \le t\}. \qquad \square$$

Using the preceding result we can compute the expected time spent in a particular state by the process Y. Let $U(i, j, t)$ be the expected time spent in j during $[0, t]$ by the process starting at i; that is,

$$(5.8) \qquad U(i, j, t) = E_i \left[\int_0^t 1_j(Y_s) \, ds \right], \quad i, j \in E, \, t \ge 0,$$

where $1_j(k) = 1$ or 0 according as $k = j$ or $k \ne j$. Passing the expectation inside the integral, noting that $E_i[1_j(Y_s)] = P_s(i, j)$, and using Proposition (5.4), we obtain the following

(5.9) PROPOSITION. For any $i, j \in E$ and $t \ge 0$,

$$U(i, j, t) = \int_0^t R(i, j, ds) \int_0^{t-s} h(j, u) \, du.$$

We will call U the *potential* function of Y. To see its relation to the potentials introduced for Markov processes in (8.4.15) we note the following: Let $f \in \mathbb{B}_+$, and define

$$(5.10) \qquad Uf(i, t) = E_i \left[\int_0^t f(Y_s, s) \, ds \right], \quad i \in E, \, t \ge 0$$

with the usual convention that $f(\Delta, s) = 0$ for all s. If we interpret $f(j, t)$ as the rate of reward being received at an instant t at which the state is j, then $Uf(i, t)$ becomes the expected value of the total reward received during $[0, t]$ starting at state i. In particular, if the rate of reward depends only on the state, namely, if $f(j, t) = g(j)$ independent of t, then

$$(5.11) \qquad U^\alpha g(i) = E_i \left[\int_0^\infty e^{-\alpha t} g(Y_t) \, dt \right], \quad \alpha \ge 0,$$

is exactly the α-*potential* of g starting at i. Moreover, for $\alpha = 0$, if we drop the superscript, (5.11) becomes the *potential* of g starting at i.

To compute Uf for $f \in \mathbb{B}_+$, we first pass the expectation inside the integral and note that $P_s(i, j) \, ds$ is simply $U(i, j, ds)$. So

$$(5.12) \qquad Uf(i, t) = \sum_j \int_0^t U(i, j, ds) f(j, s)$$

$$= \sum_j \int_0^t P_s(i, j) f(j, s) \, ds$$

$$= \sum_j \int_0^t R(i, j, du) \int_0^{t-u} h(j, s) f(j, u + s) \, ds.$$

For the α-potential of g we have

$$(5.13) \qquad\qquad U^\alpha g(i) = \sum_j U^\alpha(i, j) g(j),$$

where for any $i, j \in E$ and $\alpha > 0$,

$$(5.14) \qquad U^\alpha(i, j) = \int_0^\infty e^{-\alpha t} U(i, j, dt)$$

$$= \int_0^\infty e^{-\alpha t} P_t(i, j) \, dt = R_\alpha(i, j) h_\alpha(j)$$

with R_α defined by (2.11), and (see also (2.10))

$$(5.15) \qquad h_\alpha(j) = \int_0^\infty e^{-\alpha t} h(j, t) \, dt = \frac{1}{\alpha} [1 - \sum_k Q_\alpha(j, k)].$$

We next relate the recurrence properties of (X, T) to those of Y by considering the total amount of time spent by Y in various states. Starting at i, the total time spent in j has the expectation

$$(5.16) \qquad U(i, j) = U(i, j, +\infty) = E_i \left[\int_0^\infty 1_j(Y_s) \, ds \right], \quad i, j \in E.$$

Using Proposition (5.4) to evaluate $P_t(i, j)$ and changing the order of integration in the expression obtained, we have

$$(5.17) \qquad U(i, j) = \int_0^\infty P_t(i, j) \, dt$$

$$= \int_0^\infty dt \int_0^t R(i, j, ds) h(j, t - s)$$

$$= \int_0^\infty R(i, j, ds) \int_s^\infty h(j, t - s) \, dt = R(i, j, +\infty) m(j)$$

where $m(j) = \int h(j, u)\, du$ is the mean sojourn time in state j. It follows that if j is recurrent, then $U(i, j) = +\infty$ unless $F(i, j) = 0$. If $F(i, j) = 0$, then $U(i, j) = 0$. And if j is transient, then $U(i, j) = R(i, j)\, m(j)$, which is finite or infinite according as $m(j) < \infty$ or $m(j) = +\infty$ unless $F(i, j) = 0$.

In the recurrent case we may obtain somewhat more precise information by considering certain ratios:

(5.18) THEOREM. Let X be irreducible recurrent and let v be an invariant measure, namely, let v satisfy $v = vP$. Then

$$(5.19) \qquad \lim_{t \to \infty} \frac{U(i, j, t)}{R(i, j, t)} = m(j), \qquad \lim_{t \to \infty} \frac{U(i, j, t)}{U(h, k, t)} = \frac{v(j)m(j)}{v(k)m(k)}$$

for all $i, j, h, k \in E$ provided that at least one of $m(j)$ and $m(k)$ be finite. Moreover, if at least one of $m(j)$ and $m(k)$ is finite, then

$$(5.20) \qquad \lim_{t \to \infty} \frac{\int_0^t 1_j(Y_s)\, ds}{\int_0^t 1_k(Y_s)\, ds} = \frac{v(j)m(j)}{v(k)m(k)}$$

almost surely.

Proof. The first limit in (5.19) follows from the form of $U(i, j, t)$ given in Proposition (5.9) and Lemma (4.25). The second limit in (5.19) follows from (5.9) again and Proposition (4.26). We omit the proof of (5.20).   □

In the case where $vm = \sum_j v(j)\, m(j) < \infty$, we may sum over all j in the second limit in (5.19) and (5.20). This yields

(5.21) COROLLARY. Suppose X is irreducible recurrent, v satisfies $v = vP$, and $vm < \infty$. Then

$$\lim_{t \to \infty} \frac{1}{t} U(i, k, t) = \frac{1}{vm} v(k)m(k).$$

Moreover, almost surely,

$$\lim_{t \to \infty} \frac{1}{t} \int_0^t 1_k(Y_s)\, ds = \frac{1}{vm} v(k)m(k).$$

The preceding corollary provides an intuitive explanation for the following limit theorem, whose proof is immediate from the main limit theorem (4.17), Proposition (5.4), and the fact that $h(k, \cdot)$ is monotone non-increasing with integral $m(k)$.

(5.22) Theorem. Suppose X is irreducible recurrent, v is a solution of $v = vP$, and $m(k) < \infty$. Then for any $i \in E$,

$$\lim_{t \to \infty} P_i\{Y_t = k\} = \frac{1}{vm} v(k) m(k)$$

in the aperiodic case, and

$$\lim_{n \to \infty} P_i\{Y_{x+n\delta} = k\} = \frac{\delta}{vm} v(k) \sum_n h(k, x - \delta_{ik} + n\delta)$$

in the case where the states are periodic with period δ; here δ_{ik} is the first jump point of the distribution $F(i, k, \cdot)$, and the summation on the right is over all n for which $x + n\delta \geq \delta_{ik}$.

In the case of a regular Markov process Y, the sojourn distributions are all exponential, and therefore all states are aperiodic. Moreover,

$$m(j) = \int_0^\infty h(j, t)\, dt = \int_0^\infty e^{-\lambda(j)t}\, dt = \frac{1}{\lambda(j)},$$

so the limit claimed in (5.22) is precisely (8.5.10), and we have thus also proved Theorem (8.5.11).

We end this section with a comment on the definition of a semi-Markov process. Starting with a Markov renewal process (X, T), if we define the process Y by (5.1), then Δ plays the role of an absorbing state. In some situations, however, one has to start with a process Y which is defined for all $t \in [0, \infty)$ and which is similar to a semi-Markov process, but Δ is not absorbing. Namely, if $T_1, T_2, \ldots$ are the successive jump times of Y and if $X_0, X_1, X_2, \ldots$ are the successive states visited by Y, then (X, T) is a Markov renewal process, but the process Y is defined beyond $L = \lim T_n$ so that Δ is not absorbing. This is the case, for example, if Y is a Markov process with no instantaneous states, and the jump times $T_1, T_2, \ldots$ are such that $L = \sup T_n < +\infty$ with a positive probability. Then Δ behaves worse than an instantaneous state, and the analysis becomes much harder. In such a case, the number $P_t(i, j)$ we have in (5.4) is the probability of $\{Y_t = j \text{ and } L > t\}$ given $Y_0 = i$, and the relation between $P_i\{Y_t = j\}$ and the semi-Markov kernel Q is more complex. In this case, the Markov renewal equation (5.5) still holds, but it has more than one solution. To obtain all possible solutions, by Theorem (3.6), we are led to examine all possible solutions to $h = Q * h$, which, as we have seen in (3.14)ff., has to do with the behavior of the Markov chain X on transient states near the "infinity" Δ. The problem, which is usually referred to as the *boundary theory*, is a generalization of the boundary theory for Markov processes. At this writing, there are no good results known.

It is possible, on the other hand, to obtain a number of limiting results under suitable regularity conditions even if Y has infinitely many jumps in a finite interval. In the next section we introduce a class of such processes (see Example (6.6) in particular).

6. Semi-Regenerative Processes

A stochastic process $Z = \{Z_t; t \geq 0\}$ was called regenerative, roughly speaking, if there existed certain random times forming a renewal process at each of which the future of Z became a probabilistic replica of the process itself. Such random times were then called times of regeneration for Z. We will now weaken this concept by letting the future after a time of regeneration depend also on the "state" of a Markov renewal process at that time. Just as the real importance of renewal theory came from its applications to regenerative processes, the importance of the Markov renewal theory comes from applications to semi-regenerative processes.

A random variable T taking values in $[0, +\infty]$ is called a stopping time for Z provided that for any $t < \infty$, the occurrence or non-occurrence of the event $\{T \leq t\}$ can be determined once the history $\{Z_u; u \leq t\}$ of Z before t is known. If T is a stopping time for Z, we denote by $\{Z_u; u \leq T\}$ the history of Z before the stopping time T.

(6.1) DEFINITION. Let $Z = \{Z_t; t \geq 0\}$ be a stochastic process with a topological state space F, and suppose that the function $t \longrightarrow Z_t(\omega)$ is right continuous and has left-hand limits for almost all ω. The process Z is said to be *semi-regenerative* if there exists a Markov renewal process (X, T) with infinite lifetime satisfying the following:

    (a)   for each $n \in \mathbb{N}$, T_n is a stopping time for Z;

    (b)   for each $n \in \mathbb{N}$, X_n is determined by $\{Z_u; u \leq T_n\}$;

    (c)   for each $n \in \mathbb{N}$, $m \geq 1$, $0 \leq t_1 < t_2 < \cdots < t_m$, and function f defined on F^m and positive,

$$E_i[f(Z_{T_n+t_1}, \ldots, Z_{T_n+t_m}) \mid Z_u; u \leq T_n] = E_j[f(Z_{t_1}, \ldots, Z_{t_m})] \quad \text{on } \{X_n = j\}.$$

                 □

In this definition E_i and E_j refer to the expectations given the initial state for the Markov chain X. Condition (a) is the same as in (9.2.18); without it neither (b) nor the left-hand side of (c) would have any meaning. Condition (b) means that an observer of the process Z can tell X_n if he has watched Z until the time T_n. Condition (c) is the most important; it considers a random variable $W_n = f(Z_{T_n+t_1}, \ldots, Z_{T_n+t_m})$ which is a function of the values that Z takes at the future times $T_n + t_1, \ldots, T_n + t_m$. The left-hand side of (c) is the conditional expectation of W_n as computed by an observer who has watched Z until the "present" time T_n. To a second observer who takes T_n as the time origin, the same random variable will appear as $\hat{W}_0 = f(\hat{Z}_{t_1}, \ldots, \hat{Z}_{t_m})$ where $\hat{Z}_t = Z_{T_n+t}$, and therefore, if he is told that the "state" at time $\hat{T}_0 = T_n$ is $\hat{X}_0 = X_n = j$ (which is the initial state as far as he is concerned), his expectation of the random variable in question will be the quantity on the right-hand side of (c). The process is semi-regenerative if the two observers come up with the same answer; in other words, if the extra information that the first observer has concerning the past of Z is worthless as far as predicting the future is concerned.

(6.2) **Example.** *Regenerative processes.* This is the case where in the condition (6.1c) the right-hand side does not depend on the state X_n at all. This is much stronger than what we require of semi-regenerative processes. $\square$

(6.3) **Example.** *Semi-Markov processes.* Let (X, T) be a Markov renewal process with infinite lifetime, and let Y be the semi-Markov process associated with it (defined by (5.1)). Then Y is semi-regenerative, and we have already seen how Markov renewal theory is used to study the process Y. Let us further define

$$(6.4) \qquad\qquad V_t = T_{n+1} - t \quad \text{if } T_n \leq t < T_{n+1},$$

for all $t \geq 0$, and put $Z_t = (Y_t, V_t)$, $t \geq 0$. Then Z is a semi-regenerative process with state space $F = E \times \mathbb{R}_+$. We show this first for the case $T_0 = 0 < T_1 < T_2 < \cdots$. Then $t \to V_t$ is continuous and decreasing everywhere except at $T_0, T_1, T_2, \ldots$, at which it jumps from the left-hand value of zero to the right-hand values $T_1 - T_0, T_2 - T_1, T_3 - T_2, \ldots$. Hence, $T_1, T_2, \ldots$ are the first, second, $\ldots$ jump times of V, and as such each one is a stopping time of V, and therefore of Z. For each n, $X_n = Y_{T_n}$, and therefore X_n is determined by the history of Z until T_n. Finally, the check for condition (c) follows from the Markov renewal property of (X, T). So Z is semi-regenerative. More generally, when $0 = T_0 \leq T_1 \leq T_2 \leq \cdots$, we define $\hat{T}_0 = 0, \hat{T}_1$ as the first jump time of V, $\hat{T}_2$ as the second jump time of V, and so on, and set $\hat{X}_n = Y_{\hat{T}_n}$. Then Z is semi-regenerative provided $(\hat{X}, \hat{T})$ is a Markov renewal process. But the latter follows from the strong Markov property applied to (X, T) after noting that $\hat{T}_n = T_{N_n}, \hat{X}_n = X_{N_n}$, where $N_0 = 0, N_1 = \inf\{n > 0: T_n > 0\}, N_2 = \inf\{n > N_1 : T_n > T_{N_1}\}$, etc.

Without more explanations we point out a number of other semi-regenerative processes connected with a given Markov renewal process (X, T) with infinite lifetime. Let Y be the semi-Markov process associated with (X, T), let V be as defined in (6.4), and define

$$(6.5) \qquad\qquad U_t = t - T_n \quad \text{if } T_n \leq t < T_{n+1}.$$

Then U_t is the time since the last transition of Y before the instant t. The two-dimensional process $Z = (Y, U)$ is semi-regenerative; its state space is $F = E \times \mathbb{R}_+$. Still another semi-regenerative process is (Y, U, V), whose state space is $E \times \mathbb{R}_+ \times \mathbb{R}_+$. □

(6.6) EXAMPLE. *General semi-Markov processes.* Instead of starting with a Markov renewal process and then defining Y in terms of it, we now define Y directly. Let $E = \{0, 1, 2, \ldots\}$, and suppose $Y = \{Y_t; t \geq 0\}$ is a right continuous stochastic process taking values in $E \cup \{\infty\}$. Suppose that for any $i \in E$, the time set $H_i = \{t : Y_t = i\}$ is a union of countably many disjoint intervals whose lengths add up to infinity; the right continuity of Y implies that the component intervals of H_i are closed on the left and open on the right. For the special state $+\infty$ we suppose that the time set $H_\infty = \{t : Y_t = \infty\}$ contains no open intervals and that its length $\int_0^\infty 1_\infty(Y_s)\, ds$ is well defined and equal to zero. Moreover, we suppose that no jump of Y is of infinite magnitude. In other words, the process Y may go to and come from the state $+\infty$ but the manner in which it can do so is severely restricted.

Concerning its probabilistic behavior we suppose the following. For any finite set $A \subset E$, if S is the time of first entrance to A by Y, then we have

(6.7) $$E_i[W_S \mid Y_u; u \leq S] = E_j[W_0] \quad \text{on } \{Y_S = j\}$$

for any random variable $W_S = f(Y_{S+t_1}, \ldots, Y_{S+t_m})$; here $W_0 = f(Y_{t_1}, \ldots, Y_{t_m})$, $f \geq 0$ is defined on E^m, $0 \leq t_1 < \cdots < t_m$, and E_i is the expectation given $Y_0 = i, i \in E$.

The process Y we have defined is very close to the semi-Markov processes examined in the preceding section. If $T_1, T_2, \ldots$ are the successive times of jumps of Y, and if $\sup T_n = \infty$ almost surely, then Y is precisely a simple semi-Markov process associated with (X, T), where the T_n are the jump times and $X_n = Y_{T_n}$. But if $L = \sup T_n < \infty$ with positive probability, what we have is much more complex than the minimal semi-Markov process associated with this (X, T). Note that now $Y_L = +\infty$, but the process Y does not stay there for any length of time, and Y is already defined beyond L. It can be shown that every state $i \in E$ is transient for the Markov chain X, but the process Y spends infinite time in state i. In other words, though the number of visits to i is finite in the interval $[0, L)$, the process Y comes back to i after L again and again. The times $S_1, S_2, \ldots$ of successive visits to that fixed state i form, as before, a renewal process; and by our hypothesis on the time set $H_i = \{t : Y_t = i\}$, this renewal process S is recurrent.

To see that Y is semi-regenerative we take a fixed finite subset $D \subset E$ which contains the initial state; let $T_1, T_2, \ldots$ be the successive times of visits to that fixed set D, and set $X_0 = Y_0, T_0 = 0, X_n = Y_{T_n}$ (more precisely, $T_1 = \inf\{t : Y_s \neq Y_0$ for some $s \leq t$ and $Y_t \in D\}, T_2 = \inf\{t > T_1 : Y_s \neq X_1$

for some $s \in (T_1, t]$ and $Y_t \in D\}$, and so on). Each T_n is a stopping time, X_n clearly belongs to the history until T_n, and condition (6.1c) follows from (6.7) by a repeated application of the latter. In particular, (6.7) implies that (X, T) is a Markov renewal process.   □

Other examples of semi-regenerative processes are the Markov processes studied in Chapter 8, and some queueing processes which we will take up specially in the next section.

When Z is semi-regenerative, the Markov renewal process (X, T) figuring in Definition (6.1) is called its *embedded* process. It is possible to have more than one Markov renewal process embedded in the same semi-regenerative process. In such a situation, we may use different embedded ones to study the different aspects of Z.

The following are the main theorems of this section.

(6.8) THEOREM. Let Z be a semi-regenerative process with state space F, let (X, T) be a Markov renewal process imbedded in Z, and let Q and R be the semi-Markov kernel and the Markov renewal kernel corresponding to (X, T). Define, for any suitable set $A \subset F$,

(6.9) $$K_t(i, A) = P_i\{Z_t \in A, T_1 > t\}, \quad i \in E, t \geq 0;$$

and

(6.10) $$P_t(i, A) = P_i\{Z_t \in A\}, \quad i \in E, t \geq 0.$$

Then for any $i \in E$ and $t \geq 0$, we have

(6.11) $$P_t(i, A) = \sum_{j \in E} \int_0^t R(i, j, ds) K_{t-s}(j, A).$$

Proof. We can write

$$P_t(i, A) = P_i\{Z_t \in A, T_1 > t\} + P_i\{Z_t \in A, T_1 \leq t\}$$
$$= K_t(i, A) + E_i[I_{[0, t]}(T_1) P_i\{Z_t \in A \mid Z_u; u \leq T_1\}].$$

Now, by the regeneration property (6.1c),

$$P_i\{Z_t \in A \mid Z_u; u \leq T_1\} = P_{t-T_1}(X_1, A) \quad \text{on } \{T_1 \leq t\}.$$

So

$$P_t(i, A) = K_t(i, A) + E_i[I_{[0, t]}(T_1) P_{t-T_1}(X_1, A)]$$
$$= K_t(i, A) + \sum_{j \in E} \int_0^t Q(i, j, ds) P_{t-s}(j, A).$$

This is a Markov renewal equation, and by Theorem (3.6), (6.11) is a solution of it. That this is the only solution follows from (3.10), (3.14), and the hypothesis that (X, T) has infinite lifetime (see Definition (6.1)).   □

(6.12) THEOREM. In the hypothesis of Theorem (6.8), suppose further that (X, T) is an irreducible aperiodic recurrent process, let v be an invariant measure for X, let $m(j) = E_j[T_1]$, and suppose $vm < \infty$. Then

$$(6.13) \qquad \lim_{t \to \infty} P_t(i, A) = \frac{1}{vm} \sum_j v(j) \int_0^\infty K_t(j, A)\, dt$$

provided that $t \to K_t(j, A)$ be Riemann integrable for each $j \in E$.

Proof. The function $t \to h(j, t) = P_j\{T_1 > t\}$ is monotone decreasing and integrable with $\int h(j, t)\, dt = m(j)$. Thus, it is in $\mathbb{D}$. Since $K_t(j, A) \le h(j, t)$ and since $t \to K_t(j, A)$ is integrable by hypothesis, this implies that $t \to K_t(j, A)$ is also in $\mathbb{D}$. Now (6.13) follows from Proposition (4.9) provided that E be finite.

Suppose next that E is not finite. We can write

$$(6.14) \qquad P_t(i, A) = \sum_{j \in E} P_i\{Z_t \in A, Y_t = j\},$$

where Y is the semi-Markov process associated with (X, T). The process (Z, Y) is semi-regenerative with the same imbedded Markov renewal process (X, T), and using Theorem (6.8) above, we have

$$(6.15) \qquad P_i\{Z_t \in A, Y_t = j\} = \int_0^t R(i, j, ds) K_{t-s}(j, A).$$

Since $t \to K_t(j, A)$ is directly Riemann integrable, from the key renewal theorem we obtain

$$(6.16) \qquad \lim_{t \to \infty} P_i\{Z_t \in A, Y_t = j\} = \eta(j) \int_0^\infty K_t(j, A)\, dt,$$

where $\eta(j) = v(j)/vm$ by Theorem (4.3).

On the other hand, let D be a finite subset of E which includes the initial state i, and define the Markov renewal process $(\hat{X}, \hat{T})$ as in Theorem (1.13). Then $(\hat{X}, \hat{T})$ is an irreducible aperiodic recurrent Markov renewal process which is also imbedded in Z. Using the arguments of the first paragraph of this proof with $(\hat{X}, \hat{T})$ replacing (X, T), we see that the limit, as $t \to \infty$, of $P_t(i, A)$ exists. This fact, along with (6.14) and (6.16), implies

$$(6.17) \qquad \lim_{t \to \infty} P_t(i, A) \ge \sum_j \lim_t P_i\{Z_t \in A, Y_t = j\}$$

$$= \sum_j \eta(j) \int_0^\infty K_t(j, A)\, dt.$$

If the strict inequality held for some A, then using the inequality for the complement of A and noting that $P_t(i, A) + P_t(i, A^c) = 1$, we would have

$$1 > \sum_j \eta(j) \int K_t(j, A) \, dt + \sum_j \eta(j) \int K_t(j, A^c) \, dt$$

$$= \sum_j \eta(j) \int [K_t(j, A) + K_t(j, A^c)] \, dt$$

which is absurd since $K_t(j, A) + K_t(j, A^c) = P_j\{T_1 > t\}$, and the integral of this is $m(j)$, and $\sum \eta(j) m(j) = 1$. Hence, in (6.17) the equality must hold for every A. $\square$

The following are some applications of the preceding two theorems.

(6.18) EXAMPLE. *Semi-Markov processes.* Let (X, T) be an irreducible recurrent Markov renewal process and let Y, V, U be as defined in Example (6.3). Since (Y, V) is semi-regenerative, we have

$$(6.19) \qquad P_i\{Y_t = j, V_t > y\} = \int_0^t R(i, j, ds) h(j, t + y - s)$$

by Theorem (6.8), since the kernel K in (6.9) now is

$$K_t(i, \{j\} \times (y, \infty)) = P_i\{Y_t = j, V_t > y, T_1 > t\}$$
$$= I(i, j) P_i\{T_1 > t + y\} = I(i, j) h(i, t + y)$$

with $h(i, t) = P_i\{T_1 > t\} = 1 - \sum_j Q(i, j, t)$ as before. Applying the limit theorem (6.12), we obtain, when (X, T) is irreducible recurrent aperiodic and $m(j) < \infty$,

$$(6.20) \qquad \lim_{t \to \infty} P_i\{Y_t = j, V_t > y\} = \frac{1}{vm} v(j) \int_y^\infty h(j, s) \, ds.$$

Similarly, noting that

$$P_i\{Y_t = j, U_t > x, V_t > y, T_1 > t\} = I(i, j) I_{(x, \infty)}(t) h(i, t + y),$$

we have, from Theorem (6.8) applied now to the semi-regenerative process (Y, U, V),

$$(6.21) \quad P_i\{Y_t = j, U_t > x, V_t > y\} = \int_{[0, t-x)} R(i, j, ds) h(j, t + y - s).$$

Using Theorem (6.12) next, when (X, T) is recurrent aperiodic and $m(j) < \infty$, we get

$$(6.22) \qquad \lim_{t\to\infty} P_i\{Y_t = j, U_t > x, V_t > y\} = \frac{1}{vm} v(j) \int_{x+y}^{\infty} h(j, s) \, ds.$$

Various other quantities of interest concerning (Y, U, V) can be obtained by similar means. □

(6.23) EXAMPLE. *System availability.* Consider a piece of equipment with a finite number of components; suppose that the failure of any one component is a failure for the equipment itself. Let $T_0 = 0, T_1, T_2, \ldots$ be the times of successive failures, and let X_n be the type of the component causing the nth failure. The time $T_{n+1} - T_n$ between two failures is the sum of the repair time of the component which failed at T_n and a failure-free interval following the repair. We suppose all components have exponential lifetimes, with the component j having the parameter $\lambda(j)$; and suppose that the repair time of the component j has distribution $\varphi(j, \cdot)$. Under these assumptions, (X, T) is a Markov renewal process with semi-Markov kernel Q given by

$$(6.24) \qquad Q(i, j, t) = \int_0^t \lambda(j) e^{-\lambda u} \varphi(i, t - u) \, du$$

where $\lambda = \sum \lambda(i)$. Note that

$$P(i, j) = Q(i, j, \infty) = \frac{\lambda(j)}{\lambda}$$

independent of i, so $\{\lambda(j)\}$ is also the invariant measure.

Now let Y_t be the component which caused the last failure before t, and define W_t as 1 or 0 according as the equipment is working or under repair at time t. Then (Y, W) is a semi-regenerative process. Since

$$P_i\{Y_t = j, W_t = 0, T_1 > t\} = I(i, j)[1 - \varphi(i, t)],$$

Theorem (6.8) yields

$$(6.25) \qquad P_i\{Y_t = j, W_t = 0\} = \int_0^t R(i, j, ds)[1 - \varphi(j, t - s)],$$

and Theorem (6.12) gives, supposing that the mean repair times $b(j)$ are finite,

$$(6.26) \qquad \lim_{t\to\infty} P_i\{Y_t = j, W_t = 0\} = \frac{\lambda(j)b(j)}{1 + \sum_i \lambda(i)b(i)}.$$

In this example the times T_n of failures form a renewal process with the distribution of $T_{n+1} - T_n$ being $F = \varphi * \psi$, where $\varphi(t) = \sum \varphi(i, t) \lambda(i)/\lambda$ and

$\psi(t) = 1 - e^{-\lambda t}$. Because of this, the above results can also be obtained by using renewal theory. But it is easier to obtain them by the present methods.

□

7. Applications to Queueing Theory

In Examples (1.19) and (1.20) we showed the existence of Markov renewal processes embedded in the queue size processes of M/G/1 and G/M/1 queueing systems. In Chapter 6 we studied the underlying Markov chains. We now bring together these results to obtain the time-dependent behavior of the queue size processes. The methods we are using also apply to more general queueing systems.

We start with the M/G/1 queueing system. We use the notations introduced in Example (1.19) and Section 5 of Chapter 6. We denote by Y_t the number of customers present in the system at time t. $Y = \{Y_t; t \geq 0\}$ is then a semi-regenerative process with state space $E = \{0, 1, 2, \ldots\}$; its embedded Markov renewal process is (X, T), which was described in (1.19). We let R denote the Markov renewal kernel corresponding to Q in (1.19). For each $i, k \in E, t \geq 0$, define

$$(7.1) \qquad\qquad P_t(i, k) = P_i\{Y_t = k\}.$$

Using Theorem (6.8) to compute this, we obtain the following

(7.2) Theorem. For any $i, k \in E$ and $t \geq 0$,

$$(7.3) \qquad\qquad P_t(i, k) = \sum_{j=0}^{k} \int_0^t R(i, j, ds) K(t - s, j, k),$$

where

$$(7.4)$$

$$K(t, j, k) = \begin{cases} e^{-at} & j = 0, k = 0; \\[2mm] \displaystyle\int_0^t a\, e^{-a(t-s)} ds[1 - \varphi(s)] \frac{e^{-as}(as)^{k-1}}{(k-1)!} & j = 0, k > 0; \\[2mm] \dfrac{[1 - \varphi(t)]e^{-at}(at)^{k-j}}{(k-j)!} & j > 0, k \geq j; \\[2mm] 0 & \text{otherwise.} \end{cases}$$

Proof. In view of (6.8) we need only to show that $K(t, j, k) = P_j\{Y_t = k, T_1 > t\}$ is as claimed in (7.4). For $j > 0$, the first departure time T_1 is the same as the first service time; therefore, $T_1 > t$ with probability $1 - \varphi(t)$, and given that, $Y_t = k$ if and only if there are $k - j$ arrivals during $[0, t]$. This explains

(7.4) for $j > 0$. For $j = 0$ and $k = 0$, $K(t, j, k)$ is merely the probability that there are no arrivals in $[0, t]$. Suppose, finally, that $j = 0$ and $k > 0$; then in order that $Y_t = k$ and $T_1 > t$, there must be an arrival at some time $t - s$ before t, the service time starting at $t - s$ must last at least s units, and during that time interval $[t - s, t]$ there must be $k - 1$ arrivals. This explains $K(t, j, k)$ for $j = 0, k > 0$. $\square$

As before, let b be the mean service time (which we assume is finite). Then the recurrence properties of (X, T) are determined by the traffic intensity $r = ab$. It follows from Proposition (6.5.14) and Theorem (6.5.34) that the underlying Markov chain X is recurrent non-null if $r < 1$, recurrent null if $r = 1$, and transient if $r > 1$. In the recurrent non-null case $r < 1$, the limiting distribution π of the Markov chain X was computed in Theorem (6.5.20). The following theorem shows that the limiting behavior of Y is the same as that of X.

(7.5)   THEOREM.   The limits

$$\pi(j) = \lim_{t \to \infty} P_t(i, j), \quad j \in E,$$

exist and are independent of i. If $r \geq 1$, then $\pi(j) = 0$ for all $j \in E$. If $r < 1$, then the $\pi(j)$ are given by (6.5.21)–(6.5.23).

Proof. (a)   *Transient case.* Suppose $r > 1$; then (X, T) is transient, and we have from (7.3)

$$\lim_{t \to \infty} P_t(i, k) = \sum_{j=0}^{k} R(i, j, \infty) K(\infty, j, k)$$

provided that $K(\infty, j, k) = \lim_{t \to \infty} K(t, j, k)$ exist for all j, k. In fact, these limits exist and are all zero; to see this we use (7.4) and the approximations $K(t, 0, 0) = e^{-at}$, $K(t, 0, k) \leq \psi * (1 - \varphi)(t)$, and $K(t, j, k) \leq 1 - \varphi(t)$ for $j, k > 0$; here, and later,

$$\psi(t) = 1 - e^{-at}, \quad t \geq 0.$$

(b)   *Recurrent case.* Suppose $r \leq 1$; then X is irreducible recurrent. Moreover, the sojourn time in state 0 has an exponentially distributed component; this implies that $Q(0, j, t)$ is not a step function, and by Proposition (2.25), all states are aperiodic in (X, T).

From (7.4) we have $K(t, 0, 0) = e^{-at}$, $K(\cdot, 0, k) \leq \psi * (1 - \varphi)$ for any $k > 0$, and $K(\cdot, j, k) \leq 1 - \varphi$ for $j, k > 0$. It is clear that e^{-at} is directly Riemann integrable; $1 - \varphi$ is directly Riemann integrable since it is monotone and is integrable (integral equal to b), and $\psi * (1 - \varphi)$ is directly integrable by (9.2.16d). These facts show that $t \to K(t, j, k)$ is directly Riemann integrable for every j, k.

Therefore, the key renewal theorem applies to each one of the terms in (7.3), and we have

$$(7.6) \qquad \lim_{t \to \infty} P_t(j, k) = \sum_{j=0}^{k} \eta(j) \int_0^\infty K(t, j, k) \, dt,$$

where $\eta(j)$ is the mean recurrence time of j in (X, T). By Theorem (4.3) we have $\eta(j) = v(j)/\sum v(i)m(i)$, where v is some invariant measure for X. We now have $m(i) = b$ for all $i > 0$ and $m(0) = b + 1/a$. So if $r = 1$, we have $\sum v(i) \, m(i) = v(0)/a + \sum v(i) = \infty$, which implies that $\eta(j) = 0$ for all j, which in turn implies via (7.6) that $\lim P_t(j, k) = 0$ in that case. In case $r < 1$, we may choose v to be the limiting distribution π of X given in Theorem (6.5.20). Then we have, noting that $\sum \pi(i) \, m(i) = 1/a$,

$$(7.7) \qquad \eta(j) = a\pi(j), \quad j \in E.$$

To complete the proof, there remains to show that putting (7.7) into (7.6) yields $\pi(k)$. We put this computational result as a separate lemma. $\qquad \Box$

(7.8) LEMMA. Let π be the limiting distribution of X. Then for any $k \in E$,

$$a \sum_{j=0}^{k} \pi(j) \int_0^\infty K(t, j, k) \, dt = \pi(k).$$

Proof. Let the quantity on the left be denoted by $v(k)$. To show that $v(k) = \pi(k)$ for all k it is sufficient to show, by the uniqueness theorems for power series, that the two series

$$H(z) = \sum_{k=0}^{\infty} v(k)z^k, \qquad G(z) = \sum_{k=0}^{\infty} \pi(k)z^k$$

are equal for all $z \in [0, 1)$. Changing the order of summations and the integration, using (7.4), with $\lambda = a - az$ for short, we have

(7.9)

$$\begin{aligned}
(1 - z)H(z) &= \lambda \sum_{j=0}^{\infty} \pi(j) \int_0^\infty dt \sum_{k=j}^{\infty} K(t, j, k)z^k \\
&= \pi(0)\left[\int_0^\infty \lambda e^{-at} \, dt + z \int_0^\infty dt \int_0^t \lambda e^{-\lambda s}[1 - \varphi(s)]ae^{-a(t-s)} \, ds \right] \\
&\quad + \sum_{j=1}^{\infty} \pi(j)z^j \int_0^\infty \lambda e^{-\lambda t}[1 - \varphi(t)] \, dt \\
&= \pi(0)[1 - z + z(1 - F(z))] + \sum_{j=1}^{\infty} \pi(j)z^j(1 - F(z)) \\
&= G(z)[1 - F(z)] + \pi(0)(1 - z)F(z),
\end{aligned}$$

where

$$(7.10) \qquad F(z) = \int_0^\infty e^{-\lambda t} \varphi(dt), \quad \lambda = a - az.$$

On the other hand, using the system of linear equations $\pi = \pi P$, we obtain

$$(7.11) \qquad zG(z) = \sum_i \pi(i) \sum_j P(i, j) z^{j+1}$$

$$= [\pi(0)z + \pi(1)z + \pi(2)z^2 + \cdots]F(z)$$

$$= G(z)F(z) - \pi(0)(1 - z)F(z),$$

after noting that $F(z)$ defined by (7.10) is also $\sum q_j z^j$.

Solving for $\pi(0)(1 - z)F(z)$ in (7.11), putting the result in (7.9), we obtain

$$(1 - z)H(z) = G(z)[1 - F(z)] + G(z)F(z) - zG(z) = (1 - z)G(z).$$

Thus, $H(z) = G(z)$ as claimed for all $z < 1$.   □

Finally, we consider the cumulative time spent in particular states. Suppose a "penalty" cost of $f(k)$ units per unit time is applicable whenever the process is in state k. Then the total cost during $[0, t]$ has the expected value

$$(7.12) \qquad Uf(i, t) = E_i \left[\int_0^t f(Y_s) \, ds \right]$$

if the initial number of customers was i.

(7.13) THEOREM. For any $i \in E$ and $t \geq 0$

$$Uf(i, t) = \sum_k U(i, k, t) f(k),$$

where, in the notations of Theorem (7.2),

$$(7.14) \qquad U(i, k, t) = \sum_{j=0}^k \int_0^t R(i, j, ds) \int_0^{t-s} K(u, j, k) \, du.$$ □

For the limit, as $t \to \infty$, of the average cost we have the following

(7.15) THEOREM. Suppose the traffic intensity $r = ab < 1$. Then

$$\lim_{t \to \infty} E_i \left[\frac{1}{t} \int_0^t f(Y_s) \, ds \right] = \pi f$$

independent of the initial state i; here π is the limiting distribution given in (6.5.20), and f is non-negative and bounded.

Proof. The limit of interest is the limit of $Uf(i, t)/t$ as $t \to \infty$. Noting that $\sum_k U(i, k, t)/t$ is bounded by one, we see by the bounded convergence theorem that

$$(7.16) \qquad \lim_t Uf(i, t)/t = \sum_k f(k) \lim_t U(i, k, t)/t.$$

Writing $U(i, k, t)/t = [U(i, k, t)/R(i, k, t)] \cdot [R(i, k, t)/t]$, and applying (4.27) to the term $U(i, k, t)/R(i, k, t)$, we obtain, from the form of $U(i, k, t)$ given in (7.14),

$$\lim_{t \to \infty} \frac{1}{t} U(i, k, t) = \sum_{j=0}^{k} \frac{1}{\pi(k)} \pi(j) \int_0^\infty K(u, j, k)\, du \lim_{t \to \infty} \frac{1}{t} R(i, k, t)$$

$$= \sum_{j=0}^{k} \pi(j) \int_0^\infty K(u, j, k)\, du \frac{1}{\pi(k)} a\pi(k).$$

By Lemma (7.8) the last term is $\pi(k)$, and together with (7.16), this completes the proof. ☐

We next consider the dual queueing system G/M/1. We denote the embedded Markov renewal process by (X^*, T^*), and the queue size process by Y^*, that is, Y_t^* is the number of customers, at time t, in the system of Example (1.20). Then Y^* is a semi-regenerative process with (X^*, T^*) embedded in it. It is worth mentioning that $t \to Y_t$ is taken to be right continuous, so at the time T_n^* of the nth arrival we have $Y^*(T_n^*) = X_n^* + 1$. Since X_0^* is different from Y_0^*, we mention once more that, below, $P_i\{\cdot\} = P\{\cdot \mid X_0^* = i\}$.

We denote by R^* the Markov renewal kernel corresponding to Q^* described in (1.20), and let

$$(7.17) \qquad P_t^*(i, k) = P_i\{Y_t^* = k\}.$$

The proof of the following theorem parallels that of (7.2), and we will omit it.

(7.18) THEOREM. For any $i, k \in E$ and $t \ge 0$,

$$P_t^*(i, k) = \sum_{j=0}^{\infty} \int_0^t R^*(i, j, ds) K^*(t - s, j, k),$$

where

$$(7.19) \qquad K^*(t, j, k) = \begin{cases} [1 - \varphi(t)]\dfrac{e^{-at}(at)^{j+1-k}}{(j + 1 - k)!} & \text{if } k > 0,\, j \ge k - 1; \\[3mm] [1 - \varphi(t)] \displaystyle\sum_{n=j+1}^{\infty} \dfrac{e^{-at}(at)^n}{n!} & \text{if } k = 0,\, j \ge 0; \\[3mm] 0 & \text{otherwise.} \end{cases}$$

Let $b = \int (1 - \varphi(t))\, dt$; that is, b is the mean interarrival time. By Theorem (6.6.12), the underlying Markov chain X^* is transient if $r = ab < 1$ and recurrent otherwise. In the recurrent case, Theorem (6.6.5) shows that X^* is non-null if $r = ab > 1$ and null otherwise. If $r > 1$, the limiting distribution of X^* is given by

$$(7.20) \qquad \pi^*(j) = (1 - \beta)\beta^j, \quad j \in E,$$

where $\beta < 1$ is the number satisfying

$$(7.21) \qquad \beta = \sum q_j \beta^j = \int_0^\infty e^{-(a-a\beta)t} \varphi(dt).$$

Finally, when $r = 1$, the proof of (6.6.5) shows that an invariant measure is

$$(7.22) \qquad v(j) = 1, \quad j \in E.$$

Before listing the main limit theorem, we note that the process (X^*, T^*) is aperiodic if and only if the arrival process is aperiodic. Otherwise, if the arrivals can occur only at times $0, \delta, 2\delta, 3\delta, \ldots$, then (X^*, T^*) is periodic with period δ. This follows immediately from the definition of periodicity for a state in (X^*, T^*) and the fact that the T_n^* are the times of arrivals.

(7.23) THEOREM. *If* $ab < 1$, *or if* $ab = 1$ *and the variance* d^2 *of an interarrival time is finite, then*

$$\lim_{t \to \infty} P_t^*(i, k) = 0$$

for all $i, k \in E$. *If* $ab > 1$ *and the arrival process is aperiodic, then*

$$(7.24) \qquad \lim_{t \to \infty} P_t^*(i, k) = \begin{cases} 1 - \dfrac{1}{ab} & \text{if } k = 0, \\[2mm] \dfrac{1}{ab}(1 - \beta)\beta^{k-1} & \text{if } k > 0. \end{cases}$$

If $ab > 1$ *and the arrival process is periodic with period* δ, *then for any* $x \in [0, \delta)$ *we have*

$$(7.25) \qquad \lim_{n \to \infty} P_{x+n\delta}^*(i, k) = \begin{cases} 1 - c(x) & \text{if } k = 0, \\ c(x)(1 - \beta)\beta^{k-1} & \text{if } k > 0, \end{cases}$$

where

$$(7.26) \qquad c(x) = \delta(1 - \beta)e^{-(a-a\beta)x}[b - be^{-(a-a\beta)\delta}]^{-1}.$$

Proof. (a)   *Case $ab < 1$.* Let Y_t^+ denote the number of customers in the system just before the first arrival after t. Since $Y_t^* \geq Y_t^+$, we have

$$(7.27) \qquad P_i\{Y_t^* = k\} = \sum_{j=0}^{k} P_i\{Y_t^* = k, \, Y_t^+ = j\} \leq \sum_{j=0}^{k} P_i\{Y_t^+ = j\}.$$

Let S be the last time the process Y^* was in a state $j \leq k$, that is, let $S = T_N^*$, where $N = \sup\{n: X_n^* \leq k\}$. If $ab < 1$, then X^* is transient. Therefore, when $ab < 1$, N is finite with probability one, and thus $S = T_N^*$ is finite with probability one, and we have

$$(7.28) \qquad\qquad\qquad \lim_t P_i\{S \geq t\} = 0.$$

On the other hand, the last term in (7.27) is $P_i\{Y_t^+ \leq k\} \leq P_i\{S \geq t\}$, since $Y_t^+ \leq k$ implies that $S \geq t$. Hence, when $ab < 1$, (7.27) and (7.28) complete the proof.

(b) *Case $ab = 1$.* We still have (7.27) holding, and to show that $P_t^*(i, k) \rightarrow 0$ we need only show that

$$(7.29) \qquad\qquad \lim_t P_i\{Y_t^+ = j\} = 0, \quad i, j \in E.$$

The process Y^+ is semi-regenerative, and using Theorem (6.8), we have (for fixed k)

$$(7.30) \qquad\qquad P_i\{Y_t^+ = k\} = \sum_j \int_0^t R^*(i, j, ds) g(j, t - s)$$

where

$$(7.31) \qquad\qquad g(j, t) = P^*(j, k) - Q^*(j, k, t).$$

Since $ab = 1$, the invariant measure v is given by (7.22). We now show that g defined by (7.31) is in $\mathbb{D}_v$. It is clear that $g(j, \cdot)$ is right continuous and decreasing; furthermore,

$$vg(0) \leq \sum v(j)P^*(j, k) = v(k) = 1 < \infty,$$

and

$$\int vg(t)\, dt = \begin{cases} \sum_j \int_0^\infty (q_j - q_j(t))\, dt = b < \infty & \text{if } k > 0, \\ \sum_j \int_0^\infty (r_j - r_j(t))\, dt = a(b^2 + d^2) < \infty & \text{if } k = 0. \end{cases}$$

It follows from Proposition (4.15c) that g is in $\mathbb{D}_v$. We now may apply Theorem (4.17) to (7.30); this shows that (7.29) holds, since $vm = \sum v(i)b = \infty$ in the present case.

(c)   *Case $ab > 1$, T^* aperiodic.* We may now take v to be the limiting distribution π^* given in (7.20). Since $\sum \pi(i)m(i) = b$, Theorem (6.12) applies to give

(7.32) $$\lim_t P_t^*(i, k) = \frac{1}{b} \sum_j (1 - \beta)\beta^j \int K^*(t, j, k)\, dt.$$

Now, in view of (7.21),

$$\int_0^\infty (1 - \varphi(t))e^{-(a - a\beta)t}\, dt = \frac{1}{a - a\beta}\left[1 - \int e^{-(a - a\beta)t}\varphi(dt)\right] = \frac{1}{a}$$

so that

(7.33) $$\int_0^\infty dt \sum_j (1 - \beta)\beta^j K^*(t, j, k) = \begin{cases} \dfrac{(1 - \beta)\beta^{k-1}}{a} & \text{if } k > 0, \\[2mm] \dfrac{1}{a} & \text{if } k = 0. \end{cases}$$

Putting (7.33) in (7.32) yields the desired result.

(d)   *Case $ab > 1$, T^* periodic,* follows similarly by a direct application of Theorem (6.12) (except that the computations involved are somewhat longer).

□

A special case of some interest is the one where the arrivals are scheduled to occur at times $1, 2, 3, \ldots$; that is, $\varphi(t) = 0$ for $t < 1$ and $\varphi(t) = 1$ for $t \geq 1$. If $a > 1$, then we have $ab = a > 1$, and the limit result (7.25) is applicable. Now, $\beta = e^{-(a - a\beta)}$, $\delta = 1$, $c(x) = \beta^x$.

8. Exercises

(8.1)   Let (X, T) be a Markov renewal process with state space $E = \{a, b\}$ and semi-Markov kernel Q given as

$$Q(t) = \begin{bmatrix} 0.6(1 - e^{-5t}) & 0.4 - 0.4e^{-2t} \\ 0.5 - 0.2e^{-3t} - 0.3e^{-5t} & 0.5 - 0.5e^{-2t} - te^{-2t} \end{bmatrix}.$$

(a)   Compute the transition matrix P for the Markov chain X.
(b)   Compute the conditional distributions $G(i, j, \cdot)$ of the sojourn time in state i given that the next state is j for all i, j.

(8.2)   Let (X, T) be as in (8.1). Supposing that $X_0 = a$, $X_1 = a$, $X_2 = b$, $X_3 = b$, compute the conditional distributions
(a) $P\{T_1 \leq x, T_2 - T_1 \leq y, T_3 - T_2 \leq z \,|\, X_0, X_1, X_2, X_3\}$,
(b) $P\{T_3 \leq t \,|\, X_0, X_1, X_2, X_3\}$.

(8.3) In Example (1.17) of traffic flow, suppose $T_1, T_2, \ldots$ are the positions of successive vehicles along a highway at a fixed instant. There are two types of vehicles: cars and trucks, which will be shortened to *cr* and *tr* respectively. Suppose the successive types form a Markov chain X with transition probabilities $P(cr, cr) = 0.8$ and $P(tr, tr) = 0.1$; and suppose the headways, measured in feet, between different types have the distributions

$$G(cr, cr, x) = 1 - e^{-(x-50)/20}, \quad x \geq 50;$$

$$G(cr, tr, x) = 1 - e^{-(x-20)/10}, \quad x \geq 20;$$

$$G(tr, cr, x) = 1 - e^{-(x-100)/30}, \quad x \geq 100;$$

$$G(tr, tr, x) = 1 - e^{-(x-80)/20}, \quad x \geq 80.$$

(a)   Compute the semi-Markov kernel Q for (X, T).

(b)   Compute the distribution $F(tr, tr, \cdot)$ of the distance between two trucks.

(c)   Find an expression for the expected number of trucks in the interval $[0, x]$ of the highway.

(d)   Find the ratio of the number of trucks to the number of vehicles on the highway.

(8.4) A machine has two components a and b whose lifetimes are exponential with respective parameters 0.01 and 0.04. The machine fails if either component fails, and the repair times are random with respective distributions σ and ψ for a and b. Let Y_t be a, b, or c according as, at time t, the component a is being repaired, the component b is being repaired, or the machine is working.

(a)   Show that Y is a semi-Markov process with the semi-Markov kernel

$$Q = \begin{bmatrix} 0 & 0 & \sigma \\ 0 & 0 & \psi \\ 0.2\varphi & 0.8\varphi & 0 \end{bmatrix},$$

where $\varphi(t) = 1 - e^{-0.05t}$.

(b)   Compute $P_c\{Y_t = j\}$ for all j.

(8.5) *Clustered renewal processes.* This provides a model for situations where a failure might go undetected for some time, thereby causing secondary failures. Suppose the lifetime of an item has the distribution φ. When it fails, it may cause secondary failures with distribution ψ for the times between. The probability of detection at a time of failure is p; and once detected, the failed item is replaced by an identical one. Let X_n be 1 or 2 according as the nth failure is a primary or secondary failure, and let T_n be the time of the nth failure.

(a)   Show that (X, T) is a Markov renewal process with semi-Markov kernel

$$Q = \begin{bmatrix} p\varphi & q\psi \\ p\varphi & q\psi \end{bmatrix}.$$

(b)   Find the distribution $F(1, 1, \cdot)$ of the time between two primary failures.

(c)   Find the distribution of the number of secondary failures between two primary ones.

(8.6) *Continuation.* Suppose each failure costs one dollar and the replacement cost for the item is eight dollars. Find the limit, as $t \to \infty$, of the average expected cost for the interval $[0, t]$; in other words, find

$$\lim_{t \to \infty} \frac{1}{t} \sum_{j=1}^{2} R(i, j, t)[1 + 8P(j, 1)].$$

(8.7) Consider the machine in Exercise (8.4). Suppose repairing a costs 18 dollars per unit time, repairing b costs 4 dollars per unit time, and the machine earns 10 dollars per unit time when it is working. Compute the limit, as $t \to \infty$, of the average net earnings (here $g = (-18, -4, 10)$)

$$\frac{1}{t} \int_{0}^{t} g(Y_s)\, ds,$$

assuming that the mean repair times are $m(a) = 2$ and $m(b) = 1$ for a and b.

(8.8) *Counters of type* I. Particles arriving at a counter can be separated into N types (either by their physical natures or by their energy levels or by some other characteristic). Let X_n and T_n be the type and time of arrival, respectively, of the nth particle. Suppose (X, T) is a Markov renewal process with a semi-Markov kernel Q (and state space $E = \{1, 2, \ldots, N\}$). An arriving particle which finds the counter free gets registered and locks it for some random time with distribution $\psi(i, \cdot)$ if its type is i. Arrivals during a locked period have no effects whatsoever.

    (a) Let $\hat{T}_0 = 0$, $\hat{T}_1$, $\hat{T}_2, \ldots$ be the times of successive registrations and $\hat{X}_0, \hat{X}_1, \hat{X}_2, \ldots$ the types of the particles registered. Show that $(\hat{X}, \hat{T})$ is a Markov renewal process.

    (b) Show that the semi-Markovian kernel $\hat{Q}$ corresponding to $(\hat{X}, \hat{T})$ is

$$\hat{Q}(i, k, t) = \int_{0}^{t} \psi(i, ds) \sum_{j \in E} \int_{0}^{s} R(i, j, du)[Q(j, k, t - u) - Q(j, k, s - u)].$$

    (c) Show that when X is irreducible recurrent with invariant measure v, $\hat{X}$ is irreducible recurrent with an invariant measure $\hat{v}$ satisfying

$$v(j) = \sum_{i} \hat{v}(i) \int_{0}^{\infty} \psi(i, ds) R(i, j, s).$$

(8.9) *Continuation.* Let $P(t, i, j)$ be the probability, starting with a type i registration, that the counter is locked at time t with a type j registration as the last one.

    (a) Compute $P(t, i, j)$.

    (b) Compute $\lim_{t} P(t, i, j)$.

(8.10) *Continuation.* Compute the ratio of the number of registrations to the number of arrivals for the type j particles in the long run.

(8.11) *Counters of type* II. An arriving particle which finds the counter free gets registered and locks the counter for some random length of time. A particle which finds the counter locked does not get registered, but erases the influence of all the previous particles, and locks the counter for some random time. Let X_n be the type and T_n the time of arrival for the nth particle. Suppose (X, T) is a Markov renewal process with a semi-Markov kernel Q with state space $E = \{1, \ldots, N\}$, $N < \infty$; and let $\psi(i, \cdot)$ be the distribution of the locked period effected by a type i particle.

(a) Let Y_t be the type of the last particle arriving before t, and let W_t be 1 or 0 according as the counter is locked or free at time t. Show that

$$f(i, t) = P_i\{Y_t = k, W_t = 1\}$$

satisfies a Markov renewal equation

$$f = g + Q * f$$

with

$$g(i, t) = I(i, k)[1 - \sum_j Q(i, j, t)][1 - \psi(i, t)].$$

(b) Compute $P_i\{Y_t = k, W_t = 1\}$.

(c) Supposing that X is irreducible recurrent with limiting distribution π, compute the limit as $t \longrightarrow \infty$ of the average time spent locked during $[0, t]$, that is, compute

$$\lim_{t \to \infty} \frac{1}{t} \int_0^t W_s \, ds.$$

(8.12) *Continuation.* Let $\hat{T}_0 = 0$, $\hat{T}_1$, $\hat{T}_2$, ... be the times of successive registrations and $\hat{X}_0$, $\hat{X}_1$, $\hat{X}_2$, ... the corresponding particle types.

(a) Show that $(\hat{X}, \hat{T})$ is a Markov renewal process.

(b) Let $\hat{Q}$ be the semi-Markov kernel corresponding to $(\hat{X}, \hat{T})$. Show that for fixed k, $f(i, t) = \hat{Q}(i, k, t)$ satisfies a "Markov renewal equation"

$$f(i, t) = g(i, t) + \sum_j \int_0^t \bar{Q}(i, j, ds) f(j, t - s),$$

where $\bar{Q}(i, j, ds) = Q(i, j, ds)(1 - \psi(i, s))$.

(c) Let $\bar{Q}^n$ be defined by (2.3) and (2.4) by replacing Q in those expressions by $\bar{Q}$, and let $\bar{R} = \sum_n \bar{Q}^n$. Show that the solution for $\hat{Q}$ is

$$\hat{Q}(i, k, t) = \sum_j \int_0^t \bar{R}(i, j, ds)[Q(j, k, t - s) - \bar{Q}(j, k, t - s)].$$

(d) Find the limit of the ratio $\hat{R}(i, j, t)/R(i, j, t)$ of the expected number of registrations to the expected number of arrivals, during the interval $[0, t]$ starting with a registration of type i, for the particles of type j.

(8.13) *Generalized Markov renewal equations.* A function $f \in \mathbb{B}$ is said to satisfy a general Markov renewal equation if $f = g + Q * f$ for some function $g \in \mathbb{B}$ and generalized semi-Markov kernel Q: each $Q(i, j, \cdot)$ is a mass function as before, but $\sum_j Q(i, j, +\infty) \leq 1$ (instead of being equal to one). Suppose the state space E is finite, and let $R = \sum Q^n$ as before, with the Q^n defined by (2.3) and (2.4).

(a)   Show that $f = R * g$ is the unique solution of $f = g + Q * f, f, g \in \mathbb{B}$.

(b)   Show that each $R(i, j, \cdot)$ is still a renewal function (possibly delayed).

(c)   Show that if Q is irreducible and $\sum_j Q(i, j, +\infty) < 1$ for some i, then $R(i, j, +\infty) < \infty$ for all i, j.

(8.14)   *Pedestrian delay problem.* Consider the traffic flow model of Exercise (8.3). Suppose the speeds of the vehicles are almost all equal, and by choosing the time unit properly, we take this speed to be one foot per unit of time. Suppose a pedestrian arrives at the highway, to cross it, just as a car is passing by. To cross it, he accepts a gap of x units with probability $\varphi(x)$ or $\psi(x)$ according as the approaching vehicle is a car or a truck. He evaluates the gaps independent of his previous decisions, and he does not become impatient.

(a)   Find the distribution of the delay W before he is able to start crossing.

(b)   Compute the expected value of W.

(8.15)   *General semi-Markov processes.* Let Y be a semi-Markov process in the sense of Example (6.6). Let T be the time of its first jump, define $Q(i, j, t) = P_i\{Y_T = j, T \leq t\}$, and let $\hat{R}(i, j, t)$ be the expected number of visits to j by Y during $[0, t]$ given the initial state i.

(a)   Show that $\hat{R} = I + Q * \hat{R}$.

(b)   Show that $\hat{R} = R \equiv \sum Q^n$ if and only if

$$\lim_n Q^n * 1 = 0.$$

(8.16)   *Continuation.* Let $D \subset E$ be a finite set; let $\hat{T}_0 = 0, \hat{T}_1, \hat{T}_2, \ldots$ be the successive times Y visits D, and let $\hat{X}_n = Y_{T_n}$.

(a)   Show that $(\hat{X}, \hat{T})$ is a Markov renewal process embedded in the semi-regenerative process Y.

(b)   If $\hat{Q}$ is the semi-Markovian kernel corresponding to $(\hat{X}, \hat{T})$, and if $\hat{R}$ is as defined in (8.15), then show that $\hat{R}(i, j, t) = \sum \hat{Q}^n(i, j, t)$ for all $i, j \in D$.

(c)   Show that for $i, j \in D$,

$$P_i\{Y_t = j\} = \int_0^t \hat{R}(i, j, ds)h(j, t - s),$$

where

$$h(j, t) = 1 - \sum_k Q(j, k, t) = P_j\{T > t\}.$$

(8.17)   *Continuation.* Let U be as defined by (5.8), but for the present process Y.

(a)   Show that Proposition (5.9) holds with R there replaced by $\hat{R}$. Similarly for (5.12), (5.14), and (5.17).

(b)   Show that (5.19), (5.20), (5.21), and (5.22) all hold provided that $i, j, h, k \in D$, that R there be replaced by $\hat{R}$, that $v(j)/v(k)$ there be replaced by the expected number of visits to j by $\hat{X}$ between two visits to k, and that $vm/v(j)$ there be replaced by the expected value of the time between two visits to j.

(8.18)   *Virtual waiting time in G/M/1 queue.* Let V_t be the amount of time the server needs to serve all the customers who are in the system at time t; then V_t is called the backlog at t, or the virtual waiting time at t.

(a) Show that

$$P_i\{V_t \le x | Y_t^* = k\} = \sum_{n=k}^{\infty} \frac{e^{-ax}(ax)^n}{n!}, \quad x \ge 0.$$

(b) Show that when $ab \le 1$, $\lim_t P_i\{V_t \le x\} = 0$ for all $x < \infty$.

(c) Show, when $ab > 1$ and the arrivals are aperiodic, that

$$\lim_t P_i\{V_t \le x\} = 1 - \frac{1}{ab}e^{-(a-a\beta)x}.$$

(8.19) M/G/1 *queue with finite capacity.* Consider the queueing system described in Exercise (6.8.23). Give a treatment of this system in continuous time by using the methods of Section 7 for the corresponding infinite capacity system.

(8.20) G/M/1 *queue with finite capacity.* Consider the system described in Exercise (6.8.24). Give a treatment, paralleling that of Section 7, of the time-dependent behavior of this system.

To those readers who think this book could have done just as well without this last chapter, I should point out the special reason for its inclusion: This chapter is based on ÇINLAR [1].

Semi-Markov processes were introduced by LÉVY [1] *and* SMITH [1] *at the same time, independent of each other. The term* Markov renewal process *is due to* PYKE [1], *who gave an extensive treatment of many aspects of such processes in a series of papers. Our treatment of Sections 3 and 4 parallels* FELLER [4], [5], *and* [6]. *For generalizations in several directions and further references to recent research see* ÇINLAR [2], [3].

Afterword

The two glaring omissions alluded to in the foreword are the theories of Brownian motion and martingales. In an effort to keep the mathematical level low, we had to work without the tools of measure theory and *real* real analysis. In the case of martingales and Brownian motion, this restriction would allow only a shallow, superficial treatment. That being at variance with our tastes, it seemed best to omit them altogether. The interested reader will find a good introduction to martingales in NEVEU [1]. LAMPERTI [1] gives a lucid introduction to Brownian motion; and DYNKIN and YUSHKEVICH [1] discuss some related problems of modern interest. The definitive work on Brownian motion and other Markov processes with continuous paths is the book by ITÔ and MCKEAN [1].

This should satisfy most readers. To the remaining insistent few, I own myself a debtor for those two items. I shall pay off this debt in due time, along with my other book-debts, which, in addition to a chapter on martingales, include one on the horses wearing them, another on the people who play the horses, and their chamber-maids, green-gowns, and old hats. For now, this is the end of our

INTRODUCTION TO STOCHASTIC PROCESSES

Non-Negative Matrices

Throughout the following all matrices are finite-dimensional. They are all square unless otherwise specified. We shall avoid declaring their dimensions unless it is necessary. In particular, in writing AB, it will be understood that the number of columns of A is equal to the number of rows of B, so that their product makes sense. Matrices will be denoted by capital letters such as A, B, C; column vectors by small letters such as f, g, h; and row vectors by small Greek letters such as π, ν. The identity matrix is denoted by I, the column vector of ones by 1, of zeros by 0. Our purpose here is to supplement the computational aspects of the theory of Markov chains with finitely many states.

1. Eigenvalues and Eigenvectors

We assume that the reader is either familiar with most of the following summary or willing to accept the stated results without proofs.

Given a matrix A, the determinant

$$(1.1) \qquad \varphi(x) = det(xI - A)$$

defines a polynomial φ on the field of complex numbers. Then φ is called the *characteristic polynomial* of A. If A is $n \times n$, then φ is a polynomial of degree n, and the *characteristic equation* $\varphi(x) = 0$ has exactly n roots $\lambda_1, \ldots, \lambda_n$. The complex numbers $\lambda_1, \ldots, \lambda_n$ so obtained are called the *eigenvalues* of A. In other words, λ is an eigenvalue of A if and only if $\varphi(\lambda) = det(\lambda I - A) = 0$. An eigenvalue λ is said to be *simple* if λ appears only once among the

$\lambda_1, \ldots, \lambda_n$. If λ appears $m \geq 2$ times among $\lambda_1, \ldots, \lambda_n$, then the eigenvalue λ is said to have multiplicity m.

If λ is an eigenvalue of A, then there exists at least one column vector $f \neq 0$ satisfying

$$(1.2) \qquad\qquad\qquad Af = \lambda f;$$

then f is called an *eigenvector* of A corresponding to the eigenvalue λ. Conversely, if (1.2) holds for some scalar λ and vector $f \neq 0$, then λ is an eigenvalue of A and f is an eigenvector corresponding to it.

If f is an eigenvector, so is cf for any constant $c \neq 0$. Up to this multiplication by a constant, there exists exactly one eigenvector corresponding to λ when λ is simple. When λ has multiplicity m, it has at most m linearly independent eigenvectors corresponding to it.

Since the determinant of a matrix is the same as that of its transpose, A and its transpose A^T have the same characteristic polynomial and, therefore, the same eigenvalues. If g is an eigenvector of A^T corresponding to λ, then its transpose $\pi = g^T$ satisfies

$$(1.3) \qquad\qquad\qquad \pi A = \lambda \pi.$$

Then the row vector π is called a row eigenvector of A corresponding to λ. Since π^T is an eigenvector of A^T, everything that is said about eigenvectors is true also for row eigenvectors.

If f is an eigenvector corresponding to an eigenvalue λ, then f is linearly independent of all the eigenvectors corresponding to eigenvalues distinct from λ. Thus, if A is $n \times n$ and the eigenvalues $\lambda_1, \ldots, \lambda_n$ are all distinct, then the eigenvectors $f_1, \ldots, f_n$ corresponding to $\lambda_1, \ldots, \lambda_n$ form a linearly independent set of vectors. Therefore, the matrix $N = [f_1 \ldots f_n]$ whose jth column is f_j is a non-singular matrix and has an inverse N^{-1}. Noting that $Af_j = \lambda_j f_j$, we can write $AN = ND$ by letting D to be the diagonal matrix whose diagonal entries are $\lambda_1, \ldots, \lambda_n$. This shows that we have

$$(1.4) \qquad\qquad\qquad A = NDN^{-1}.$$

In the converse direction, suppose there is a diagonal matrix D and a non-singular matrix N such that (1.4) holds. Then each diagonal entry of D is an eigenvalue of A, the jth column of N is an eigenvector corresponding to the eigenvalue appearing as the jth diagonal entry of D, and the ith row of N^{-1} is a row eigenvector corresponding to the eigenvalue appearing as the ith diagonal entry of D.

A matrix A for which the representation (1.4) is possible with D diagonal is said to be *diagonalizable*. If all the eigenvalues of A are distinct, then A is diagonalizable. But the distinctness of eigenvalues is not necessary for A

to be diagonalizable; all that is needed is the existence, corresponding to each eigenvalue which has multiplicity greater than one, of as many linearly independent eigenvectors as the multiplicity of that eigenvalue.

The form (1.4) is also useful in computations of the powers of a given matrix, etc. If $A = NDN^{-1}$, then $A^2 = NDN^{-1}NDN^{-1} = ND^2N^{-1}$, and by induction

$$(1.5) \qquad\qquad A^k = ND^kN^{-1}, \quad k = 0, 1, \ldots .$$

Since D^k is just the diagonal matrix whose diagonal entries are the kth powers of the eigenvalues $\lambda_1, \ldots, \lambda_n$, it is easy to compute. Moreover, (1.5) can be used to compute

$$(1.6) \qquad\qquad e^{tA} = \sum_{k=0}^{\infty} \frac{t^k}{k!} A^k = Ne^{tD}N^{-1},$$

where

$$(1.7) \qquad\qquad e^{tD} = \begin{bmatrix} e^{\lambda_1 t} & & & 0 \\ & \cdot & & \\ & & \cdot & \\ & & & \cdot \\ 0 & & & e^{\lambda_n t} \end{bmatrix}.$$

In computing eigenvalues there is no need to take any determinants (which is computationally clumsy). The easiest relation to use is the following: *the sum of the eigenvalues of A is equal to the sum of the diagonal entries of A.* The latter sum is called the *trace* of A, and what was just said is that

$$(1.8) \qquad\qquad tr(A) = \sum_i \lambda_i$$

where $\lambda_1, \ldots, \lambda_n$ are the eigenvalues of A. More generally, since the *eigenvalues of A^k are the kth powers of the eigenvalues of A,*

$$(1.9) \qquad\qquad tr(A^k) = \sum_i \lambda_i^k, \quad k = 0, 1, \ldots .$$

If A is $n \times n$, there are n eigenvalues to compute; the n equations needed can be obtained from (1.9) for $k = 1, \ldots, n$.

To see why these relations hold we first note that whether A is diagonalizable or not, there is a non-singular matrix N and a *lower-triangular* (all entries above the diagonal are zero) matrix D such that $A = NDN^{-1}$. Since $\varphi(x) = det(xI - A) = det(N(xI - D)N^{-1}) = det(N)det(xI - D)det(N^{-1})$ $= det(xI - D)$, the eigenvalues of A are the same as those of D. But since D is lower-triangular, $det(xI - D) = (x - \lambda_1)\cdots(x - \lambda_n)$ where $\lambda_1, \ldots, \lambda_n$

are the diagonal entries of D. Hence, the eigenvalues of A are the diagonal entries of D. Next, note that $A^k = NDN^{-1}NDN^{-1} \cdots NDN^{-1} = ND^kN^{-1}$, and the matrix D^k is lower-triangular since D is. Hence, the eigenvalues of A^k are the diagonal entries of D^k. But the diagonal entries of D^k are the kth powers of the corresponding diagonal entries of D, and the latter are the eigenvalues of A. Thus, the eigenvalues of A^k are $\lambda_1^k, \ldots, \lambda_n^k$ if the eigenvalues of A are $\lambda_1, \ldots, \lambda_n$. This shows that (1.9) holds if (1.8) holds. To show (1.8), we first observe that $tr(AB) = tr(BA)$:

$$tr(AB) = \sum_i (AB)(i, i) = \sum_i \sum_j A(i, j)B(j, i)$$
$$= \sum_j \sum_i B(j, i)A(i, j) = \sum_j (BA)(j, j) = tr(BA).$$

We apply this to $A = NDN^{-1}$: $tr(A) = tr(N(DN^{-1})) = tr(DN^{-1}N) = tr(D)$. Since the diagonal entries of D are the eigenvalues of A, the trace of D is just the sum of them. Hence (1.8) holds.

2. Spectral Representations

We limit ourselves to diagonalizable matrices. Accordingly, suppose we can write $A = NDN^{-1}$, where D is diagonal. Let the diagonal entries of D be $\lambda_1, \ldots, \lambda_n$; let the *columns* of N be $f_1, \ldots, f_n$; and let the *rows* of N^{-1} be $\pi_1, \ldots, \pi_n$; that is,

$$(2.1) \quad N = \begin{bmatrix} f_1(1) & \cdots & f_n(1) \\ \cdot & \cdots & \cdot \\ \cdot & \cdots & \cdot \\ \cdot & \cdots & \cdot \\ f_1(n) & \cdots & f_n(n) \end{bmatrix}, \quad N^{-1} = \begin{bmatrix} \pi_1(1) & \cdots & \pi_1(n) \\ \cdot & \cdots & \cdot \\ \cdot & \cdots & \cdot \\ \cdot & \cdots & \cdot \\ \pi_n(1) & \cdots & \pi_n(n) \end{bmatrix},$$

$$D = \begin{bmatrix} \lambda_1 & & 0 \\ & \cdot & \\ & & \cdot \\ 0 & & \lambda_n \end{bmatrix}.$$

Since $N^{-1}N = I$, we have, for the scalar $\pi_j f_k$,

$$(2.2) \qquad \pi_j f_k = \begin{cases} 0 & \text{if } j \neq k, \\ 1 & \text{if } j = k. \end{cases}$$

We next define B_k to be the matrix obtained by multiplying the column vector f_k (looked upon as an $n \times 1$ matrix) with the row vector π_k (looked upon as a $1 \times n$ matrix); that is, $B_k = f_k \pi_k$, or more explicitly,

$$(2.3) \qquad B_k = \begin{bmatrix} f_k(1)\pi_k(1) & \cdots & f_k(1)\pi_k(n) \\ \vdots & \ddots & \vdots \\ f_k(n)\pi_k(1) & \cdots & f_k(n)\pi_k(n) \end{bmatrix}.$$

It follows from (2.2) that

$$(2.4) \qquad B_j B_k = f_j \pi_j f_k \pi_k = \begin{cases} 0 & \text{if } j \neq k, \\ B_j & \text{if } j = k. \end{cases}$$

Finally, the representation $A = NDN^{-1}$ with (2.1) means that

$$\begin{aligned} A(i,j) &= \sum_k \sum_{k'} N(i,k)D(k,k')N^{-1}(k',j) \\ &= \sum_k N(i,k)D(k,k)N^{-1}(k,j) \\ &= \sum_k f_k(i)\lambda_k \pi_k(j) = \sum_k \lambda_k B_k(i,j); \end{aligned}$$

in other words,

$$(2.5) \qquad A = \lambda_1 B_1 + \cdots + \lambda_n B_n.$$

This representation of A in terms of the eigenvalues λ_k and the matrices B_k defined by (2.3) is called the *spectral representation* of A. In addition to displaying all the eigenvalues and eigenvectors in an explicit fashion, the representation (2.5) can also facilitate certain computations. For example, it follows from (2.4) that for the kth power of A we have

$$(2.6) \qquad A^k = \lambda_1^k B_1 + \cdots + \lambda_n^k B_n;$$

and (2.6) in turn yields

$$(2.7) \qquad e^{tA} = \sum_{k=0}^{\infty} \frac{t^k}{k!} A^k = e^{t\lambda_1} B_1 + \cdots + e^{t\lambda_n} B_n.$$

The expressions (2.6) and (2.7), in the case of Markov matrices or generators, provide the means for computing the k-step transition probabilities and the transition functions. Furthermore, they may be used to obtain the limits of A^k, as $k \to \infty$, and of e^{tA}, as $t \to \infty$, along with estimates of rates of convergence and bounds for the errors involved in certain approximations. We will return to these topics after the next section; in the meantime, we give the following examples to illustrate the concepts mentioned.

(2.8) EXAMPLE. Let

$$P = \begin{bmatrix} 0.8 & 0.2 \\ 0.3 & 0.7 \end{bmatrix}.$$

It follows from the observation $P1 = 1$ that $\lambda_1 = 1$ is an eigenvalue and $f_1 = 1$ is an eigenvector corresponding to it. Since the trace of P is $0.8 + 0.7 = 1.5$, and since the trace is equal to the sum of the eigenvalues, the other eigenvalue must be $\lambda_2 = 0.5$. The row eigenvector π_1 corresponding to $\lambda_1 = 1$ and satisfying $\pi_1 f_1 = 1$ is $\pi_1 = (0.6, 0.4)$. Hence, $B_1 = f_1 \pi_1$ is

$$B_1 = \begin{bmatrix} 0.6 & 0.4 \\ 0.6 & 0.4 \end{bmatrix};$$

and since (2.6) gives $P^0 = I = B_1 + B_2$ for $k = 0$, we have

$$B_2 = \begin{bmatrix} 0.4 & -0.4 \\ -0.6 & 0.6 \end{bmatrix}.$$

The spectral representation for P^k becomes

$$P^k = \begin{bmatrix} 0.6 & 0.4 \\ 0.6 & 0.4 \end{bmatrix} + (0.5)^k \begin{bmatrix} 0.4 & -0.4 \\ -0.6 & 0.6 \end{bmatrix}, \quad k = 0, 1 \ldots.$$

In the limit, as $k \rightarrow \infty$, $(0.5)^k$ approaches zero, and we get

$$P^\infty = \lim_k P^k = \begin{bmatrix} 0.6 & 0.4 \\ 0.6 & 0.4 \end{bmatrix}.$$

To obtain an estimate of the speed of convergence, note that

$$|P^k(i, j) - P^\infty(i, j)| \le (0.6)(0.5)^k, \quad k = 0, 1, \ldots$$

for all i, j. This shows that P^k converges to P^∞ geometrically fast, and even for moderate values of k, say $k = 10$, P^∞ approximates P^k up to the third decimal place. $\qquad \square$

(2.9) EXAMPLE. Consider the matrix A which is the generator of a Markov process with three states; let

$$A = \begin{bmatrix} -2 & 2 & 0 \\ 1 & -2 & 1 \\ 0 & 6 & -6 \end{bmatrix}.$$

One eigenvalue is $\lambda_1 = 0$, corresponding to which we have an eigenvector $f_1 = 1$ and a row eigenvector $\pi_1 = (0.3, 0.6, 0.1)$. The trace of A is $-2 - 2 - 6 = -10$, and the trace of A^2 is $(4 + 2) + (2 + 4 + 6) + (6 + 36) = 6 + 12 + 42 = 60$. Since $\lambda_1 = 0$, these mean that $\lambda_2 + \lambda_3 = -10$, $\lambda_2^2 + \lambda_3^2 = 60$. Putting $\lambda_1 = x$ and $\lambda_2 = y$, we see that $y = -10 - x$ and $x^2 + 100 + 20x + x^2 = 60$; that is, $x^2 + 10x + 20 = 0$. Solving it, we obtain $\lambda_2 = -5 + \sqrt{5}, \lambda_3 = -5 - \sqrt{5}$. An eigenvector corresponding to λ_2 is $f_2 = (3 + \sqrt{5}, -2, 3 - 3\sqrt{5})^T$, and a row eigenvector for the same λ_2 is $\pi_2 = \frac{1}{40}(3 + \sqrt{5}, -4, 1 - \sqrt{5})$, which is normalized to have $\pi_2 f_2 = 1$. It follows that

$$
B_1 = \begin{bmatrix} 0.3 & 0.6 & 0.1 \\ 0.3 & 0.6 & 0.1 \\ 0.3 & 0.6 & 0.1 \end{bmatrix},
$$

$$
B_2 = \frac{1}{20} \begin{bmatrix} 7 + 3\sqrt{5} & -6 - 2\sqrt{5} & -1 - \sqrt{5} \\ -3 - \sqrt{5} & 4 & -1 + \sqrt{5} \\ -3 - 3\sqrt{5} & -6 + 6\sqrt{5} & 9 - 3\sqrt{5} \end{bmatrix},
$$

and since $B_1 + B_2 + B_3 = A^0 = I$,

$$
B_3 = \frac{1}{20} \begin{bmatrix} 7 - 3\sqrt{5} & -6 + 2\sqrt{5} & -1 + \sqrt{5} \\ -3 + \sqrt{5} & 4 & -1 - \sqrt{5} \\ -3 + 3\sqrt{5} & -6 - 6\sqrt{5} & 9 + 3\sqrt{5} \end{bmatrix}.
$$

Hence,

$$
P_t = e^{tA} = B_1 + e^{-(5 - \sqrt{5})t}B_2 + e^{-(5 + \sqrt{5})t}B_3
$$

with B_1, B_2, and B_3 as computed above. Taking limits, as $t \to \infty$, we obtain

$$
\lim_t P_t = P_\infty = B_1,
$$

and we have, for any t,

$$
|P_t(i, j) - P_\infty(i, j)| \le e^{-(5 - \sqrt{5})t}(|B_2(i, j)| + |B_3(i, j)|)
$$

$$
\le \frac{12\sqrt{5}}{20} e^{-(5 - \sqrt{5})t}
$$

$$
\le 1.35 e^{-2.7t}.
$$

In other words, P_t converges to P_∞ exponentially fast.   □

3. Positive Matrices

The following theorem, which was proved by Perron in 1907, is the basic result on non-negative matrices. Its applications to non-negative matrices will be found in the next section.

(3.1) THEOREM. Let P be a strictly positive matrix, that is, $P(i,j) > 0$ for all i, j. Then P has an eigenvalue α which is real, positive, simple, and strictly greater than the absolute value of any other eigenvalue. To this maximal eigenvalue there corresponds a strictly positive eigenvector.

(3.2) REMARK. Define

$$a = \inf_i P1(i), \quad b = \sup_i P1(i);$$

that is, a and b are the smallest and the largest row sums. If $a = b$, then the maximal eigenvalue $\alpha = a$; otherwise, if $a < b$, then $a < \alpha < b$. If P is a Markov matrix, then $P1 = 1$, which shows that $\alpha = 1$ is the maximal eigenvalue. $\qquad\square$

Proof of (3.1). Let S be the set of all $\lambda \geq 0$ such that $\lambda f \leq Pf$ for some vector f with $0 \leq f \leq 1$. Define b as in (3.2). Consider an element λ of S; for some f we have $\lambda f \leq Pf$; this implies that if $c = \sup_j f(j)$,

$$(3.3) \qquad \lambda f(i) \leq \sum_j P(i,j)f(j) \leq c\sum_j P(i,j) \leq cb,$$

which implies (upon choosing i so that $f(i) = c$) that $\lambda \leq b$. Hence the set S is a subset of $[0, b]$ and, being bounded, must have its supremum α in $[0, b]$. We will show that

$$(3.4) \qquad\qquad\qquad \alpha = \sup S$$

is the desired eigenvalue.

Choose $\alpha_n \in S$ such that $\lim \alpha_n = \alpha$. For each α_n, there is $0 \leq f_n \leq 1$ such that $\alpha_n f_n \leq Pf_n$. Since $0 \leq f_n \leq 1$ for all n, we may choose a subsequence which converges to some vector f, $0 \leq f \leq 1$. Then we have $\alpha f \leq Pf$. In fact, we must have $\alpha f = Pf$. For otherwise, $Pf - \alpha f \geq 0$ and $Pf - \alpha f \neq 0$ imply $P(Pf - \alpha f) > 0$ by the strict positivity of P. This means that, for $\hat{f} = Pf$, we have $\alpha\hat{f} < P\hat{f}$, which implies that we can pick $\varepsilon > 0$ small enough to have $(\alpha + \varepsilon)\hat{f} \leq P\hat{f}$. Thus $\alpha + \varepsilon \in S$, which contradicts the definition (3.4) of α. Hence, $\alpha f = Pf$; α is an eigenvalue; f is an eigenvector; and $f \geq 0$ implies $Pf > 0$, which implies $f > 0$ and $\alpha > 0$.

Let λ be an eigenvalue of P; let $\lambda g = Pg$. Then, taking absolute values on both sides, we get $|\lambda| h \leq Ph$ for $h(i) = |g(i)|$. This implies that $|\lambda| \in S$ and hence $|\lambda| \leq \alpha$. In fact, if $|\lambda| = \alpha$, then $\lambda = \alpha$. For if $|\lambda| = \alpha$, then $\alpha h \leq Ph$ implies $\alpha h = Ph$, which together with $\lambda g = Pg$ implies that $g = ch$ for some constant c. Then, we must have $\lambda h = Ph$, which shows that λ is real and therefore $\lambda = |\lambda| = \alpha$.

Let g be an eigenvector corresponding to α; define $h = |g|$. Then $\alpha g = Pg$ implies $\alpha h \leq Ph$, which implies $\alpha h = Ph$, and therefore $h > 0$. So no eigenvector g corresponding to α can have any zero entries. This implies that to the eigenvalue α there corresponds only one eigenvector (up to multiplication by a constant). Otherwise, if there were two linearly independent eigenvectors f and g corresponding to α, then we could choose the numbers c and d such that the eigenvector $cf + dg$ has at least one zero entry.

Finally, to show that α is a simple root of $\varphi(x) = 0$, we will show that the derivative $\varphi'(\alpha)$ of φ at α is non-zero. We have $\varphi'(\alpha) = \sum_j R(j,j)$ with $R = R_\alpha$, where R_x is the adjoint matrix of $(xI - P)$; that is, R_x satisfies

$$R_x(xI - P) = (xI - P)R_x = \varphi(x)I.$$

Since $\varphi(\alpha) = 0$, $R = R_\alpha$ satisfies $\alpha R = PR$. Hence, every column of R is an eigenvector corresponding to α, and by the uniqueness of f satisfying $\alpha f = Pf$, we have

(3.5) $$R(i,j) = \pi(j)f(i),$$

for some numbers $\pi(j)$. Conversely, R also satisfies $RP = \alpha R$, which shows that every row of R is a row eigenvector corresponding to α. Since a row eigenvector of P is a column eigenvector of the transpose P^T of the matrix P, we see by applying the preceding statements to the positive matrix P^T that any row eigenvector of P corresponding to α must be strictly positive (up to multiplication by a constant). Hence, π figuring in (3.5) above is a constant $c \neq 0$ times $v > 0$ satisfying $\alpha v = vP$. Hence $\sum_j R(j,j) \neq 0$. This completes the proof of (3.1).   □

(3.6) REMARK. Continuing the last paragraph of the above proof, we note that the derivative φ' must be positive at α since α is the largest root of $\varphi(x) = 0$. So the constant c there must be positive, and we must have $R(i,j) > 0$ for all i, j.   □

(3.7) REMARK. It might be worth proving the remark made about the maximal eigenvalue of a strictly positive Markov matrix (see Remark (3.2)). Since, then, $P1 = 1$, the number 1 is an eigenvalue. On the other hand, it was shown within the first paragraph of the preceding proof that the set S is contained in $[0, b]$ where $b = \sum_j P(i,j) = 1$. So we must have $\alpha = 1$.

4. Non-Negative Matrices

Perron's theorem generalizes to non-negative matrices in an essentially unchanged form. This section is devoted to showing how to reduce the general case to Perron's case.

A matrix P is called *non-negative* if $P(i,j) \geq 0$ for all i, j. The concepts of irreducibility and periodicity for Markov chains depended only on the locations of positive entries rather than on the actual values of the entries. Therefore these concepts remain the same for arbitrary non-negative matrices. For the sake of completeness we will give the definitions here in matrix language.

By a *permutation* of a square matrix P we mean a permutation of the columns of P combined with the same permutation of the rows (this corresponds to relabeling the states in the Markov chain case). A matrix P is called *reducible* if there exists a permutation which puts it in the form

$$(4.1) \qquad \bar{P} = \begin{bmatrix} P_1 & 0 \\ T & Q_1 \end{bmatrix}$$

where P_1, Q_1 are square. Otherwise, P is called *irreducible*. If P is reducible to the form (4.1), the matrices P_1 and Q_1 may be further reducible. Carrying out this process a number of times, P can finally be put in the form

$$(4.2) \qquad P = \begin{bmatrix} P_1 & 0 & \cdots & 0 & 0 & \cdots & 0 \\ 0 & P_2 & \cdots & 0 & 0 & \cdots & 0 \\ \cdot & \cdot & \cdot & \cdot & \cdot & & \cdot \\ \cdot & \cdot & \cdot & \cdot & \cdot & & \cdot \\ \cdot & \cdot & \cdot & \cdot & \cdot & & \cdot \\ 0 & 0 & \cdots & P_k & 0 & \cdots & 0 \\ T_{11} & T_{12} & \cdots & T_{1,k} & Q_1 & \cdots & 0 \\ \cdot & \cdot & & \cdot & \cdot & \cdot & \cdot \\ \cdot & \cdot & & \cdot & \cdot & \cdot & \cdot \\ \cdot & \cdot & & \cdot & \cdot & \cdot & \cdot \\ T_{m1} & T_{m2} & \cdots & T_{mk} & T_{m,k+1} & \cdots & Q_m \end{bmatrix},$$

where all the diagonal blocks are square and the matrices on the diagonal, namely $P_1, P_2, \ldots, P_k, Q_1, \ldots, Q_m$, are all irreducible. This is called the *normal form* of a reducible matrix.

If P is in the normal form (4.2), then its characteristic polynomial is

$$(4.3) \qquad \varphi(x) = \varphi_1(x)\varphi_2(x) \cdots \varphi_k(x)\psi_1(x) \cdots \psi_m(x),$$

where $\varphi_1, \varphi_2, \ldots, \varphi_k, \psi_1, \ldots, \psi_m$ are the respective characteristic polynomials of $P_1, P_2, \ldots, P_k, Q_1, \ldots, Q_m$. Hence the eigenvalues of P are the eigenvalues of $P_1, \ldots, P_m, Q_1, \ldots, Q_k$ put together. Therefore, there is no loss of generality in restricting ourselves to irreducible matrices below.

Among irreducible matrices P we need to distinguish two types: those for which $P^k > 0$ for some k (and therefore for all k large), and those for which there is no such k. In the case of a transition matrix P, these are respectively the aperiodic and periodic ones. Accordingly, an irreducible non-negative matrix P is called *aperiodic* if $P^k > 0$ for some k. Otherwise, it is called periodic with period $\delta \geq 2$ if there exists a permutation which puts it in the form

$$(4.4) \qquad P = \begin{bmatrix} 0 & A_1 & 0 & \cdot & \cdot & 0 \\ 0 & 0 & A_2 & \cdot & \cdot & 0 \\ \cdot & \cdot & \cdot & \cdot & & \cdot \\ \cdot & \cdot & \cdot & & \cdot & \cdot \\ \cdot & \cdot & \cdot & & & \cdot \\ 0 & 0 & 0 & \cdot & \cdot & A_{\delta-1} \\ A_\delta & 0 & 0 & \cdot & \cdot & 0 \end{bmatrix}$$

where the diagonal blocks are square (but $A_1, \ldots, A_\delta$ probably are *not* square). Then the powers of P are all of a specific form, and in particular,

$$(4.5) \qquad P^\delta = \begin{bmatrix} P_1 & & & 0 \\ & P_2 & & \\ & & \cdot & \\ & & & \cdot \\ 0 & & & P_\delta \end{bmatrix},$$

where the matrices $P_1, \ldots, P_\delta$ are all irreducible aperiodic and

$$(4.6) \qquad P_1 = A_1 \cdots A_\delta; \quad P_2 = A_2 \cdots A_\delta A_1; \quad \ldots ;$$
$$P_\delta = A_\delta A_1 \cdots A_{\delta-1}.$$

The following theorem was proved by Frobenius in a series of three papers published in 1908, 1909, 1912. As we shall see, it can be reduced to Perron's theorem fairly easily.

(4.7) THEOREM. Let P be an irreducible non-negative matrix. Then P has an eigenvalue α which is real, positive, and simple. For any other eigenvalue λ of P we have $|\lambda| \leq \alpha$. To this maximal eigenvalue α there corresponds a strictly positive eigenvector.

If P is aperiodic, then for any other eigenvalue λ of P we actually have $|\lambda| < \alpha$. If P is periodic with period δ, then there are exactly δ eigenvalues

$\lambda_1, \ldots, \lambda_\delta$ with absolute values exactly equal to α. These eigenvalues are all distinct and are given by

$$(4.8) \qquad \lambda_1 = \alpha, \quad \lambda_2 = \alpha c, \quad \lambda_3 = \alpha c^2, \quad \ldots, \quad \lambda_\delta = \alpha c^{\delta-1},$$

where c is the complex number $c = e^{2\pi i/\delta}$, $i = \sqrt{-1}$.

Proof. (a) *Aperiodic case.* Suppose P is irreducible aperiodic. Then there is some integer k such that P^k is strictly positive. By Perron's theorem (3.1), P^k has an eigenvalue β which is real, positive, simple and which is greater than $|\lambda|$ for any other eigenvalue λ of P^k. But the eigenvalues of P^k are the kth powers of the eigenvalues of P. Hence, there is an eigenvalue α of P such that $\alpha^k = \beta$ and $|\alpha| > |\lambda|$ for any other eigenvalue λ of P. Next, let f be an eigenvector of P corresponding to α. Then $Pf = \alpha f$ implies that $P^k f = \alpha^k f = \beta f$. By Perron's theorem, the eigenvector f corresponding to β can be taken to be strictly positive. Taking $f > 0$, we see that $Pf > 0$ and hence $\alpha f > 0$, which means α is a real positive number.

(b) *Periodic case.* Suppose P is irreducible and is periodic with period δ, and assume that it is already in the form (4.4). Then P^δ is in the form (4.5) with $P_1, \ldots, P_\delta$ irreducible aperiodic. Applying the preceding paragraph to each P_i separately, we see that P_i has an eigenvalue β_i which is real, positive, and simple and exceeds all other eigenvalues in absolute value, and to which there corresponds a strictly positive (unique up to a constant factor) eigenvector f_i. Note that by (4.6),

$$P_{i-1}A_{i-1} = A_{i-1}P_i$$

for all $i = 1, \ldots, \delta$ (we are identifying 0 with δ here and below. Since $A_{i-1} \neq 0$ and $f_i > 0$, the vector $A_{i-1}f_i \neq 0$. Moreover, it satisfies

$$(4.9) \qquad \beta_i(A_{i-1}f_i) = A_{i-1}(\beta_i f_i) = A_{i-1}P_i f_i = P_{i-1}(A_{i-1}f_i);$$

that is, the maximal eigenvalue β_i of P_i is an eigenvalue of P_{i-1}. Therefore, we must have $\beta_i \leq \beta_{i-1}$, which implies, since i is arbitrary, that

$$(4.10) \qquad \beta_1 = \cdots = \beta_\delta \equiv \beta.$$

Putting (4.10) together with (4.9), we see that we can take

$$(4.11) \qquad f_1 = A_1 f_2, \quad f_2 = A_2 f_3, \quad \ldots, \quad f_{\delta-1} = A_{\delta-1}f_\delta,$$

in which case

$$(4.12) \qquad A_\delta f_1 = \beta f_\delta.$$

Finally, define $\alpha = \beta^{1/\delta}$, $c = e^{2\pi i/\delta}$, and put

(4.13)
$$D = \begin{bmatrix} I & 0 & \cdot & \cdot & 0 \\ 0 & cI & \cdot & \cdot & 0 \\ \cdot & \cdot & \cdot & & \cdot \\ \cdot & \cdot & & \cdot & \cdot \\ \cdot & \cdot & & & \cdot \\ 0 & 0 & \cdot & \cdot & c^{\delta-1}I \end{bmatrix}, \quad f = \begin{bmatrix} f_1 \\ \alpha f_2 \\ \cdot \\ \cdot \\ \cdot \\ \alpha^{\delta-1}f_\delta \end{bmatrix},$$

where the partitioning of D and f is in accord with that of P as given in (4.4). Then, defining $\lambda_1, \ldots, \lambda_\delta$ as in (4.8), we see that (4.11), (4.12), and (4.13) imply

(4.14) $\lambda_1 f = Pf, \quad \lambda_2 Df = PDf, \quad \ldots, \quad \lambda_\delta D^{\delta-1}f = PD^{\delta-1}f.$

In other words, each one of the numbers $\lambda_1, \ldots, \lambda_\delta$ is an eigenvalue; and, in particular, $\lambda_1 = \alpha$ is real, positive, and simple and has the strictly positive eigenvector f corresponding to it. This completes the proof.   □

The following summarizes the situation in the case of irreducible Markov matrices. The proof follows by using (3.2) and (3.7) together with Theorem (4.7).

(**4.15**) COROLLARY.   Let P be an irreducible Markov matrix. Then the number 1 is a simple eigenvalue of P. For any other eigenvalue λ of P we have $|\lambda| \leq 1$. If P is aperiodic, then $|\lambda| < 1$ for all other eigenvalues of P. If P is periodic with period δ, then there are δ eigenvalues with absolute value one. These are all distinct and are

$$\lambda_1 = 1, \quad \lambda_2 = c, \quad \ldots, \quad \lambda_\delta = c^{\delta-1}, \quad c = e^{2\pi i/\delta}.$$

To the eigenvalue 1 there corresponds a unique (up to a constant factor) strictly positive row eigenvector.

(4.16) EXAMPLE.   Suppose P has rows $(0.6, 0.4)$, $(0.2, 0.8)$. The maximal eigenvalue is 1; the other eigenvalue is $tr(P) - 1 = 0.4$.

(4.17) EXAMPLE.   Suppose

$$P = \begin{bmatrix} 0.2 & 0.3 & 0.5 \\ 0.2 & 0.5 & 0.3 \\ 0 & 0.4 & 0.6 \end{bmatrix}.$$

The maximal eigenvalue is 1. Since $tr(P) = 0.2 + 0.5 + 0.6 = 1.3$ and $tr(P^2) = (0.04 + 0.06) + (0.06 + 0.25 + 0.12) + (0.12 + 0.36) = 1.01$, the other two eigenvalues $\lambda_2 = x$, $\lambda_3 = y$ satisfy

$$1 + x + y = 1.30, \qquad 1 + x^2 + y^2 = 1.01.$$

solving for x, we obtain $y = 0.3 - x$ and $2x^2 - 0.6x + 0.08 = 0$. So $x^2 - 0.3x + 0.04 = 0$, which yields $x = 0.15 + i\sqrt{0.0175}$, $y = 0.15 - i\sqrt{0.0175}$. Hence the eigenvalues of P are

$$1, \quad 0.15 + i\sqrt{0.0175}, \quad 0.15 - i\sqrt{0.0175}.$$

Note that the absolute values of the eigenvalues distinct from 1 are both equal to $\sqrt{0.04} = 0.2$, which is definitely smaller than 1.

To the maximal eigenvalue 1 there corresponds the column eigenvector 1 and the row eigenvector $(4, 16, 17)$. ☐

(4.18) EXAMPLE. Let

$$P = \begin{bmatrix} 0 & 0 & 0.2 & 0.3 & 0.5 \\ 0 & 0 & 0.5 & 0.5 & 0 \\ 0.4 & 0.6 & 0 & 0 & 0 \\ 1 & 0 & 0 & 0 & 0 \\ 0.2 & 0.8 & 0 & 0 & 0 \end{bmatrix}.$$

The matrix P is irreducible and periodic with period 2. Now,

$$P^2 = \begin{bmatrix} 0.48 & 0.52 & 0 & 0 & 0 \\ 0.70 & 0.30 & 0 & 0 & 0 \\ 0 & 0 & 0.38 & 0.42 & 0.20 \\ 0 & 0 & 0.20 & 0.30 & 0.50 \\ 0 & 0 & 0.44 & 0.46 & 0.10 \end{bmatrix}.$$

The eigenvalues of the 2×2 matrix on the upper left-hand corner are, by inspection, 1 and -0.22. This is sufficient to compute all the eigenvalues of P: We note that in the proof of Frobenius' theorem, in the periodic case, (4.9) holds for any eigenvalue β_i of P_i. This implies the following

(**4.19**) REMARK. The set of eigenvalues of a periodic irreducible matrix P, regarded as a system of points in the complex plane, goes over into itself under a rotation of the plane by the angle $2\pi/\delta$.

Returning to our example, since 1 and -0.22 are eigenvalues for P^2, their square roots 1, -1, $i\sqrt{0.22}$, and $-i\sqrt{0.22}$ will have to be eigenvalues for P. There remains one eigenvalue to guess; by Remark (4.19), that

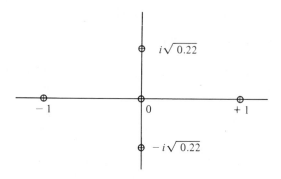

Figure A.4.1   Eigenvalues of the matrix P with period 2.

eigenvalue will have to go into itself under a rotation of the plane by the angle of 180 degrees. So that remaining eigenvalue must be 0.   □

(4.20) EXAMPLE.   Suppose P is a 5×5, irreducible, periodic Markov matrix with period 3. Then its eigenvalues are

$$1, \quad e^{2\pi i/3}, \quad e^{4\pi i/3}, \quad 0, \quad 0$$

(because, by (4.19), for any eigenvalue $\lambda \neq 0$, the image $\lambda e^{2\pi i/3}$ of λ under a rotation of the plane by 120 degrees is again an eigenvalue).   □

5.   Limits and Rates of Convergence

In this section we will limit ourselves to consideration of irreducible aperiodic Markov matrices. If P is irreducible but periodic with period δ, then what we are saying holds for each one of the irreducible aperiodic matrices $P_1, \ldots, P_\delta$ on the diagonal of P^δ. If P is reducible, then again, what we have to say here applies to each irreducible block separately. The following is the basic theorem.

(5.1) THEOREM.   Let P be an irreducible aperiodic Markov matrix. Then for all i, j,

(5.2) $$\lim_{k \to \infty} P^k(i, j) = \pi(j) > 0$$

independent of i. The row vector π is the unique solution of

(5.3) $$\pi P = \pi, \quad \pi 1 = 1.$$

Moreover, the convergence in (5.2) is geometric: there exist constants $\alpha > 0$

and $0 \leq \beta < 1$ such that

(5.4) $$|P^k(i,j) - \pi(j)| \leq \alpha\beta^k, \quad k = 1, 2, \ldots$$

for all i, j. Here β can be taken to be the largest number amongst the absolute values of the eigenvalues of P excluding the eigenvalue 1.

Proof. We will prove this only in the case where P is diagonalizable. By Corollary (4.15), the largest eigenvalue of P is 1; to this there corresponds a column eigenvector, which we can take to be $f = 1$, and a row eigenvector π, which we can take to be that satisfying (5.3). Then in the spectral representation (2.5) for P we have

(5.5) $$\lambda_1 = 1, \quad B_1(i,j) = \pi(j),$$

for all i, j. Now, (2.6) becomes

(5.6) $$P^k = B_1 + \lambda_2^k B_2 + \cdots + \lambda_n^k B_n, \quad k = 0, 1, \ldots.$$

Since P is aperiodic, the absolute values $|\lambda_2|, \ldots, |\lambda_n|$ are all strictly less than 1; therefore, as $k \to \infty$, $\lambda_2^k, \ldots, \lambda_n^k \to 0$. This shows that

$$\lim_k P^k = B_1,$$

which is the same as (5.2) in view of (5.5).

Going back to (5.6) we have, for any i, j,

(5.7) $$|P^k(i,j) - \pi(j)| = |\lambda_2^k B_2(i,j) + \cdots + \lambda_n^k B_n(i,j)|$$
$$\leq |\lambda_2|^k |B_2(i,j)| + \cdots + |\lambda_n|^k |B_n(i,j)|.$$

Now let

$$\beta = \max\{|\lambda_2|, \ldots, |\lambda_n|\}.$$

Now $\beta < 1$, and we have, in (5.7),

(5.8) $$|P^k(i,j) - \pi(j)| \leq \beta^k(|B_2(i,j)| + \cdots + |B_n(i,j)|).$$

Finally, let

(5.9) $$\alpha = \sup_{i,j} [|B_2(i,j)| + \cdots + |B_n(i,j)|].$$

Then (5.8) yields

$$|P^k(i,j) - \pi(j)| \leq \alpha\beta^k,$$

which is the desired expression (5.4). $\qquad\square$

(5.10) REMARK. Note that the estimate in (5.4) can be replaced by better estimates such as (5.8). However, (5.4) is quite good for many purposes, for it gives qualitative assurance that $P^k(i, j)$ converges to $\pi(j)$ very fast.

In fact, it is easy to show that the smallest entry $\inf_i P^k(i, j)$ on the jth column *increases* to $\pi(j)$ as k increases; and the largest entry $\sup_i P^k(i, j)$ in the jth column *decreases* to $\pi(j)$. Therefore, if the powers P^k of P are computed for a few k, by comparing the smallest and the largest entries on a fixed column, we can tell how close the entries on that column are to their common limiting value.   □

(5.11) EXAMPLE. Suppose

$$P = \begin{bmatrix} 0.6 & 0.4 \\ 0.5 & 0.5 \end{bmatrix}.$$

The eigenvalues are 1 and 0.1. The limiting probabilities $\pi(j)$ are easily computed to be $\pi = (\tfrac{5}{9}, \tfrac{4}{9})$. So

$$B_1 = \begin{bmatrix} \tfrac{5}{9} & \tfrac{4}{9} \\ \tfrac{5}{9} & \tfrac{4}{9} \end{bmatrix}, \quad B_2 = \begin{bmatrix} \tfrac{4}{9} & -\tfrac{4}{9} \\ -\tfrac{5}{9} & \tfrac{5}{9} \end{bmatrix}.$$

We have

$$|P^k(i, j) - \pi(j)| \le (\tfrac{5}{9})(0.1)^k.$$

In the case of 2×2 matrices it is easily seen that α is the largest of the $\pi(j)$ and $\beta = |\operatorname{tr}(P) - 1|$.   □

(5.12) EXAMPLE. Consider the matrix P of Example (4.17). The limiting distribution π is $\pi = (\tfrac{4}{37}, \tfrac{16}{37}, \tfrac{17}{37})$. So

$$\lim_k P^k = \begin{bmatrix} \tfrac{4}{37} & \tfrac{16}{37} & \tfrac{17}{37} \\ \tfrac{4}{37} & \tfrac{16}{37} & \tfrac{17}{37} \\ \tfrac{4}{37} & \tfrac{16}{37} & \tfrac{17}{37} \end{bmatrix};$$

and

$$|P^k(i, j) - \pi(j)| \le \alpha(0.2)^k, \quad k = 1, 2, \dots$$

and we can take α to be 2.   □

(5.13) EXAMPLE. Consider the periodic matrix P of Example (4.18). Note that we have

$$P = \begin{bmatrix} 0 & A \\ B & 0 \end{bmatrix}, \quad P^2 = \begin{bmatrix} AB & 0 \\ 0 & BA \end{bmatrix},$$

so that $\pi_1 AB = \pi_1$, $\pi_2 BA = \pi_2$ can be satisfied by taking π_1 to be a solution of $\pi_1 AB = \pi_1$ and setting $\pi_2 = \pi_1 A$. Using the 2×2 matrix AB on the upper left-hand corner of P^2, we obtain $\pi_1 = (70, 52)$ as a solution to $\pi_1 AB = \pi_1$. Then $\pi_2 = \pi_1 A$ becomes $\pi_2 = (40, 47, 35)$. Suitable normalizations yield

$$\lim_{k \to \infty} P^{2k} = \frac{1}{122} \begin{bmatrix} 70 & 52 & 0 & 0 & 0 \\ 70 & 52 & 0 & 0 & 0 \\ 0 & 0 & 40 & 47 & 35 \\ 0 & 0 & 40 & 47 & 35 \\ 0 & 0 & 40 & 47 & 35 \end{bmatrix} = P',$$

$$\lim_{k \to \infty} P^{2k+1} = \frac{1}{122} \begin{bmatrix} 0 & 0 & 40 & 47 & 35 \\ 0 & 0 & 40 & 47 & 35 \\ 70 & 52 & 0 & 0 & 0 \\ 70 & 52 & 0 & 0 & 0 \\ 70 & 52 & 0 & 0 & 0 \end{bmatrix} = P''.$$

Since the eigenvalues of P^2 were $1, -0.22, 1, -0.22$, and 0, we see that

$$|P^{2k}(i, j) - P'(i, j)| \leq \alpha(0.22)^k,$$

and similarly

$$|P^{2k+1}(i, j) - P''(i, j)| \leq \alpha(0.22)^k. \qquad \square$$

*   *   *

A complete treatment of non-negative matrices is given in Chapter 13 of GANTMACHER [1]. It is usual for books on matrices to prove FROBENIUS' theorem directly without making use of the notion of periodicity. Such proofs are long and difficult. By reversing the historical order of things, that is, by using Markov-chain-theoretic thinking to prove FROBENIUS' theorem rather than vice versa, we have obtained a much simpler proof.

References

J. Azema, M. Kaplan-Duflo, and D. Revuz
[1] Mésure invariante sur les classes récurrentes des processus de Markov. *Z. Wahrscheinlichkeitstheorie verw. Geb.* **8**, 157–181 (1967).

R. M. Blumenthal and R. K. Getoor
[1] *Markov Processes and Potential Theory.* Academic Press, New York, 1968.

K. L. Chung
[1] *Markov Chains with Stationary Transition Probabilities*, second edition. Springer-Verlag, Berlin, 1967.
[2] *A Course in Probability Theory.* Harcourt, Brace & World, Inc. New York, 1968.
[3] Crudely stationary counting processes. *Amer. Math. Monthly*, **79**, 867–877 (1972).

E. Çinlar
[1] Markov renewal theory. *Adv. Appl. Prob.* **1**, 123–187 (1969).
[2] Markov additive processes, I. *Z. Wahrscheinlichkeitstheorie verw. Geb.* **24**, 85–93 (1972).
[3] Markov additive processes, II. *Z. Wahrscheinlichkeitstheorie verw. Geb.* **24**, 95–121 (1972).

C. Dellacherie
[1] *Capacités et Processus Stochastiques.* Springer-Verlag, Berlin, 1972.

E. B. Dynkin and A. A. Yushkevich
[1] *Markov Processes: Theorems and Problems.* Plenum Press, New York, 1969.

W. FELLER
[1] On the integro-differential equations of purely discontinuous Markov processes. *Trans. Amer. Math. Soc.* **48**, 488–515 (1940).
[2] Fluctuation theory of recurrent events. *Trans. Amer. Math. Soc.* **67**, 98–119 (1949).
[3] *An Introduction to Probability Theory and Its Applications*, **1**. Wiley, New York, 1950. (Second edition, 1957; third edition, 1967).
[4] Boundaries induced by non-negative matrices. *Trans. Amer. Math. Soc.* **83**, 19–54 (1956).
[5] On boundaries and lateral conditions for the Kolmogorov differential equations. *Ann. Math.* **65**, 527–570 (1957).
[6] On semi-Markov processes. *Proc. Nat. Acad. Sci.* **51**, 653–659 (1964).
[7] *An Introduction to Probability Theory and Its Applications*, **2**, second edition. Wiley, New York, 1971.

M. FRÉCHET
[1] *Méthode des Fonctions Arbitraires. Théorie des Événements en Chaîne dans le cas d'un Nombre Fini d'États Possibles.* Gauthier-Villars, Paris, 1938. (Second edition, 1952).

F. R. GANTMACHER
[1] *The Theory of Matrices*, 2 volumes. Chelsea, New York, 1959.

B. V. GNEDENKO and A. N. KOLMOGOROV
[1] *Limit Distributions for Sums of Independent Random Variables*. Addison Wesley, Cambridge, 1954.

T. E. HARRIS
[1] *The Theory of Branching Processes*. Springer-Verlag, Berlin, 1963.

K. ITÔ and H. P. MCKEAN
[1] *Diffusion Processes and Their Sample Paths*. Springer-Verlag, Berlin, 1965.

J. G. KEMENY, J. L. SNELL, and A. W. KNAPP
[1] *Denumerable Markov Chains*. Van Nostrand, New York, 1966.

D. G. KENDALL
[1] Stochastic processes occurring in the theory of queues and their analysis by the method of the imbedded Markov chain. *Ann. Math. Statist.* **24**, 338–354 (1953).

A. N. KOLMOGOROV
[1] *Foundations of the Theory of Probability*. Chelsea, New York, 1956. (Second English edition of *Grundbegriffe der Wahrscheinlichkeitsrechnung*, Berlin, 1933.)
[2] Anfangsgründe der Theorie der Markoffschen Ketten mit unendlichen vielen möglichen Zustanden. *Mat. Sbornik N. S. Ser.* 607–610 (1936).

A. Y. KHINTCHINE
[1] *Mathematical Methods in the Theory of Queueing*. Griffin, London, 1960.

J. LAMPERTI
[1] *Probability: a Survey of the Mathematical Theory.* Benjamin, New York, 1966.

P. LÉVY
[1] Processus semi-Markoviens. *Proc. Intern. Congr. Math.* (Amsterdam) **3**, 416–426 (1954).

M. NAGASAWA and K. SATO
[1] Some theorems on time change and killing of Markov processes. *Kodaï Math. Sem. Reports* **15**, 195–219 (1963).

J. NEVEU
[1] *Mathematical Foundations of the Calculus of Probability.* Holden-Day, San Francisco, 1965.

S. OREY
[1] *Limit Theorems for Markov Chain Transition Probabilities.* Van Nostrand, New York, 1971.

R. PYKE
[1] Markov renewal processes: definitions and preliminary properties. *Ann. Math. Statist.* **32**, 1231–1242 (1961).

A. RÉNYI
[1] Remarks on the Poisson process. *Studia Sci. Math. Hungar.* **2**, 119–123 (1967).

W. L. SMITH
[1] Regenerative stochastic processes. *Proc. Roy. Soc. London Ser. A* **232**, 6–31 (1955).
[2] Renewal theory and its ramifications. *J. Roy. Statist. Soc. B* **20**, 243–302 (1958).

S. WATANABE
[1] On discontinuous additive functionals and Lévy measures of a Markov process. *Japan J. Math.* **34**, 53–70 (1964).

Answers to Selected Exercises

Chapter 1

(4.2) (a) $\Omega = \{(\omega_1, \omega_2, \omega_3): \omega_i = H \text{ or } T, i = 1, 2, 3\}$.

(b) $\Omega = \{(\omega_1, \omega_2, \ldots): \omega_i = H \text{ or } T, i = 1, 2, \ldots\}$.

(c) $\Omega = \{(\omega_1, \omega_2, \ldots): -\infty < \omega_i < +\infty, i = 1, 2, \ldots\}$.

(d) $\Omega = \{(\omega_1, \ldots, \omega_{20}): \omega_i \in \{0, 1, \ldots, 100\}, i = 1, 2, \ldots, 20\}$.

(e) $\Omega = \{(\omega_1, \omega_2, \ldots): -100 \le \omega_i \le 150\}$.

(f) $\Omega = \{(\omega_1, \omega_2, \ldots): 0 \le \omega_i < \infty, i = 1, 2, \ldots\}$.

(4.4) (a) 0.64×0.2^5; $0.64 \times 0.2^{k-2}$, $m \in \{1, \ldots, k-1\}$, $k \in \{2, 3, \ldots\}$.

(b) $\frac{1}{6}$.

(c) 0.6.

(d) $\frac{1}{13}$.

(e) $\frac{1}{7}$.

(f) 1.

(4.6) $P\{X = n\} = \dfrac{e^{-14}14^n}{n!}$, $P\{Y = m\} = \dfrac{e^{-7.14}7.14^m}{m!}$, $m, n \in \mathbb{N}$.

$P\{X = n \mid Y = m\} = \dfrac{e^{-6.86}(6.86)^{n-m}}{(n-m)!}$, $n \ge m$.

$P\{Y = m \mid X = n\} = \dbinom{n}{m}0.51^m \cdot 0.49^{n-m}$, $m \le n$.

$P\{X - Y = k \mid X = n\} = \dbinom{n}{k}0.49^k 0.51^{n-k}$, $k \le n$.

$P\{X - Y = k \mid Y = m\} = e^{-14}\dfrac{7.14^m 6.86^k}{m!\, k!}$.

(4.8) $e^{-1.32}$.

(4.10) (a) $p^2\left[p^2 + \dfrac{(1-p)^2}{n-1}\right]^{-1}$.

(b) $p(1 - p)\left[1 - p^2 - \dfrac{(1 - p)^2}{n - 1}\right]^{-1}.$

(c) Same as (b).

(d) $p(1 - p)\left[2p(1 - p) + (n - 2)\left(\dfrac{1 - p}{n - 1}\right)^2\right]^{-1}.$

Chapter 2

(3.2) $-0.2, 2.8, 13.4.$

(3.4) (a) $\dfrac{a + b}{2}, \dfrac{(b - a)^2}{12}, \dfrac{1}{2}.$

(b) $P\{Y \le t\} = t, 0 \le t \le 1.$

(3.6) $\beta^2.$

(3.12) (a) $aN, b^2N + a^2N^2.$

(b) $ac, a^2(d^2 + c^2) + b^2c, a^2d^2 + b^2c.$

Chapter 3

(5.2) $0.5^3; 3 \times 0.05 \times 0.95^2.$

(5.4) $1.04.$

(5.6) $10.2.$

(5.10) $0.19.$

(5.14) (a) p^3q^{n-3} for $k \ge 1, m \ge k + 1, n \ge m + 1.$

(b) $pq^{n-m-1}, n \ge m + 1.$

(c) $m + \dfrac{1}{p}.$

(d) $\dfrac{p\alpha}{1 - q\alpha}\alpha^{T_2}.$

(5.16) $1 - 0.04 \times 0.96^6.$

(5.18) (a) $[(1 - p_1)(1 - p_2)]^{12}[p_1 + p_2 - p_1p_2].$

(b) $\dfrac{1 - p_2}{(p_1 + p_2 - p_1p_2)}.$

(c) $[(1 - p_1)(1 - p_2)]^{12}[1 - p_2]p_1.$

(5.20) $110; 470.5.$

Chapter 4

(8.2) (a) $2t, 2t.$

(b) $N_t + 2s.$

(8.4) (a) $\dfrac{\lambda(\lambda t)^{12}e^{-\lambda t}}{12!}, t \ge 0.$

(b) $\displaystyle\sum_{n=13k}^{13k+12} \frac{e^{-\lambda t}(\lambda t)^n}{n!}, k \in \mathbb{N}.$

(8.8) (a) 28.56.

(b) $\exp[-14 + 6.72\alpha + 3.12\alpha^2 + 2.08\alpha^3 + 1.04\alpha^4 + 1.04\alpha^5].$

(8.12) $E[Z] = \lambda \displaystyle\int_0^\infty f(x)\, dx; \ Var(Z) = \lambda \displaystyle\int_0^\infty f(x)^2\, dx.$

Chapter 5

(4.2) (a) 0.096.

(b) 14.41.

(c) $F_k(w, b) = 0.4^{k-1}\, 0.6, k \ge 1; \ F_1(b, b) = 0.2, F_2(b, b) = 0.8.$

(4.4) (a) 1 and 3 are recurrent non-null aperiodic; 2 and 4 are transient.

(b) 1, 3, and 4 are recurrent non-null aperiodic; 2 and 5 are transient.

(c) All states are recurrent non-null aperiodic.

(d) 4 and 7 are transient; others are recurrent non-null aperiodic.

(e) 2 is transient; others are recurrent non-null aperiodic.

(f) 1, 2, 3, 4, and 6 are recurrent non-null periodic with period 3; other states are transient.

(4.8) (a) 1 if $i = 0$, $k = 0$; 0.45 if $i \in \{1, 2, 3\}$, $k = 1$; 0.55 if $i \in \{1, 2, 3\}$, $k = -1$; 0.25 if $i \ge 4$, $k = 2$; 0.30 if $i \ge 4$, $k = 1$; 0.25 if $i \ge 4$, $k = -2$; 0 otherwise.

(c) Rows are $(1, 0, 0, \ldots)$; $(0.55, 0, 0.45, 0, \ldots)$; $(0, 0.55, 0, 0.45, 0, \ldots)$; $(0, 0, 0.55, 0, 0.45, 0, \ldots)$; $(0, 0, 0.45, 0, 0, 0.30, 0.25, 0, \ldots)$; $(0, 0, 0, 0.45, 0, 0, 0.30, 0.25, 0, \ldots)$; \ldots.

(d) 0 is absorbing; all others are transient.

Chapter 6

(8.2) (a) $\begin{bmatrix} \frac{23}{30} & 1 & 1 \\ \frac{20}{30} & \frac{4}{5} & 1 \\ 0 & 0 & 1 \end{bmatrix}$

(b) $\begin{bmatrix} 1 & 1 & 0 & 0 \\ 1 & 1 & 0 & 0 \\ 1 & 1 & \frac{4}{5} & \frac{5}{7} \\ 1 & 1 & 1 & \frac{6}{7} \end{bmatrix}$

(c) $\begin{bmatrix} \frac{23}{30} & 1 & 1 & 1 & 1 \\ \frac{20}{30} & \frac{4}{5} & 1 & 1 & 1 \\ 0 & 0 & 1 & 1 & 1 \\ 0 & 0 & 1 & 1 & 1 \\ 0 & 0 & 1 & 1 & 1 \end{bmatrix}$

(d) $\begin{bmatrix} 1 & 1 & 0 & 0 & 0 \\ 1 & 1 & 0 & 0 & 0 \\ \frac{1}{3} & \frac{1}{3} & \frac{2}{5} & \frac{2}{3} & 0 \\ 0 & 0 & 0 & 1 & 0 \\ 1 & 1 & 0 & 0 & \frac{1}{2} \end{bmatrix}$

(8.4) (a) $(\frac{52}{93}, \frac{21}{93}, \frac{20}{93})$.
 (b) $(\frac{3}{8}, \frac{5}{8})$.

(8.6) $(\frac{1}{5}, \frac{1}{5}, \frac{1}{5}, \frac{1}{5}, \frac{1}{5})$

(8.8) $\frac{1}{m}(1 - p_1 - \cdots - p_j)$, where $m = \sum k p_k$.

(8.10) (a) 0.7.
 (b) $\frac{0.7}{0.3} = \frac{7}{3}$.

(8.12) (a) $\begin{bmatrix} a & b & c & 0 & 0 & 0 & 0 & 0 \\ a & b & c & 0 & 0 & 0 & 0 & 0 \\ a & b & c & 0 & 0 & 0 & 0 & 0 \\ 0 & 0 & 0 & d & e & 0 & 0 & 0 \\ 0 & 0 & 0 & d & e & 0 & 0 & 0 \\ a & b & c & 0 & 0 & 0 & 0 & 0 \\ a & b & c & 0 & 0 & 0 & 0 & 0 \\ a & b & c & 0 & 0 & 0 & 0 & 0 \end{bmatrix}$
 $a = \frac{20}{89}, b = \frac{45}{89}, c = \frac{24}{89}, d = \frac{4}{9}, e = \frac{5}{9}$.
 (b) $\frac{89}{20}, \frac{89}{45}, \frac{89}{24}, \frac{9}{4}, \frac{9}{5}, \infty, \infty, \infty$.
 (c) $\frac{1}{9}$ if $i \in \{4, 5\}$, $\frac{3}{89}$ otherwise.

(8.14) (b) $\frac{9}{8}$.

(8.16) (a) $v = (232, 363, 237, 58, 300, 333, 262)$.
 (c) $\frac{2283}{1785} = \frac{(vf)}{(v1)}$.

(8.18) $\pi_0 = 1 - ab$; $\pi_1 = (1 - ab)(e^{ab} - 1)$; $\pi_2 = (1 - ab)(e^{2ab} - e^{ab} - abe^{ab})$

(8.22) $\frac{10}{58}(\frac{48}{58})^j, j \in \mathbb{N}$; 27.84.

(8.28) For $p = 0.4$, $\eta = 0.823$.

Chapter 7

(5.2) (a) $\frac{1}{844}(5900, 6500, 0)$
 (b) $\frac{1}{110}(0, 0, -40, -376)$
 (c) $\frac{1}{1301}(4000, 4700, 0, 0, 0)$

(5.4) (a) $\frac{135}{7}$.
 (b) $\frac{10}{7}$.

(5.14) (a) 6.
  (b) 0.
  (c) 5.
  (d) 5.

(5.16) (a) $v = (3, 5, 3, 4\frac{1}{3}, 3)$; $T_0 = \inf\{n \geq 0: X_n \in \{b, e\}\}$.
  (b) $\frac{2}{3}, \frac{1}{3}, 0$ for $k = 5, 3$, other.
  (c) $\frac{2}{3} \cdot 5 + \frac{1}{3} \cdot 3 = \frac{13}{3}$.

(5.18) $T_0 = 0$; $v = (5, 3)$.

(5.20) $v = (5, 4)$; $T_0 = \inf\{n \geq 0: X_n = a\}$.

(5.22) $v = (7, 4, 3)$; $T_0 = \inf\{n: X_n \in \{a, c\}\}$.

(5.24) (a) $v = (4, 2, 2)$; $T_0 = \inf\{n: X_n \in \{a, b\}\}$.
  (b) $v = (4, 2, \frac{7}{6})$; $T_0 = \inf\{n: X_n \in \{a, b\}\}$.

(5.26) $v = (5, \frac{31}{9})$.

(5.30) $T_0(\omega) = \begin{cases} 0 & \text{if } X_0(\omega) \in \{b, c\}, \\ 1 & \text{if } X_0(\omega) \notin \{b, c\} \text{ and } X_1(\omega) \in \{a, b\}, \\ 2 & \text{otherwise.} \end{cases}$

Chapter 8

(7.8) (c) $P_t = \begin{bmatrix} p + qe^{-ct} & q - qe^{-ct} \\ p - pe^{-ct} & q + pe^{-ct} \end{bmatrix}$.

  (d) $U^\alpha = \dfrac{1}{\alpha(\alpha + c)} \begin{bmatrix} \alpha + cp & cq \\ cp & \alpha + cq \end{bmatrix}$.

(7.10) $\dfrac{1}{\alpha(\alpha^2 + 10\alpha + 18)} \begin{bmatrix} \alpha^2 + 8\alpha + 6 & 2\alpha + 6 & 6 \\ 2\alpha + 6 & \alpha^2 + 5\alpha + 6 & 3\alpha + 6 \\ 6 & 3\alpha + 6 & \alpha^2 + 7\alpha + 6 \end{bmatrix}$.

(7.12) (a) $\frac{1}{6}g(X_0) + \frac{1}{4}g(X_1)1_b(X_1)$.
  (b) $\frac{55}{6}$.

(7.14) $84U^\alpha g = \dfrac{1}{\alpha}(24, 24, 24) + \dfrac{1}{\alpha + 6}(-21, 42, 168) + \dfrac{1}{(\alpha + 14)}(-3, 18, -24)$.

(7.16) (a) $\pi = \frac{1}{6}(1, 2, 3)$.
  (b) $\pi g = \frac{80}{3}$.

(7.18) $\frac{1}{9}(1, 3, 5)$.

(7.20) $\dfrac{1 - r}{1 - r^{m+1}}(1, r, \ldots, r^m)$ where $r = \dfrac{a}{b}$.

(7.22) $\pi(0) = \frac{1}{11}$; $\pi(i) = \frac{2}{11}(0.8)^{i-1}$, $i \geq 1$.

Chapter 9

(**4.2**)   (c)   Limits are $\frac{3}{2}\tau$, $\frac{3}{2}$, $\frac{3}{4}$ respectively.

(**4.10**) (a) $1 - \int_{[0,t]} R(ds)(1 - \varphi(t - s))(1 - \psi(t - s))$ where $R = \sum F^n$.

        (b) $G(t) = 1 - \int_{[0,t]} R(ds)(1 - \varphi(t - s))$.

        (c) $\dfrac{\left[\int\int_{[0,\infty]} \varphi(ds)\psi(s)\right]}{\left[\int\int_{[0,\infty]} (1 - \varphi(s))\, ds\right]}$.

(**4.12**) (a) $1 - G(t) = 1 - \psi(t) + \int_{[0,t]} \psi(du) \int_{[0,t-u]} R(ds)(1 - \varphi(t - u - s))$.

        (b) $\int_{[0,t]} \hat{R}(du)(1 - \psi(t - u))$, where $\hat{R} = \sum G^n$.

        (c) $\dfrac{M_t}{N_t} \longrightarrow \dfrac{1}{\int_{[0,\infty]} \psi(ds)R(s)}$.

(**4.14**) (b) $\dfrac{1}{a + b} \int_{(x,\infty)} (1 - \varphi(u))\, du$.

(**4.20**) $1 - F = (1 - \varphi)(1 - \psi)$.

Chapter 10

(**8.2**)   (a) $(1 - e^{-5x})(1 - e^{-2y})(1 - e^{-2z} - 2ze^{-2z})$.

        (b) $1 - (\frac{35}{27} + \frac{10}{9}t + \frac{10}{3}t^2)e^{-2t} + (\frac{8}{27} + \frac{10}{3})e^{-5t}$.

(**8.4**)   (b) $\int_0^t R(c, j, ds)h(j, t - s)$ where $R = \sum Q^n$, and

           $h(j, \cdot) = 1 - \sigma, 1 - \psi, 1 - \varphi$ respectively for $j = a, b, c$.

(**8.6**) $(1 + 8p)m$ with $m = p \int (1 - \varphi) + q \int (1 - \psi)$.

(**8.10**) $\dfrac{\hat{v}(j)}{v(j)}$ where $\hat{v}\hat{P} = \hat{v}$, $v = vP$; $\hat{v}$ and v satisfy (8.9c).

Index of Notations

$A(i, j)$	254
$\mathbb{B}, \mathbb{B}_+$	324
$\mathbb{D}$	295
$\mathbb{D}_\pi$	332
$F(i, j)$	122
$F(t), F^n(t)$	285
$F*f(t)$	284
$F(i, j, t)$	319
$G(i, j)$	149
$\eta(i)$	329
$I(i, j)$	108
$I_A(x)$	7
$K(t, A)$	299
$K_t(i, A)$	346
$\lambda(i)$	247
m	289
$m(j)$	329
$\mathbb{N}$	6
$P(i, j)$	108
$P^n(i, j)$	110
$P_t(i, j)$	233
$Q(i, j)$	247
$Q(i, j, t)$	314
$Q^n(i, j, t)$	318
$Q*f(i, t)$	323

r	171
$\mathbb{R}, \mathbb{R}_+$	6
$R(i, j)$	123
$R^\alpha(i, j), R^\alpha g(i), Rg(i)$	198
$R(t), Rf$	286
$R(i, j, t)$	319
$R * f(i, t)$	324
$S(i, j)$	145
$U^\alpha(i, j), U^\alpha g(i)$	256
$U(i, j, t), Uf(i, t)$	339
$1_j(i)$	108
$i \longrightarrow j$	127
$\binom{n}{k}$	48
$\Omega, \varnothing$	2

Subject Index

A

Absorbing states, 127, 243
Additive functionals, 269
    ratio limit theorems, 270
Age processes, 141
Almost all, almost surely, 3
Aperiodicity; *see* periodicity
Arrival processes, 71
Arrival rate, 75

B

Bayes' formula, 15
Bernoulli-Laplace model of diffusion, 141
Binomial coefficients, 48
Binomial distribution, 48
Birth and death processes, 271
Bounded convergence theorem, 33
Branching processes, 183
Busy period in queueing, 188

C

Central limit theorem, 64
Certain event, 2

Chapman-Kolmogorov equation:
   for Markov chains, 110
   for Markov processes, 234
Chebyshev's inequality, 30
Closed sets, 127
Clustered renewal processes, 358
Complement of an event, 2
Compound Poisson processes, 91
Conditional distributions, 33
Conditional expectations, 34
Conditional independence, 39
Conditional probability, 14, 34
Convolutions, 284
Counters:
   of type I, 309, 316, 359
   of type II, 292, 308, 359

D

Direct integrability, 332
Direct Riemann integrability, 295
Discrete random variables, 6
Discrete topology, 240
Disjoint events, 2
Distribution function, 9
   arithmetic, 289
   empirical, 64
   estimation of, 63
Doubly Markov matrices, 114

E

Electron multipliers, 184
Empty event, 2
Erlang-n distribution, 83
Event, 1
Excessive functions, 204
Expected value, 22 ff.
Exponential distribution, 80
   memorylessness of, 17, 80
Extinction of families, 184, 193

F

Failure rate function, 309
Fee function, 222
Forward recurrence times, 85 ff.

G

Gamma distribution, 83
Generating function, 30
Generators, 256
Geometric distribution, 56

H

Harmonic functions, 205
Hypergeometric distribution, 66

I

Identity matrix, 108
Independence, 12, 39
Independent increments, 50, 61
Independent trials process, 113
Indicator function, 7
Instantaneous states, 243
Intersection, 2
Interval reliability, 309
Invariant measures:
    for Markov chains, 160
    for Markov processes, 265
Inventory theory, 115, 140
Irreducibility, 127

J

Joint distribution, 12

K

Key renewal theorem, 295
Kolmogorov's backward equation, 255

L

Laplace transform, 30
Laws of large numbers, 62

M

Machine repair problem, 276
Markov chains:
    invariant distribution, 154
    invariant measure, 160
    irreducible, 127
    K-dependent, 142
    limiting distribution, 152
    non-time-homogeneous, 142
    potentials, 196
    ratio limit theorems, 159
    recurrent, 152
    subordinated to Poisson processes, 236
    time homogeneity, 107
Markov matrix, 108
Markov processes:
    generators, 256
    irreducibility, 263
    limiting probabilities, 264
    modification of, 241
    potentials, 256
    recurrence, 263
    regular, 251
    time-homogeneity, 233
    transition functions, 233
Markov renewal equations, 324 ff.
    generalized, 360
Markov renewal functions, 319
Markov renewal kernel, 319
Markov renewal processes, 313 ff.
    classification of states, 321
    embedded, 346
    lifetime, 327
Mixtures of Poisson processes, 103
Moments, 30
Monotone convergence theorem, 33

N

Normal density function, 64
Normal distribution, 27, 45, 64
    table for, 65
Nuclear chain reactions, 194
Null states, 125

O

Optimal replacement policy, 311
Optimal stopping time, 208
Optimal strategies, 210
    ε-optimal strategies, 219, 221

P

Parameter set, 7
Payoff function, 210
    time-dependent, 230
Pedestrian delay problem, 291, 361
Periodic states, 125
Periodicity:
    for Markov chains, 160
    for renewal processes, 289
    for Markov renewal processes, 321
Poisson distribution, 76
Poisson processes, 71 ff.
    decompositions, 88
    mixtures, 103
    non-stationarity, 94
    superpositions, 87
Potential matrix, 123, 144, 197
Potential operators, 256
Potentials:
    in renewal theory, 286
    of Markov chains, 196
    of Markov processes, 256
    of semi-Markov processes, 339
Probability density function, 10
Probability distribution, 10
Probability measure, 3
Pure birth processes, 252

Q

Queueing system G/M/1, 178, 317
    finite waiting room, 193
    time dependence, 354
    virtual waiting time, 361
    waiting times, 181
Queueing system M/G/1, 169, 317
    busy period, 188
    finite waiting room, 192
    time-dependence, 350
    traffic intensity, 171
    waiting times, 175, 192
Queueing system M/M/1, 182, 238
Queueing system M/M/1/m, 274
Queueing system M/M/1/∞, 273
Queueing system M/M/s/∞, 274
    non-identical servers, 274
    with balking, 275
Queueing system M/M/∞, 275

R

Random numbers, 10
Random variables, 6
Random walks, 135, 156, 168
Ratio limit theorems:
    for Markov chains, 160
    for Markov processes, 269
    in Markov renewal theory, 336
Recurrent states, 125, 152
Regenerative processes, 298
    delayed, 302
Reliability theory, 103
    age replacement, 310, 311
    interval reliability, 309
    see also Replacements
Remaining lifetimes, 115, 157
Renewal equation, 293
Renewal function, 286
Renewal processes, 68, 284 ff.
    decomposition of, 312
    delayed, 302

lifetime, 290
periodicity, 289
recurrence, 288
stationary, 305
superposition of, 312
Renewal times, 284
Replacements, 80, 82, 92, 311, *see also* Reliability theory
Resolvent, 256
Riesz decomposition theorem, 205

S

Sample space, 1
Selling an asset, 221
Semi-Markov processes, 337 ff.
    generalized, 345
    minimal, 316, 337
    potentials, 339
Semi-Markovian kernel, 314
Semi-regenerative processes, 343
Simulation, 10, 98
Sojourn interval, 247
Space-time processes, 142
Stable states, 243
Stationary and independent increments, 50, 61, 72
Stochastic continuity, 240
Stochastic processes, 7
    independence of, 39
    parameter set of, 7
    with stationary and independent increments, 50, 61
Stopping times, 84, 118, 209, 239
Strong law of large numbers, 62
    for Markov chains, 160
    for Poisson processes, 79
    for renewal processes, 69
Strong Markov property, 117, 239
Superposition, 87
Support set, 215
System availability, 349

T

Time inverse, 95
Total probability, theorem of, 15

Traffic flow, 83, 85, 87, 89, 316
Traffic intensity, 273
Transient states, 125
Transition functions, 234
    standard, 240
Transition graph, 128
Transition matrix, 107
Transition probabilities, 107

U

Uniform distribution, 41
Union of events, 2

V

Value of a game, 210, 221
Variance, 30

W

Wald's lemma, 307
Weak law of large numbers, 62
Weibull distribution, 310